SPACE, KNOWLEDGE AND POWER

Space, Knowledge and Power
Foucault and Geography

Edited by

JEREMY W. CRAMPTON
Georgia State University, USA

and

STUART ELDEN
Durham University, UK

Routledge
Taylor & Francis Group

LONDON AND NEW YORK

First published 2007 by Ashgate Publishing

Published 2016 by Routledge
2 Park Square, Milton Park, Abingdon, Oxon OX14 4RN
711 Third Avenue, New York, NY 10017, USA

Routledge is an imprint of the Taylor & Francis Group, an informa business

British Library Cataloguing in Publication Data
Space, knowledge and power : Foucault and geography
   1. Foucault, Michel - Knowledge - Geography 2. Foucault,
   Michel - Influence 3. Spatial behavior 4. Geography -
   Philosophy
   I. Foucault, Michel II. Crampton, Jeremy W. III. Elden,
   Stuart, 1971-
   304.2'3

Library of Congress Cataloging-in-Publication Data
Space, knowledge, and power : Foucault and geography / edited by Jeremy W. Crampton and Stuart Elden.
      p. cm.
   Includes bibliographical references and index.
   ISBN: 978-0-7546-4654-9 (hardback) -- ISBN: 978-0-7546-4655-6 (pbk.)
   1. Geography--Philosophy. 2. Foucault,
Michel, 1926-1984. I. Crampton, Jeremy W. II. Elden, Stuart,
1971-

   G70.S673 2007
   910.01--dc22

                                                        2006025030

ISBN 9780754646549 (hbk)
ISBN 9780754646556 (pbk)

Contents

PART 6   Development

List of Figures and Tables

Figures

Tables

List of Contributors

John A. Agnew is Professor at the University of California, Los Angeles

Jean-Michel Brabant was a member of the *Hérodote* editorial staff

Mathew Coleman is Assistant Professor, Ohio State University

Jeremy W. Crampton is Associate Professor at Georgia State University

Stuart Elden is Reader at Durham University

Juliet J. Fall is an Academic Fellow at The Open University

Thomas Flynn is Professor at Emory University

Colin Gordon is the editor of *The Foucault Effect* and *Power/Knowledge*, and Director, NHSIA Disease Management Systems Programme, Royal Brompton Hospital (London)

Matthew Hannah is Associate Professor, the University of Vermont

David Harvey is Distinguished Professor at City University of New York Graduate Center

Philip Howell is a Senior Lecturer and Fellow of Emmanuel College, University of Cambridge

Margo Huxley is Senior Lecturer at the University of Sheffield

Alain Joxe is Director of Studies at Ecole des Hautes Etudes en Sciences Sociales, Paris

Gerry Kearns is a University Senior Lecturer and Fellow of Jesus College, University of Cambridge

William J. King is at the University of Hawaii

Edgar Knowlton Jr. is Professor Emeritus, the University of Hawaii

Stephen Legg is Lecturer at the University of Nottingham

Sara Mills is Professor at Sheffield Hallam University

Gerald Moore is in the PhD program, the University of Cambridge

Chris Philo is Professor at the University of Glasgow

Jean-Bernard Racine is Professeur ordinaire at Université de Lausanne

Claude Raffestin is Professeur honoraire at Université de Genève

Michel Riou was a contributor to *Hérodote*

Nigel Thrift is Vice-Chancellor, the University of Warwick

David Murakami Wood is Lecturer at the University of Newcastle

Acknowledgements

As editors we have incurred debts to a number of people. We are especially grateful to Claude Raffestin for allowing us to translate his text for this volume, and to Colin Gordon for permission to use his translation of 'Questions on Geography'. Éditions Gallimard granted permission to translate material from *Dits et écrits*; and Editions de la Découverte, who now control the rights to the material from *Hérodote*, granted permission for Foucault's questions and the responses. Jaime Sepulveda, former editor of *Revista centro-americana de Ciencias de la Salud*, allowed us to translate the third Rio lecture.

Previously published texts originally appeared as:

* 'Des questions de Michel Foucault à Hérodote.' *Hérodote*, No. 3, juillet–septembre 1976, 9–10.
* 'Des réponses aux questions de Michel Foucault.' *Hérodote*, No. 6, 2ᵉ trimestre, 1977, pp. 3–4, 12–15, 17–19, 21–3.
* Claude Raffestin, 'Foucault aurait-il pu révolutionner la géographie?' In *Au risque de Foucault*. Paris: Éditions de Centre Pompidou, 1997, 141–9.
* Michel Foucault, 'L'incorporation de l'hôpital dans la technologie modern.' *Revista centro-americana de Ciencias de la Salud*, No. 10, 1978, 93–104.
* Michel Foucault, 'Les mailles du pouvoir.' *Dits et écrits 1954–1988*, edited by Daniel Defert and François Ewald. Paris: Gallimard, Four Volumes, 1994, Vol. IV, pp. 182–94.
* Michel Foucault, 'Le langage de l'espace.' *Dits et écrits*, Vol. I, 407–12.
* Michel Foucault, 'La force de fuir.' *Dits et écrits*, Vol. II, 401–5.
* Michel Foucault, 'Questions on Geography.' Translated by Colin Gordon, in Gordon (Ed.) *Power/Knowledge: Selected Interviews and Other Writings, 1972–1977*. Brighton: Harvester Wheatsheaf, 1980, 63–77.

We are also deeply grateful to all our contributors for their enthusiasm for the project, despite its long gestation; to our translators Edgar Knowlton Jr., William J. King and Gerald Moore, and to our editors at Ashgate, Valerie Rose and Ellie Rivers. Catherine Alexander helped enormously with the proofs and Tom Norton provided the index.

A Note on References

Foucault's work has appeared in many forums as well as in a wide variety of reprints. We have adopted a policy of leaving the references preferred by each contributor rather than standardizing on one particular set of references. They are printed at the end of each chapter.

Introduction

Space, Knowledge and Power: Foucault and Geography

Stuart Elden and Jeremy W. Crampton

From architectural plans for asylums, hospitals and prisons; to the exclusion of the leper and the confinement of victims in the partitioned and quarantined plague town; from spatial distributions of knowledge to the position of geography as a discipline; to his suggestive comments on heterotopias, the spaces of libraries, of art and literature; analyses of town planning and urban health; and a whole host of other geographical issues, Foucault's work was always filled with implications and insights concerning spatiality. Many geographers, philosophers and social scientists have developed these issues in their own work, either through a sustained analysis of Foucault's own work, or in application in a range of other areas. Taking these debates to a new level, this book provides a series of challenges, appreciations, critiques and developments concerning the relation(s) between Foucault and geography.

In its sustained and in-depth encounter between Foucault and questions of space, place and geography *Space, Knowledge and Power: Foucault and Geography* has two main aims. The first is to provide a comprehensive overview of Foucault's engagement with geographical concerns and geography's engagement with Foucault; the second is to begin to open up a new range of themes and questions for the continuation of that engagement. That continued engagement is, it seems to us, two-way, since there is much of Foucault's own work still to discover: either in untranslated material published in his lifetime, or in the fascinating lecture courses that are appearing in French for the first time, and slowly being translated into English. As the French series editors have noted, because Foucault used these lectures as explorations rather than outlines for books, they do not reduplicate the studies he published in his lifetime. Foucault therefore still lives for us through these posthumous publications.

Questions and Responses

In 1976 Foucault took part in an interview with the geographers of the radical French journal *Hérodote*. It appeared in English translation in the 1980 collection *Power/Knowledge* as 'Questions on Geography', and has proved to be one of the most cited pieces concerning Foucault's relation to questions of space and power. Following shortly after the translation of *Discipline and Punish* in 1977, and the first volume of *The History of Sexuality* in 1978, the *Power/Knowledge* collection was designed to contextualize and make available a representative sample of his shorter writings from this period of Foucault's work, in much the same way as the *Language, Counter-Memory, Practice* collection had for an earlier period (Foucault 1977a).

Although it took a few years, Foucault found a ready audience for his geographical analyses, with some key figures of Anglophone geography part of this first wave of interest and appropriation. Writing in 1985, Felix Driver (Driver 1985) discussed the implications of *Discipline and Punish* for institutions, the law and the state. Driver was stimulated by the spatial implications of the historical transformation of power described in that book. Driver notes that for Foucault, power is both productive and negative, locally defined and yet immanent within particular fields of technology and action. Space is a vital part of the battle for control and surveillance of individuals, but it is a battle and not a question of domination. As Driver mentions this has inspired interesting discussions of governmentality (Driver 1985, 444) as the contact point between technologies of domination and technologies of the self (Foucault 1999, 162), a thematic pursued in the questions back to geographers. Chris Philo also drew on this material in work on the spatiality of madness (Philo 1986; 1989; 1992), while the work of John Pickles took up Foucault's 'insurrection of subjugated knowledges' (Pickles 1988; 2004). Derek Gregory's influential book *Geographical Imaginations* drew repeatedly on Foucault, especially the essay 'Of Other Spaces', but also displayed a sensitivity to the relations between power, knowledge and space and the 'cartographic anxiety' or crisis in geographical representation (Gregory 1994).

Hannah's contribution to this volume provides much important context to this story. It is interesting that these engagements by geographers came relatively late, mainly dating from after his 1984 death, and at some distance from Foucault's influence in other disciplines. As Hannah discusses, two points of entry proved significant – in the early 1980s at Cambridge University where Derek Gregory and his then-students (Felix Driver and Chris Philo) read Foucault, and Soja's book on postmodern geographies (Soja 1989).

Whether this is because geography in its traditional sense is not directly confronted by Foucault, or because the discipline itself did not make the 'cultural turn' is debatable. Certainly the cultural unrest of the 1960s made a number of geographers increasingly uneasy with its quantitative and law-finding focus, and demands grew for a more political account of power, race, sexuality or globalization, in which culture was a form of politics. Harvey's *Social Justice and the City* (Harvey 1973) was perhaps the first marker in this shift, and if for many during the 1970s and 1980s the old geographies were still satisfactory, there were signs of an emerging critical geography (Peet 2000).

By the early 1990s this situation was more fully in reversal, with the primary overview works taking up Foucault's work in some detail (see, for example, Cloke et al. 1991). The essay on heterotopia proved to be particularly influential, informing the work of Edward Soja (1989), and Kevin Hetherington (1997) for example.

Although this essay is an early work (it was written in 1967, but published only in Foucault 1984) it has attracted many commentaries. In it, Foucault makes a few programmatic and largely suggestive remarks that the traditional idea that time is creative and progressive, while space is static, could be reversed:

> The great obsession of the nineteenth century was, as we know, history: with its themes of development and of suspension, of crisis, and cycle, themes of the ever-accumulating past, with its preponderance of dead men ... the present epoch will perhaps be above all the epoch of space. We are in the epoch of simultaneity: we are in the epoch of juxtaposition, the epoch of the near and far, of the side-by-side, of the dispersed. We are at a moment, I believe, when our experience of the world is less that of a long life developing through time than that of a network that connects points and intersects with its own skein. (Foucault 1986, 22)

For geographers these remarks signalled something of a vindication, and despite their sketchy and possibly 'Kantian' overtones of absolute space (see Harvey, this volume), they have acted for years as a convenient standard bearer.

As Fall notes in this volume, the story within Francophone geography was rather different. Despite the *Hérodote* interview there has been little direct engagement with Foucault's relation to geography, and with the notable exception of Claude Raffestin, Francophone work which draws on Foucault is scant. While the interview with the geographers is regularly cited within Anglophone geography, what is much less well-known is that Foucault sent some questions back to the journal for a subsequent issue in 1976, which received responses the following year from some eminent Francophone scholars. It is interesting and perhaps significant that Foucault almost never went back to a journal following an interview, but as he admitted he had formed a mistaken impression about the geographers' intent and, as we further discuss below, he wanted to engage issues of space some more.

Foucault's questions back to the journal touch upon many of the issues discussed in the interview, but raise some important issues of their own. Foucault asks four related questions: What are the relations between knowledge (*savoir*), war and power? What does it mean to call spatial knowledge a science? What do geographers understand by power? and What would the geographies of medical establishments (implantations) understood as 'interventions' look like? Until now, these questions have never appeared in English translation. It is instructive to compare these questions with a parallel series that Foucault published as part of his 'Course Summary' for the 1975–1976 *Society Must Be Defended*, a lecture course on war, race, strategic knowledge, historiography and politics (Foucault 2003). While the questions cover related grounds, the latter make no mention of space or geography; and yet as the chapters by Philo and Elden demonstrate the concerns of that course can be related to geographical issues in a number of ways.

Although we reprint Foucault's interview with *Hérodote* later in this volume, the centrepiece of this book is Foucault's own questions to geographers, which are written from the perspective of an engaged and interested interlocutor, someone who is looking for answers both from himself and from others. We follow these questions with a selection of the original French language responses, revealing both in terms of an engagement with Foucault in a very different context from the Anglophone academy, but also in terms of the concerns of radical Francophone geography in the 1970s.

They are followed by newly commissioned responses, written thirty years after the initial encounter, from within the English-speaking world. These responses, which both respond to Foucault's questions and raise issues and problems of their own, are of two kinds. On the one hand we have David Harvey and Nigel Thrift, two of the most important contemporary geographers in the Anglophone academy; on the other hand two figures from outside geography who have engaged with Foucault's spatial concerns. Both Harvey and Thrift offer some incisive critiques of Foucault's work, at the same time establishing a distance between Foucauldian inspired radical geography and their own projects (see Harvey 2001; Thrift 2006 among many others). Sara Mills brings a perspective from English Studies, raising issues concerning gender; Thomas Flynn more directly answers Foucault's questions, but from the perspective of his being 'a philosopher, not a geographer'. Both these authors have dealt with Foucault's own work and its relation to spatial questions elsewhere (Flynn 1997; 2005; Mills 2003; 2005) although these analyses have largely yet to be appreciated or appropriated by geographers.

Contexts

Comparing the responses from 1977 and 2006, and from a Francophone and Anglophone perspective, raises an important issue of context. The timing of Foucault's questions is revealing, in that they were posed at the time of *Society Must be Defended* and the first volume of *The History of Sexuality*, which appeared in late 1976; are coterminous with his concern with biopower; and slightly predate his work on governmentality. They also tie back to his work on the origin of the disciplines (in *The Archaeology of Knowledge*) madness and the hospital (in *History of Madness*, *The Birth of the Clinic* and *Discipline and Punish*) and the concern for space that runs through these topics. The year 1976 was a crucial one for Foucault as it saw the collision of two projects – the analysis of discipline that he had been working on in concentrated form from the beginning of the decade; and the genealogy of the subject that occupied him until the end of his life. Stuart Elden's chapter therefore focuses on Foucault in this year, particularly looking at the concerns of strategy, medicine and the spatial politics of habitat, in order to situate the encounter with *Hérodote*. It does this by locating the questions and *Society Must Be Defended* within the various projects Foucault was engaged with at the time, both those mentioned above and lesser known collaborative works.

It is followed by two remarkable chapters that contextualize Foucault within two very different academies. Matthew Hannah's chapter 'Formations of "Foucault" in Anglo-American Geography: An Archaeological Sketch' accounts for the development of an interest in Foucault's work within British and North American geography, including material gleaned from discussions with some of the people involved with that appropriation. It is followed by a chapter by Juliet J. Fall that uncovers an encounter unknown to almost all Anglophone geographers, the way that Francophone geographers have used, critiqued or ignored Foucault. Focusing particularly on the work of the Swiss geographer Claude Raffestin, Fall shows how Foucault is read within a very different tradition. Rounding off this section is a newly translated piece by Raffestin himself, on the question of 'Could Foucault have Revolutionized Geography?' In it Raffestin demonstrates how Foucault's writings could have played a role in geography similar to the way Paul Veyne suggests they have for history, a claim that Fall's analysis helps us to understand.

Texts

Although much of Foucault's work is available in English translation, there are some striking omissions. His first major work, *Histoire de la folie à l'âge classique* has only just been translated into English in other than a heavily abridged form (Foucault 1967; 2005); and some of his collaborative projects with other scholars have never been translated at all (Farge and Foucault 1982; Foucault 1977b; Foucault et al. 1979 [though Foucault's own essay appears in 1980]). While his lecture courses are being translated into English, albeit with an understandable time-lag, the remarkable collection of his shorter writings, published in France in 1994 as the four volume, 3400 page, collection *Dits et écrits* has been only partially translated. In the three volumes of *The Essential Works of Michel Foucault* English readers are given much new material, but also a number of other essays, long available in other collections. Much material, some of it of the utmost interest, remains inaccessible to English-only speakers.

In this volume we have therefore included a very small sample of those missing texts, which we feel particularly speak to the encounter between Foucault and questions of spatiality, broadly conceived. These new translations are of two important lectures first given in Brazil; and two texts that bring questions of space to bear on literature and visual art. The first of the lectures is the final of Foucault's three 1974 Rio lectures on medicine, 'The Incorporation of the Hospital into Modern Technology' (the other two lectures appear in translation as Foucault 2000; 2004a). It only appeared in French for the first time in 1994, previously available only in Spanish and Portuguese translations. In it Foucault shows how hospital architecture and the situation of the hospital in an urban setting are crucial mechanisms in the politics of health. As such it provides a bridge between early concerns with health in *The Birth of the Clinic* and the spatial aspects of discipline and bio-power in works of the mid 1970s. Indeed it is in these lectures that bio-power first emerges as a category of Foucault's thought. Foucault argues that the 'question of the hospital at the end of the eighteenth century was fundamentally a question of space', and the medicine of this time was, through its use of space, simultaneously one of the individual and the population. Discipline, declares Foucault, is 'above all an analysis of space'. Along the way Foucault discusses both civil and military hospitals, political and military techniques of control and discipline, and mechanisms for the treatment of epidemics and the use of quarantine.

The second lecture is 'The Meshes of Power' which provides an overview analysis of Foucault's work on power in 1976. Here Foucault articulates his standard position that power is not repressive: 'I am going to try to develop, or better, to show in which direction one could better develop an analysis of power that would not simply be a negative, juridical conception of power, but a conception of a technology of power.' Given the timing of this lecture it has obvious parallels with the first volume of *The History of Sexuality*, but also with the 1973–1974 lectures on psychiatric power (Foucault 2006). Foucault draws on Marx for some ideas about the positivities of power, such as the fact that power is heterogenous: 'if we want to do an analysis of power ... we must speak of powers and try to localize them in their historical and geographical specificity'. Two objectives are underlined that prefigure their more explicit formulation in the governmentality work, especially the need for government to govern lightly and liberally. First, this discontinuous nature of power, in which the 'meshes of the net' were too wide and for example border control became problematic, and second the need: 'to find a mechanism of power that, at the same time as controlling things and people up to the finest detail, is neither onerous nor essentially predatory on society, that exercises itself in the very sense of the economic process'.

This lecture is an important addition to the governmentality literature in that it links his spatial concerns in *Discipline and Punish* (the dispersion of objects, the technologies of surveillance) to the continued focus on power in his later work. As Foucault discusses in more detail elsewhere, he identifies the second half of the 18th century as a time when politics shifted from the individual to that of populations, that is: 'living beings, traversed, commanded, ruled by processes and biological laws. A population has a birth rate, a rate of mortality, a population has an age curve, a generation pyramid, a life-expectancy, a state of health, a population can perish or, on the contrary, grow.'

The question arises however, just how *geographical* did Foucault intend the notion of governmentality to be? While it is true that the lectures of 1978–1979 were entitled *Security, Territory, and Population* the geographical analysis is somewhat underplayed in the lecture course as actually delivered. This is crucial, because governmentality and biopolitics informs the work of an increasing number of geographers (one could speak of a geo-governmentality school). Foucault also suggests, drawing on a text from the 16th century writer Guillaume de la Perrière that 'the definition of government in no way refers to territory: one governs *things*' (Foucault 1991).

But care is needed here not to read Foucault out of context; in the passage cited he is concerned with contrasting governmentality as the 'conduct of conduct' (see Huxley, this volume) with that of an earlier style of politics, that is, sovereignty. Under sovereignty (and here Foucault discusses Machiavelli at length; see Elden 2007; Holden and Elden 2005), the political focus is on defending and retaining territory, or perhaps more properly, terrain, from those enemies inside and outside the state. Machiavelli's purpose is to offer advice to the prince on how to defend territory, not just militarily but also politically. Foucault elaborates:

> I think it is not a matter of opposing things to men, but rather of showing that what government has to do with is not territory but, rather, a sort of complex composed of men and things. The things, in this sense, with which government is to be concerned are in fact men, but men in their relations, their links, their imbrication with those things that are wealth, resources, means of subsistence, the territory with its specific qualities, climate, irrigation, fertility, and so on … what counts is essentially this complex of men and things; property and territory are merely one of its variables. (Foucault 1991)

So Foucault then is shifting the emphasis from a simple retention of territorial control to a more nuanced notion of government over a 'complex' of men and things constituted as a population. And he says that a population is not just 'the sum of individuals inhabiting a territory' (Foucault 2004b) but an object itself, with birth and death rates, healthiness and so on. In Foucault's reading, Machiavelli differentiates between the object (territorial retention) and the target of the population who must be controlled. The art of government is less about geopolitics (territorial gain and retention) and more deeply geographical. This population thus had to be known in its spatial dispersion, giving rise not only to statistics but to a new form of cartography (Crampton 2004), something which Matthew Hannah develops in his study *Governmentality and the Mastery of Territory in Nineteenth-Century America* (Hannah 2000).

'The Language of Space' – a review of several novels that originally appeared in the journal *Critique* in 1964 – shows Foucault's under-appreciated qualities as a literary critic and reviewer. Foucault suggests that until the twentieth century, writing obeyed a formalism of chronology, or at least of a narrative return to its origins after a period of absence (as in the *Odyssey*). After Nietzsche and Joyce however: 'it is in space that, from the outset, language unfurls.' Here Foucault posits space as a space of freedom, as unconstrained by barriers. He adds that: 'Such is the power of language: that which is woven of space elicits space, gives itself space through an originary opening and removes space to take it back into language.' In this short piece we can see some of the same concerns that animated him in *The Order of Things*, namely the process of representation; as well as in a range of other more 'literary' texts from this period (see, for example, those collected in Foucault 1977a).

'Force of Flight' was originally published in 1973 to accompany a series of paintings by Paul Rebeyrolle (1926–2005). Rebeyrolle's work is little known outside France, but is reminiscent of that of Francis Bacon in some of its style, although with a pronounced emphasis on texture and the incorporation of objects in a blend between collage and painting. Here Foucault offers some abbreviated comments on space as power:

> In the world of prisons, as in the world of dogs ('lying down' and 'upright'), the vertical is not one of the dimensions of space, it is the dimension of power.

It dominates, rises up, threatens and flattens; an enormous pyramid of buildings, above and below; orders barked out from up high and down low; you are forbidden to sleep by day, to be up at night, stood up straight in front of the guards, to attention in front of the governor; crumpled by blows in the dungeon, or strapped to the restraining bed for having not wanted to go to sleep in front of the warders; and, finally, hanging oneself with a clear conscience, the only means of escaping the full length of one's enclosure, the only way of dying upright.

But finally the dogs in the series of paintings escape through a force of flight, a force which is not present in the paintings themselves as a representation, but which 'produces itself unspeakably *between* two canvases' will have 'left you alone in the prison where you find yourself now enclosed, high on the passing of this force which is now already far from you and whose traces you no longer see before you – the traces of one who "saves oneself"'. Spaces of captivity can be reversed: the 'inside outside'.

These four new translations are partnered by one reprinted piece, the previously mentioned interview 'Questions on Geography'. Given its importance to the posing of the encounter between Foucault and *Hérodote*, and the fact it was one of the texts from the 1980 *Power/Knowledge* collection that was not reprinted in the final volume of the *Essential Works*, devoted to 'Power', it seemed appropriate to reprint it here.

Development

Foucault ends his original *Hérodote* interview with a comment that is much cited by geographers (e.g., see Harvey and Philo, this volume): 'Geography must indeed necessarily lie at the heart of my concerns' (Foucault 1980, 77). Was Foucault just being polite? He gave many interviews throughout his life and no doubt was often asked to consider the special interests of the interlocutor. Perhaps his comment is no more than such an acknowledgement and, in fact, Foucault had just remarked on this issue himself:

> I have enjoyed this discussion with you because I've changed my mind since we started. I must admit I thought you were demanding a place for geography like those teachers who protest when an education reform is proposed, because the number of hours of natural sciences or music is being cut. So I thought 'It's nice of them to ask me to do their archaeology, but after all, why can't they do it themselves?' I didn't see the point of your objection. Now I can see that the problems you put to me about geography are crucial ones for me. (Foucault 1980, 77)

It is true in one sense that, with some signal exceptions in short works and interviews, Foucault did not write primary texts that foreground spatial concerns, but it is also true that spatiality was more than just a passing interest. It is notable for instance, as Harvey discusses in this volume, that Foucault's introduction to Kant's *Anthropology* (Foucault 1960) does not mention the latter's *Physische Geographie*, which like the *Anthropology* is based on lecture materials. We therefore feel strongly that our aim here is not to show that somehow space is the hidden explanatory lens through which Foucault's work must be seen, but rather that it is a factor, at times only in the margins or in the background, that can be found throughout much of his work.

For Foucault, space, knowledge and power were necessarily related, as he stated, 'it is somewhat arbitrary to try to dissociate the effective practice of freedom by people, the practice of social relations, and the spatial distributions in which they find themselves. If they are separated, they become impossible to understand' (Foucault 1984, 246). In many places in Foucault then, spatiality occurs as an integral part of a larger concern, and as one of us has argued extensively elsewhere, a tool of analysis rather than merely an object of it (Elden 2001).

The chapters in the last part of this book aim to develop and extend these concerns. Is Foucault's work coherent in terms of its geographical explorations? Is it possible, or necessary, for a group of commentators to speak through Foucault to each other and find common ground? The reader will certainly find in these eight chapters a number of approaches, not all of them commensurable with each other. If there is a thematic throughout this section however, it is that most of the pieces concentrate on the mid to later work, including the topics of governmentality (Huxley, Crampton, Legg, Philo), health and sexuality (Kearns and Howell), and surveillance (Wood). In our call for chapters and specific invitations we placed no time periods on possible topics and it is interesting to note this clustering. It seems to us that one reason that authors made these choices is that the mid and later period is the one most obviously seen as the political period in Foucault's work. Of these chapters Coleman and Agnew's, which engages with Gramsci, Hardt and Negri and the problem of empire, is only the most explicitly political. What these chapters share, then, is an interest in the way institutions and political rationalities 'thought out space' (Foucault 1984, 244).

This focus on the political is at first sight a puzzling one. While Foucault is rightly seen variously as a historian, philosopher, critic and activist, he has also often been characterized as a political defeatist because he did not offer foundational grounds for taking positions, or (even more problematically) that the grounds he did offer were irredeemably defeatist (in that power is seen as everywhere). This assessment has often drawn the ire of commentators interested in a more definitive – not to say Marxian – political position, as represented by the contributions of Harvey and perhaps Thrift in this volume.

We would argue forcefully that Foucault's politics is not one of defeatism: a claim that has been made before, but which bears repeating. While liberation and freedom are both desirable and achievable, they will not come about with the mere passing of certain laws or by the guarantee of rights. Rather, freedom is a practice or a process that has to be constantly undertaken. This theme runs through both the collective work on populations and the mode of their governing, and at the individual level in his later work on practices of the self. To take a particularly 'spatial' example, when Paul Rabinow interviewed him for an architecture journal in 1982 Foucault remarked:

I do not think it possible to say that one thing [architectural project] is of the order of 'liberation' and another is of the order of 'oppression.' There are a certain number of things that one can say with some certainly about a concentration camp to the effect that it is not an instrument of liberation, but one should still take into account – and this is not generally acknowledged – that, aside, from torture and execution, which preclude any resistance, no matter how terrifying a given system may be, there always remain the possibilities of resistance, disobedience, and oppositional groupings ... liberty is a practice. (Foucault 1984, 245 emphasis added)

He then made the point in more explicitly general terms:

The liberty of men is never assured by the institutions and laws that are intended to guarantee them. That is why almost all of these laws and institutions are quite capable of being turned around ... which is not to say that, after all, one may as well leave people in slums, thinking that they can simply exercise their rights there. (Foucault 1984, 245–6)

The chapters in this part develop many of these points as well as covering the major themes with which Foucault is associated: surveillance, government, power, knowledge, sexuality and mental health. The chapters bring together a geographical sensibility to these familiar questions and represent the first sustained encounter between Foucault and geography. Additionally the chapters bring in some of the concerns only just now emerging from Foucault's new translations, such as race, bio-power, and territory and populations. Throughout this part the authors do not hesitate to identify latencies and undeveloped analyses (either in Foucault's work or in that of geographers) or productive ways of going 'beyond' Foucault.

Whereas the Foucault of the early to middle periods is often seen as concentrating on dystopian aspects; surveillance and prisons for example (although see Wood's chapter, where he develops a 'post-Panoptic' critique of surveillance through actor-network theory), other contributors to this book concentrate on the overlapping period of middle to later work that emphasizes the productive side of power. Using 'diagrams' or examples Margo Huxley addresses the way that space and environment are imbricated as rationalities of government. Using the formulation of government as *le conduire des conduits* (conduct of conducts) she indicates a wide remit of government from bodily discipline to whole populations; which have nevertheless not been posed as 'substantive geographies of government' until the mid-1990s (see also Hannah 2000). These geographies are not just arranged supine for surveillance and control; rather subjectivity is 'fostered through the positive, catalytic qualities of space, places and environments' (Huxley, this volume). She identifies three such qualities (dispositional, generative and vitalist, mapping roughly onto geographies of arrangement, health, and bio-politics) that act to produce such positive geographies.

Taking up these latter two themes in more detail, Gerry Kearns similarly focuses on the shaping of subjectivity with reference to medical geography. Historically, Kearns identifies two styles of medical geography, one focused around the environment and one around spatial science. Although (or perhaps because) these histories are fragmented, they can be analyzed through three of Foucault's historical tactics: discourse/practice; subjectivation; and the politics of medical history writing. In this he is concerned to apply these historical developments to contemporary medical geographies such as AIDS. In the colonial context for example (as noted in Stoler 1995) the empire was brought back home in the sense that the dangers of the environment and the 'race experience' were not just felt overseas, but also domestically (see also Legg, this volume). A stark reminder of this is provided in the literature on the geography of AIDS. A 'discourse of blame' is constructed around an imaginary environmentalism of disease, the tropics, and race. The question then becomes one of risk groups rather than risk behaviours: the 'dangerous individual' (Foucault 1988).

Race figures again in the contribution by Crampton. He shows the way mapping 'thought out space' (Foucault 1984, 244) in the peace negotiations after World War I. 'Sites' of geographical knowledge were established to redraw Europe's borders, but Crampton traces out the previously unknown linkages to eugenics and the goal of racial partitioning. Here maps were vital in the geographic imaginary. Eugenic scientists not only contributed reams of ethnographic data to the negotiations, but also used their affiliation to push for immigration reform to exclude or sterilize the 'degenerate' stock made visible in the cartography.

One of the key themes for readers of Foucault has always been his groundbreaking analyses on surveillance and its role in the generation of knowledge. Whether it is knowledge of populations, medicine or maps, surveillance has many spatial connotations. As Wood shows, these need not be irredeemably unproductive. The disciplinary gaze attempted to ward off plague, abnormality and disorder, and its success in doing so helped inoculate it from critique. Reading through the large literature, Wood explores interpretations beyond *Discipline and Punish*, notably drawing on the work of Deleuze and actor-network theory. If the Panopticon is not the only model of power or even of surveillance, Wood argues that ANT is the only comprehensive attempt to develop a post-Foucauldian understanding of power.

This post-Foucauldian theme is expanded by Legg, who calls Foucault's silence on colonialism 'astounding'. Undoing this silence, suggests Legg, means going beyond Foucault to 'counter-discourses of modernity' that do not just focus on the imperial centre. Legg traces the positioning of geographical research in the relations between Foucault and Edward Said. For Said, Foucault lacked 'political bite' and failed to consider other spaces of existence beyond the present (Legg this volume, 271). Yet perhaps surprisingly, Said circled back to Foucault's geographical work on control of territories, not just in discourse but in material practices of government. To achieve a new emphasis beyond the imperial centre then, would suggest taking seriously Foucault's remarks on the 'insurrection of subjugated knowledges' across different scales, and that occur materially as well as discursively.

In analyzing Foucault's work on sexuality, Howell firstly clears away a number of mistaken impressions in geography and other disciplines regarding such works as the *History of Sexuality*. The lack of geographical engagement with this work – and the biased focus on power rather than sexuality – is for Howell a serious analytic oversight. Spatial orderings around sexuality or sex work such as prostitution (Howell 2004), for instance, are particularly neglected by geographers, although it is interesting to speculate how this would have been affected by Foucault's proposed book on *Woman, Mother, and Hysteric* (see Elden, this volume; 2005). Spaces of sexual penitence for prostitutes (Howell discusses the Magdalen Hospital) are particularly suggestive here as a 'site' of sexuality. Others include the family and the 'closet'. These are more than spatial metaphors but speak to the spatialization of the body, whether travelling from the suburbs to the city for sex, or the surveillance of the child's bedroom (see Brown 2000). Here Howell's appeal to biopolitics – in this instance a 'geopolitics of sexuality' – has much in common with other contributors' work emphasizing the health of the *population* (e.g., Legg on population geography, Legg 2005), Kearns on medical geographies or Crampton on the mapping of race in this volume).

Geopolitics reappears in the next chapter, in which Coleman and Agnew take on the post 9/11 'climate of fear' and the role of the United States as a (neo)imperial power. Using Hardt and Negri's *Empire* as a jumping off point (Hardt and Negri 2000) Coleman and Agnew reject its binaries of power as either centred on states or decentred in networks to argue for its re-territorialization. Interpreting Foucault as a philosopher rather than a historian – or at least as a non-periodizing historian, Coleman and Agnew argue that contra *Empire* power is more manifest in space than periodization in time. That is, power does not get exercised over 'undifferentiated blocks of subjects fixed in absolute spaces' but as a series of overlapping and discontinuous spatialities of power. Here Coleman and Agnew once again draw on Foucault's work since the mid-1970s on governmentality of populations, not as a replacement but as a supplement to disciplinary power.

If Coleman and Agnew find inspiration in this work, it falls to Philo to articulate it in the most detail. Here Philo shifts his long engagement with Foucault's work on madness and hospitals (Philo 1989; 2004) to perhaps the most political book, *Society Must be Defended*. If this book centralizes biopower, the population and the end of sovereignty, for Philo it also connects back to archaeologies of discourse and knowledge. The 'bellicose history' found therein is deeply predicated on 'the local' especially local knowledges and their 'subjugation'. It is to war and politics that Foucault looks for an elaboration of power for it is through the challenge of these counter knowledges and their local settings that alternatives can emerge. Philo underlines a key point of the book that discourses and knowledges, including histories of rightful territorial possession, battle it out to be accepted and that this results in 'an uneven geography' of knowledge about society.

The chapters in this part share a goal of going 'beyond' Foucault while at the same time seeing in a lot of his work (government of populations and territory for example) much of potential interest to geographers. While some authors offer close readings of Foucault, others use it as a jumping off point. Some of the contributors to this book have long engaged his work; for others the encounter is relatively recent. But if Foucault's questions back to geographers reveal his interests in exploring the context of strategy, tactics and conflict, then the 'developments' included here show something of geographers' own potential deployments of Foucault. Just as theory, for Foucault, is not something separate from practice, but rather a practice itself, so too is the process of critique an inherently practical tool, a mode of engaging in struggle.

Conclusion

The questions posed by Foucault to geographers are ones that spring from his own concern about issues of space, and are indicative of his thinking – both for what he asked and for what he did not pose or overlooked. This book therefore aims to cast light on Foucault's own thinking about space, power, knowledge, governmentality, and war. While Foucault may appear to be an extraordinarily familiar figure within the intellectual landscape, much remains to be known about his work. This includes the Collège de France lectures that are discussed in many of the chapters included here, but also supposedly well-known writings, and material unavailable in English.

The purpose of this collection is therefore to suggest that Foucault's position in relation to geography remains unclear. Where he does not just remain a forbidding thinker in the distance, he is too often used in an emblematic and superficial manner. The contributors to this book, although writing from a diverse series of perspectives and interests, show that it is fruitful to establish a critical encounter with Foucault's work. Such an encounter does not consist in uncritically applying Foucault's work (which itself is diverse) to close off dialogue, but rather to encounter it in a way that deliberately opens up questions, possibilities and reappraisals. It is therefore the aim of this book to go beyond the simple appropriation of Foucauldian terms such as 'heterotopia' and 'disciplinary space' and to put these terms in their larger and proper context. Foucault's work was remarkably informed by the spatial problematic at many stages of his career. Although he may not have wanted or needed to bring this problematic explicitly to the surface at all times it runs deeply. The aim is both to contextualize Foucault's work within geography and to use it as a springboard to explore these issues of spatiality in detail. As such it demonstrates not just that the relation between space, knowledge and power works in a number of ways, but equally that the relation between Foucault and geography is one that works both ways:

> These are not questions that I pose to you from the position of a knowledge that I already have. They are inquiries that I am asking myself, that I address to you, thinking that you are probably more advanced than me on this path. (Foucault 1994, III, 9; this volume, 19)

As Foucault suggested in the original interview with Hérodote, if geographers could make use of some of the 'gadgets' ['trucs'] of approach or method he had used, then he would be delighted. But, he tells them, 'if you find the need to transform my tools or use others then show what they are, because it may be of benefit to me' (Foucault 1994, III, 30). Geographers have, still, much to learn from Foucault. But as he himself acknowledged, his work, and today that of Foucault studies generally, have much to learn from geographers.

References

Brown, Michael (2000) *Closet Space: Geographies of Metaphor from the Body to the Globe*. New York: Routledge.

Cloke, Paul, Chris Philo and David Sadler (1991) *Approaching Human Geography. An Introduction to Contemporary Theoretical Debates*. London: Paul Chapman Publishing.

Crampton, Jeremy W. (2004) 'GIS and Geographic Governance: Reconstructing the Choropleth Map.' *Cartographica*, 39(1), 41–53.

Driver, Felix (1985) 'Power, Space and the Body: A Critical Assessment of Foucault's Discipline and Punish.' *Environment & Planning D: Society and Space*, 3, 425–46.

Elden, Stuart (2001) *Mapping the Present: Heidegger, Foucault, and the Project of a Spatial History*. New York: Continuum.

Elden, Stuart (2005) 'The Problem of Confession: The Productive Failure of Foucault's History of Sexuality.' *Journal for Cultural Research*, 9(1), 23–41.

Elden, Stuart (2007) 'Governmentality, Calculation, Territory.' *Environment and Planning D: Society and Space*, 25, 4.

Farge, Arlette and Michel Foucault (1982) *Le Désordre Des Familles: Lettres De Cachet Des Archives De La Bastille*. Paris: Julliard.

Flynn, Thomas R. (1997) *Sartre, Foucault, and Historical Reason, Vol. I: Toward an Existentialist Theory of History*. Chicago: University of Chicago Press.

Flynn, Thomas R. (2005) *Sartre, Foucault, and Historical Reason, Vol. II: A Poststructuralist Mapping of History*. Chicago: University of Chicago Press.

Foucault, Michel (1960) *Thèse Complémentaire Pour Le Doctorat Ès Letters: Introduction À L'anthropologie De Kant*. Unpublished manuscript.

Foucault, Michel (1967) *Madness and Civilization*. London: Routledge.

Foucault, Michel (1977a) *Language, Counter-Memory, Practice: Selected Essays and Interviews*. Edited by D.F. Bouchard. Trans. Donald F. Bouchard and Sherry Simon. Ithaca: Cornell University Press.

Foucault, Michel (1977b) *Politiques De L'habitat (1800–1850)*. CORDA.

Foucault, Michel (1980) *Power/Knowledge: Selected Interviews and Other Writings, 1972–1977*. Edited by Colin Gordon. New York: Pantheon Books.

Foucault, Michel (1984) *Space, Knowledge, and Power*. In Paul Rabinow (Ed.) *The Foucault Reader*. New York: Pantheon. 239–56.

Foucault, Michel (1986) 'Of Other Spaces.' *Diacritics*, Spring, 22–27.

Foucault, Michel (1988) 'The Dangerous Individual.' In Lawrence D. Kritzman (Ed.) *Politics, Philosophy, Culture: Interviews and Other Writings of Michel Foucault, 1977–1984*. New York: Routledge. 125–51.

Foucault, Michel (1991) 'Governmentality.' In C. Gordon, G. Burchell and P. Miller (Eds) *The Foucault Effect: Studies in Governmentality.* Chicago: University of Chicago Press. 87–104.

Foucault, Michel (1994) *Dits et écrits: 1954–1975.* Paris: Éditions Gallimard.

Foucault, Michel (1999) 'About the Beginning of the Hermeneutics of the Self.' In Jeremy Carrette (Ed.) *Religion and Culture.* New York: Routledge. 158–81.

Foucault, Michel (2000) *Power: The Essential Works of Michel Foucault 1954–1984 Volume Two.* Edited by James Faubion. Trans. Robert Hurley. London: Allen Lane.

Foucault, Michel (2003) *Society Must Be Defended: Lectures at the Collège De France, 1975–76.* New York: Picador.

Foucault, Michel·(2004a) 'Crisis of Medicine or Anti-Medicine?' *Foucault Studies*, 1(1), 5–19.

Foucault, Michel (2004b) *Sécurité, Territoire, Population: Cours Au Collège De France 1977–78.* Paris: Gallimard-Seuil.

Foucault, Michel (2005) *History of Madness.* Edited by Jean Khalfa. Trans. Jonathan Murphy.

Foucault, Michel (2006) *Psychiatric Power. Lectures at the Collège De France, 1973–74.* New York: Palgrave Macmillan.

Foucault, Michel, Blandine Barrett Kriegel, Anne Thalamy, François Beguin and Bruno Fortier (1979) *Les machines à guérir (aux origines de l'hôpital moderne).* Bruxelles: Pierre Mardaga.

Gregory, Derek (1994) *Geographical Imaginations.* Cambridge, MA and Oxford, UK: Blackwell.

Hannah, Matthew (2000) *Governmentality and the Mastery of Territory in Nineteenth-Century America.* Cambridge: Cambridge University Press.

Hardt, Michael and Antonio Negri (2000) *Empire.* Cambridge, MA: Harvard University Press.

Harvey, David (1973) *Social Justice and the City.* Baltimore: Johns Hopkins University Press.

Harvey, David (2001) *Spaces of Capital: Towards a Critical Geography.* New York: Routledge.

Hetherington, Kevin (1997) *The Badlands of Modernity: Heterotopia and Social Ordering.* London and New York: Routledge.

Holden, Adam and Stuart Elden (2005) '"It Cannot Be a Real Person, a Concrete Individual": Althusser and Foucault on Machiavelli's Political Technique.' *Borderlands*, 4(2), e-journal.

Howell, Phil (2004) 'Race, Space and the Regulation of Prostitution in Colonial Hong Kong: Colonial Discipline/Imperial Governmentality.' *Urban History*, 31(2), 229–48.

Legg, S. (2005) 'Foucault's Population Geographies: Classifications, Biopolitics and Governmental Spaces.' *Population, Space and Place*, 11(3), 137–56.

Mills, Sara (2003) *Michel Foucault.* London: Routledge.

Mills, Sara (2005) *Gender and Colonial Space*. Manchester: Manchester University Press.

Peet, Richard (2000) 'Celebrating Thirty Years of Radical Geography.' *Environment and Planning A*, 32(6), 951–3.

Philo, Chris (1986) '"The Same and the Other": On Geographies, Madness and Outsiders.' Loughborough University of Technology, Department of Geography, Occasional Paper (11).

Philo, Chris (1989) 'Enough to Drive One Mad: The Organisation of Space in Nineteenth-Century Lunatic Asylums.' In J. Wolch and Michael Dear (Eds) *The Power of Geography*. London: Unwin Hyman. 258–90.

Philo, Chris (1992) 'Foucault's Geography.' *Environment & Planning D: Society and Space*, 10, 137–61.

Philo, Chris (2004) *A Geographical History of Institutional Provision for the Insane from Medieval Times to the 1860s in England and Wales: The Space Reserved for Insanity*. Lewiston, Queenston and Lampeter: Edwin Mellen.

Pickles, John (1988) 'Knowledge, Theory, and Practice: The Role of Practical Reason in Geographical Theory.' In Reginald G. Golledge, Helen Couclelis and Peter Gould (Eds) *A Ground for Common Search*. Santa Barbara: The Santa Barbara Geographical Press. 72–90.

Pickles, John (2004) *A History of Spaces. Cartographic Reason, Mapping and the Geo-Coded World*. London: Routledge.

Soja, Edward W. (1989) *Postmodern Geographies: The Reassertion of Space in Critical Social Theory*. London: New York.

Stoler, Anne L. (1995) *Race and the Education of Desire: Foucault's 'History of Sexuality' and the Colonial Order of Things*. Durham, NC: Duke University Press.

Thrift, Nigel (2006) *Non-Representational Theories*. London: Routledge.

PART 1
Questions

Chapter 1

Some Questions from Michel Foucault to *Hérodote*

Michel Foucault, Translated by Stuart Elden

'Des questions de Michel Foucault à «*Hérodote*».' *Hérodote*, No. 3, juillet-septembre 1976, 9–10; reprinted in *Dits et écrits 1954–1988*, edited by Daniel Defert and François Ewald. Paris: Gallimard, Four Volumes, 1994, Vol. III, 94–5.

These are not questions that I pose to you from any knowledge that I might have. They are inquiries that I am asking myself, that I address to you, thinking that you are without doubt more advanced than me on this path.

1. The notion of strategy is essential if one wants to make an analysis of knowledge [*savoir*] and its relations with power. Does it necessarily imply that through the knowledge in question one wages war?
   Does strategy not allow the analysis of relations of power as techniques of *domination*?
   Or must we say that domination is only a continued form of war?
   Or alternatively, what scope would you give to the notion of *strategy*?
2. If I understand you correctly, you are aiming to constitute a knowledge of spaces [*un savoir des espaces*]. Is it important for you to constitute it as a science?
   Or do you find it acceptable to say that the break which marks the threshold of science is only a means of disqualifying certain knowledges [*savoirs*], or to make them evade examination.
   Is the division between science and non-scientific knowledge [*savoir*] an effect of power linked to the institutionalization of knowledges [*connaissances*] in the University, research centres, etc.?
3. It seems to me that you link the analysis of space or of spaces less to production and to 'resources' than to the exercise of power.
   Could you outline what you understand by power? (Through relation to the State and its apparatuses, through relation to class domination.)
   Or do you consider that the analysis of power, of its mechanisms, of its field of action, is still at the outset and that it is too soon to give general definitions?
   In particular, do you think one can reply to the question: who has power?

4. Do you think it is possible to undertake a geography – or a range of geographies – of medicine (not of illnesses, but of medical establishments along with their zone of intervention and their modality of action)?

PART 2
Francophone Responses – 1977

Chapter 2

Hérodote Editorial

Translated by Gerald Moore

'Éditorial.' *Hérodote*, No. 6, 2ᵉ trimester 1977, 3–4.

Having agreed to answer questions on geography, Michel Foucault has posed a set of questions to geographers.

Since the core of Michel Foucault's questions essentially bears on the problem of power, of domination, the responses could not be unanimous, nor could they seek to be collective.

These interrogations of power, especially of the ubiquitous vision that Foucault increasingly accords it, obviously does not only concern geography, but social practices as a whole and representations that have been made of them. The replies from geographers are therefore not specifically 'geographical', and they correspond to the idea that each one of them has, not of geography, but of society as a whole. Problems of geography have thus to a certain extent been eluded, both by Michel Foucault and by a good many of those who have endeavoured to respond to him.

When Michel Foucault asks: 'Can you outline what you understand by power?' and 'Who has power?', we think that there is no single answer, but different types of response, depending on the scale of social space that one takes into consideration: the response differs according to whether one takes planetary space (a very small scale) into consideration – in which case it concerns the role of the two superpowers and very large transnational companies; or whether one envisages a very large-scale spatial organization of the family home and the relation of power between individuals. Understanding the problem of power by systematically distinguishing between different spatial scales and different levels of analysis enables us to avoid conflating very different, but nonetheless mutually articulating, structures of power into a fluid whole and even a ubiquitous presence.

When Michel Foucault asks: 'What scope would you give to the notion of strategy?', it is here that we are more critical in respect to his whole discourse, because it tends to use the same term, strategy, to designate, on one hand, plans that are deliberate, conscious, organized, devised to attain certain objectives or defeat an adversary, by choosing means and ruses and considering the configuration of 'terrain'; on the other, dilute and unconscious tendencies, procedures in which the whole of society participates, without realizing, and which produce involuntary effects, with neither winners nor losers.

When, in his latest book, *The Will to Knowledge* (which is a crucial piece of reflection for all of us), Michel Foucault shows in substance that what he calls the 'strategy of power' proceeds not only via prohibitions regarding the essential problem of sexuality, but also – and much more – through incitements to speak about 'it', to think about 'it', this 'strategy' that he reveals is precisely one that is unconscious and involuntary, as much for those who exercise it (where are they? Everywhere) as for those who are subjected to it (who are they? All of us). It comes down to involuntary apparatuses [*dispositifs*] and unconscious collective propensities. It is essential to realize this, but [not] by saying: 'It happens as if there were a strategy and gamesmasters.' It is precisely because there is not, because there is not a conscious, deliberate strategy opposing specific adversaries for clearly perceived stakes that this tendency is so strong at the heart of our society.

When we speak of strategy and tactics, it is clearly not about these unconscious apparatuses [*dispositifs*], these collective propensities that we are thinking, but about plans, secretly or discretely constructed, devised by one of the protagonists in a relation of force, plans that take account not only of the means and characteristics of the adversary, and of the other strategy that he, too, could put to work, but also of the configuration of the 'terrain' (of topography on various scales of social space) and the relative positions that the forces present occupy. It is for this that knowing-how-to-think-space [*le savoir-penser-l'espace*] has so great an importance in all strategic reasoning.

Chapter 3

Response: Jean-Michel Brabant

Translated by Gerald Moore

Hérodote, No. 6, 2ᵉ trimester 1977, 12–14.

The Power of Scale

The notion of strategy is applicable, in current vocabulary, to a range of terms. Where it concerns us, we hold to the fact that every strategy implies a plan worked out in relation to an enemy, be it real or imaginary, concrete or potential.

The strategy with which we are occupied is that which corresponds to a practice of the domination of space, in all its forms.

Thinking about and organizing space is one of the pre-occupations of power. If every strategy of power has a spatial dimension, power also has a practice of spatial domination that is appropriate to its strategy.

This practice of spatial domination cannot be totally identified with military practice. The latter is only one aspect, one that is perhaps institutionally concentrated, of the spatial practice of power. It is situated on the plane of 'knowing-how-to-think-[the]-space [*le savoir-penser-l'espace*]' of a scarcely defined power. What characterizes power is the way that its internal complexity goes hand in hand with a multiform intervention on the plane of space.

On the stage of confined territories, weak or fragmented authorities, 'knowing-how-to-think-space' boils down to knowing how to think war. Ruling substrata are reflected on a grand-scale, just as the chief of a stronghold is essentially preoccupied with the topography of the reduced space that he is charged with defending, and not the strategic data amidst which he is situated.

When power is capable of reasoning on a smaller scale, its strategic knowledge diversifies. This is without doubt true above all of State power, where war or the threat of war is no longer the only means of extending or maintaining its hegemony over a given space. The rise of a range of forces, particularly in the sphere of economics, is based on a comprehension of the play of spaces. Developed on a small scale (we should define more rigorously the level of analysis privileged by different advisors), their strategy is often perceived only on a grand scale (or more simply, on another scale), which obscures its significance.

To decode the spatial practice of different powers is to reveal their social strategy in terms of space, it is to clarify the underlying mechanisms of the force of those who dominate, and the weakness of those who are dominated.

Strategy (as the knowledge/practice of space) can serve to subvert power itself. This knowledge/practice cannot be neutral and, if it is to be used, it must be reinvented. The evidencing of this 'knowing-how-to-think-[the]-space' of power must enable us to found, with the struggling masses, a new and efficient spatial practice.

Science and Ideology

The recognition of the scientific status of certain branches of knowledge is without doubt a way of turning these branches of knowledge into a hierarchy, which is linked to power-status and a social consensus, as well as being linked to a necessary disposition toward internal rigour.

The place of geography in this process is, without doubt, novel. Linked to power as strategic knowledge, geography is simultaneously depoliticized and 'scientified'. The establishment of this scientific status, essentially through the institution of the university, has moved geography from the domain of strategic knowledge to the rank of accessory to the ideological arsenal of power. This passage has been reinforced (in the case of French geography) by the internal epistemological evolution that, while privileging the science of places and not that of men, has refused all 'knowledge of spaces'.

This 'knowledge of spaces', obscured but in part nurtured by the 'science' of geography is, before all else, a practice at the level of power and its expert advisors. The present problem consists neither in criticizing geography on the basis of its internal epistemology, nor in putting in place a new science of space (a new or renovated geography).

Rather than placing ourselves on the level of scientific debate, it suits us to decrypt a knowledge that operates on reality, and which one can attempt to grasp at the level of practice.

The State

The notion of power must always be brought back to one's approach to the social organization of which it is the principal organizer. To avoid this reference or to skirt around this reality is to expose oneself to mistakes in analysis and to bracket the same words under different notions. The power that preoccupies us is that to which we are presently subjected in our society, not an abstract, atemporal power.

Knowledge of this power, its delimitation and the evaluation of its techniques of domination does not come down to an exclusively spatial approach. This essentially hierarchical power is identified with the power of the State, the guarantor and summit of this hierarchy. It is the armed branch of social organization, and this connotation is one of the essential objectives of our study. To map power is first to map the power of the State in all its levels [*échelons*], to define its different types of domination of space, to detect its areas of weakness and contradictions. This should be the goal of the 'knowledge of spaces' for which we are fighting.

This hierarchical and concentrated power expands and reproduces itself in various agencies of society. This hierarchization and concentration is the work of special interests who through their practice establish veritable networks of powers in which particularly dangerous zones are partitioned. From this perspective, the process of production must be at the centre of our knowledge/practice of space, because at the level of power, this knowledge and this practice are thought as a function of power.

Rather than just enumerating productions and resources – although it is not a question of under-estimating the importance of the data that power sometimes conceals – it seems to us more pertinent to situate the strategic place of these elements in space, or in the combined play of different spaces of power. The superimposition of maps of data and networks of power would certainly give us a few clues as to the space at stake in social conflicts.

Should our preoccupation limit itself, however, to the critical elucidation of present power in its mechanisms and the delimitations of its different aspects? This criticism has the function of grasping and orientating the spatial resonances of the struggles of those whom power oppresses.

In the analysis of the gestation of popular counter-powers that necessarily but not exclusively turn around the power of the State, the definition of revolutionary power as a network of power taken up and subjected to the control of different agents of the social process is at stake.

The question of knowing who has power, if we must perhaps initially try to reply to it as it is posed, is therefore foremost a question of knowing who power serves.

Chapter 4

Response: Alain Joxe

Translated by Gerald Moore

Hérodote, No. 6, 2ᵉ trimester 1977, 14–15.

Strategy: the art of making decisions conforming to the defence of an interest by taking into account a system of opposed interests and the possibilities of the decisions and defence of these other interests. (A definition 'approximating' that of game theory, rather than that of military strategy, where the notion of time is introduced straight away.)

This generalization of the notion of strategy is a semantic fact. One might deplore it, in that it ultimately means anything that one wants it to mean, starting from the moment one considers man as thinking and acting, resulting in: 'amorous strategy', the strategy of Saint-Etienne football team, 'economic strategy', government strategy, business strategy, the strategy of the board, etc.

Obviously, strategy does not mean war. Nor necessarily even conflict, but always the power of decision, which is to say power. The strategy of a leader who seeks to remain legitimate is not that of confrontation, or at least not only that. In any case – following the terminology of the games of strategy – we generally distinguish between the aspects of cooperation and struggle, the carrot and the stick, promise and threat, like the two faces of Janus, who marks a *threshold* and not a space. It is rather that strategy is the art of thinking the threshold of the passage to the act [*passage à l'acte*?]. It is an art because strategy cannot give itself as an object to be explained by the artist: it must be presupposed. One can think only the strategy of '*x*' ... and this '*x*' escapes the object of the strategic study in question, because one can only *strategically* draw the distinction between the *interest* and the *person* who has the power to make decisions. (These terms are synonymous with game theory.) Every distinction drawn between interest and person (for example, the Marxist analysis that distinguishes between the bourgeois *class* and the bourgeois *parties*) can only have strategic sense if one is equally to establish that there exist contradictions between the bourgeois class and party, and that there are accordingly two interests present.

There are thus extreme limitations to the reflection on strategy, insofar as it is not applicable to a perfectly disciplined organization whose interest is defined as a mission; one could say that it is the goal of armies to build up strategic tools, which is to say means, defined as tightly as possible, that have a monopoly over the decision to employ violence, and which tend to reduce the reality of social conflicts to matrices of calculus.

Chapter 5

Response: Jean-Bernard Racine and Claude Raffestin

Translated by Gerald Moore

Hérodote, No. 6, 2ᵉ trimester 1977, 17–19.

Geography as a Tool of Domination

We don't believe that the notion of strategy, as we have defined it, necessarily implies that through knowledge one makes war. The link with the concept of domination, on the other hand, seems to us more evident, more grounded. In geography, the knowledge linked to the 'scientific' analysis of central places has been transformed into strategy and, very precisely, into a technique of dominating and even occupying economic, political and geographic space. Is this knowledge theoretically grounded? Apparently, geographers of the new school that has come from the works of Walter Christaller have hardly doubted this, even if each one of them today recognizes that the hierarchical organization of urban groups is only one form amongst many others of modalities of relations between towns. Just as, in terms of intra-urban structure, one recognizes that the centre of theocratic towns (exactly like the 'city' in current metropolises) are forms linked to the differentiation and hierarchy of social orders totally different from those that, for example, enabled the birth of the *agora* in the Greek *polis*. If the *agora* speaks of homogeneity and equality in place of differentiation and hierarchy, it is because, for the historian of towns, 'scientific' knowledge [*savoir*] is nothing other than the knowledge [*connaissance*] of that which conforms to the logic of the development of a given society.

How will this knowledge be used? The dominant class can content itself to await the manifestations of a more or less spontaneous revolution, the biggest challenge in the system. It can also, knowing of this revolution scientifically, seek actively to utilize it. The application of the criteria of minimal energy to the division of a surface (where surfaces must be 'efficiently' divided between concurrent centres) in fact leads to cutting space up into regular polygons, with hexagons enabling the best paving of a surface, minimizing the costs of movement and limitations. It is thus that the Third Reich decided to organize 'rationally' the distribution of market towns along the planes of conquered Poland. We know that, since this aborted attempt, explicitly founded on the works of Christaller, the examples of the use of this *strategy of domination* (military, agricultural, commercial, social) have multiplied. Should this problematic not, for all that, be rejected entirely?

Popular Knowledge and Theory

We know that the discovery of the hierarchical structuring of urban systems is historically linked to a quantitative and theoretical revolution. It was in fact in order to confront Christaller's theory of central places with reality, to make the passage between the world of theory and the empirical world in a manner that would leave one assured of the faithfulness of the results obtained, that men like Chauncy Harris, William Garrison and Brian Berry, geographers seeking general explanations in the image of Christaller himself, opted to recourse to quantitative analysis. Has the knowledge that they acquired been reappropriated by power, in the service of properly ideological ends? Is it a question of contesting the scientific, theoretical principal of minimal energy, or the use that has been made of the 'knowledge' enabled by an understanding of the principle? In this respect, two positions appear to us to be possible for those who speak from the left.

The first position consists in wanting to take no other guide, no other principle of action than the democratic control of the production of our space, the place of our ethics. The geographer no longer begins with 'science', whose discourse so frequently functions to nurture and make acceptable the dominant ideology, but with 'popular knowledge', whose expression he or she can seek to shape. This, evidently, is a way of constructing – by rooting oneself, if possible, in the masses (but what would this mean for universities?) – a counter-discourse of possible alternatives, an excellent means of inciting people to call for the democratic control of the production of their space, a control that, in the final analysis, is the sole criterion of truth. With some slight nuances, some of us seem to have made this choice. Others, however – and we are amongst them – hold onto what, in their practices, could be qualified as 'a necessary bit of positivism', or a 'minimal positivism'. Obviously attentive to the fact that, in their models, the signifier does not entertain a necessary and natural relation with the signified, but is always contingent, they nonetheless agree to differentiate between strictly empirical knowledge and theoretically grounded science. The fact of being situated in a diachronic perspective forbids them from forgetting that there is not, in space, a signified linked definitively to a signifier. On the contrary, the same signifier can connote multiple and sometimes even opposed signifieds. The urban centre of the mediaeval town connotes a zone of appropriation and collective participation, whereas present centres connote appropriation by economic power. And their researches lead them to the same perspective as those who chose the first position: the counter-discourse in which scientific *information* is used as needed to prevent the effects of the dominant ideology. And why not, rightly, also the principles that underlie the theory of central places?

Democratic Control of the Production of Space

How better to ensure the democratic control of the production of space than to show the scientific possibility of a division of political space into a series of units of different sizes, in accordance with hierarchical levels of their own specific functions, following the principle put forth by Christaller and threshold theory? Since 1970 David Harvey has proposed that we examine a territorial organization that, while being hierarchical in its nature, allows maximal local participation with access, in return, for each of the units under consideration, to services close to the optimal condition. William Bunge has followed through with this idea (Bunge and Bordessa 1975), and has proposed using Christaller's work no longer for economic, but for democratic ends; no longer as a support for the centralization of power, but for its decentralization. By starting with the smallest unit viable for the exercising of a political function (a commune of 200 inhabitants in Canada), he was thus able to construct a seven-layered hierarchy in search of a model that enabled him to assign governmental functions to each of the hierarchical levels. The threshold theory used with the aim of maximizing economic profit can very easily be substituted with a threshold theory analogous in its formulation and use of mathematics, but directed not toward profit so much as democracy and maximizing the validity of expression at the heart of the city, for example, or equally the heart of a region or nation. The energy expended in scientific work can allow us to find the threshold of a unit in an apportioning of space that maximizes information and relations. The knowledge that initially and traditionally was a strategic instrument of domination can well become an instrument of freedom. This is true to the extent that William Bunge had to risk his career for this type of 'turning back' of information against the dominant power, as much in Detroit, where dealing with the optimization of apportioning schools, as in Ottawa and Toronto, where researching, as one of the conditions of the survival of humanity, the possibility of a democratic restructuring of urban governments. The 'rejection' of his discourse by the dominant power tends to authenticate it. We are a long way, here, from 'geography that serves no purpose': ultimately, it unsettles and disturbs.

Reference

Bunge, William and Robert Bordessa (1975) *The Canadian Alternative. Survival, Expeditions and Urban Change.* Geographical Monographs No. 2. New York: New York University.

Chapter 6

Response: Michel Riou

Translated by Gerald Moore

Hérodote, No. 6, 2ᵉ trimester 1977, 21–3.

The questions that Michel Foucault poses, it seems to me, are of a nature that allows the expression of several essential principles, and also are highly capable of supporting a Marxist conception of geography. With regard to Marxist geography, this will be undertaken later, as an illustration, a deepening and an enriching. One could always ask Mr Jacques Lévy to have a go at these questions: that would prevent him from passing his time pretending not to understand you.[1] And I express myself on his subject all the more freely in that I, too, am a member of the Communist Party, a reader of *France nouvelle* and Marxist in every possible way.

Class Struggle as a Source of Knowledge

Michel Foùcault's first question, it seems to me, can be broken down in the following way: What is war? What is strategy?

For me, war is the enduring exercise of concentrated force. War, in the contemporary era, is only one of the forms of class struggle and capitalist competition. It is just the most spectacular and most murderous manifestation of the antagonisms at work in class societies. Knowledge, obviously including geographical knowledge, but also physical, chemical, mathematical and sociological knowledge, amongst others, is necessary for war. It is not by itself the source of knowledge, however. The source of knowledge resides in permanent, everyday, hard-fought class struggle. The majority of progress in the domain of the natural sciences has been brought about within the framework of capitalist competition. The majority of knowledges in the domain of the social sciences have been developed in order better to preserve the domination of one class over another, to avoid revolts and troubles of any sort. It is not only geography that is born of combat, it is knowledge as a whole. Through the specific form that it gives to antagonisms, war by its nature accelerates the progress

---

1   Editors' Note: Jacques Lévy (1952–) is director of Laboratoire Chôros in Lausanne. He is the author of *Europe: Une Geographie*. Paris: Hachette, 1997, and the editor of *From Geopolitics to Global Politics: A French Connection*. London: Frank Cass, 1991, among many other publications.

of knowledge, but it is not the only source of it. Class struggle expresses itself through geographical knowledge, even if war has contributed and still contributes powerfully to its constitution. This, of course, by no means signifies that war has no need of geography, whichever camp one happens to belong to!

The second part of the question deals with strategy. Being a man of politics only insofar as I am a citizen, this interests me less. It seems very clear to me, however, that strategy is the art of engaging in combat in favourable conditions. Besides the military, it interests trade unionists, politicians and intellectuals. In our own domain of the combat of ideas, it seems to me extremely important to know where, when and by whom such and such an aspect of the dominant ideology will be attacked. I am not, for all that, speaking of war. But it seems to me undeniable that there is combat, which is to say precise objectives, and tactics, adversaries who are defined as such, and victory and defeat. There is nonetheless a nuance of size to be respected: political struggle, under both its pacific and military forms, has in view the destruction of an adversary, insofar as they are an adversary. Strategy has the goal, in this case, of ensuring a provisional domination with a view to definitive elimination. Turning to intellectual struggle, this does not aim at the elimination of a source of ideas, the potential enriching that would constitute the adversary. Domination, which is to say an endlessly accrued audience for its thought, is enough. Intellectual struggle can only have strategic objectives in view. This is one of the reasons for which it cannot be decisive, and for which it can no longer achieve completion. The leading role of the working class in the domain of art and spirituality does not consist in imposing silence. It consists in taking its thoughts on its allies and adversaries a little further every day, starting from, and in, the framework of its own approach. But the process is dialectical: it is only to the extent that it listens and allows speech that the proletariat is capable of providing social reality with a sufficiently just and convincing analysis for itself to be heard and followed. Strategy is therefore a means of intellectual domination; but above all it is important to know exactly what one means by domination.

Who has Power?

One cannot reply to this in a non-dynamic way. Power demonstrates, reveals itself. To the profit of whoever brings about the evolution of a society? To the profit of whoever has power. And the definition of power is just that: it is the capacity in which a person, a class or an institution finds them- or itself able to make the whole social body evolve to their or its own profit. It goes without saying that political power is not always real power, and that in the final analysis, and for a sufficiently long period, power has belonged to who holds the fundamental means of production. I believe that this much can be drawn from any slightly serious historical study. With regard to the very nature of power, it lies quite simply in armed force; all the rest is only the premises, trappings, symbols and consequences of the possession of armed force. It seems to me that, since Marx, the question has been settled...

Geography as an Instrument of Liberation

Geography must be made a means of reading the global crisis of imperialism, capitalism and centralism in all its forms. Geography is the projection of history in space. It is not about becoming a sociologist, an ethnologist, a political scientist or similar. What defines a science is the questions that it poses. What defines geography, what distinguishes it from the other merchants of social science, is the spatial analysis that it practices. At the level of the landscape, the plan and the map. Social forces manifest themselves in space. They inscribe themselves on the landscape, the plan and the map.

Space is the place where history inscribes itself, and geography should be the analysis of that which dwells and is born there. The cost of this would be that geographers become what they should be: awakeners of consciousness, educators and thereby liberators.

We will only arrive at this by continually drawing on practice, on the transformation of nature and society.

Without doubt, geography presupposes specialists, institutions and credits. But I am convinced that it cannot progress in the silence of laboratories, out of permanent contact with the masses, on pain of endlessly getting stuck in old ruts. Only a specialist can help the masses to analyse space, but only the masses live space, and know concretely what it is. Only a long-term policy [*une politique de longue durée*] can put geography (and knowledge as a whole) in the service of the masses – not professions of faith by such and such a minister or such and such a geographer.

PART 3
Anglophone Responses – 2006

Chapter 7

The Kantian Roots of
Foucault's Dilemmas

David Harvey

At the end of his interview with the editors of *Hérodote*, Foucault says:

> I have enjoyed this discussion with you because I've changed my mind since we started
> ... Now I can see that the problems you put to me about geography are crucial ones for
> me. Geography acted as the support, the condition of possibility for the passage between
> a series of factors I tried to relate. Where geography itself was concerned, I either left
> the question hanging or established a series of arbitrary connections ... Geography must
> indeed necessarily lie at the heart of my concerns. (1980, 77; this volume, 182)

Foucault's subsequent appeal for guidance from the geographers as to how this
'condition of possibility' might work focuses, it is interesting to note, more on the
concept of space than it does on geography. While most geographers would accept
that spatiality is one of their foundational concepts I doubt that there are many who
would now argue for such a narrow conception of the subject (Harvey 2004). The
further geography drifts away from the positivist grounding it assumed in the 1960s
as a uniquely spatial science, the grander and more difficult to answer become the
questions of its actual epistemological status. Foucault's request for clarification
deserves some answer. Interestingly, an answer of sorts can be fashioned out of
Foucault's own intellectual history.

Consider, for example, the phrase 'condition of possibility' that Foucault deploys
to describe the position of geography in relation to his own work. This is a very
Kantian phrase. In appealing to it, Foucault is, I suspect, harking back to his earlier
encounter with Kant's anthropology. Foucault translated Kant's text on *Anthropology
from a Pragmatic Point of View* into French (it was published in 1964). He closed
his introduction to the translation with a promise to prepare a lengthier work to take
up many of the questions that the text posed. This longer work never materialized
but Foucault did leave behind an extended unpublished 'Commentaire' that is now
available to us (Kant 1964; Foucault 2005).

In his commentary Foucault argues that Kant's *Anthropology* is a vital rather
than a marginal text in relation to Kant's overall thinking, that, in effect, Kant's three
critiques could not have been written without it. Amy Allen summarizes Foucault's
argument this way:

The *Anthropology* (perhaps unwittingly) breaks open the framework of the critical philosophy, revealing the historical specificity of our a priori categories, their rootedness in historically variable social and linguistic practices and institutions. Foucault's reading of Kant's *Anthropology* thus suggests that Kant's system contains the seeds of its own radical transformation, a transformation that Foucault will take up in his own work: namely the transformation from the conception of the a priori as universal and necessary to the historical a priori; and the related transformation from the transcendental subject that serves as the condition of possibility of all experience to the subject that is conditioned by its rootedness in specific historical, social and cultural circumstances. (Allen 2003)

It is this transition in thinking from a disembodied and transcendental to a rooted subject that is critical. What kind of theory of the species being of 'man' could be constructed through scientific enquiry? The primary vehicle for Kant to explore this question, according to Foucault, is supplied by the *Anthropology*. For Foucault, however, it was to be supplied by an archaeology, but of knowledge rather than of artefacts (1982). The question this immediately poses for us is: why could it not have been equally well supplied, for both Kant and Foucault, by geography? Geography, after all, traffics very heavily in the idea of the rootedness of the human subject in specific historical, social, cultural and, we should add, environmental circumstances. Geography, even more so than anthropology, aspires to the status of science. What Foucault seems to admit in his interview and in his subsequent questions, is that geography could well have performed a parallel function to that assumed in Kant's *Anthropology* or in his own archaeology. In Kant's case, however, the answer is even more interesting. He did indeed view geography as 'a condition of possibility' of all other forms of knowledge and he began teaching it many years before he began his enquiries into anthropology (see May 1970; Zammito 2002). If Foucault was aware of this fact, he never mentions it, perhaps because Kant's geography was never published in Kant's lifetime and when it was subsequently published from a compilation of Kant's and students' notes, it had very little impact. So the question for us is: why did the geography as a condition of possibility of all other forms of knowledge fall by the wayside until it is finally recognized, under pressure from the editors of *Hérodote*, by Foucault?

Given the strong Kantian influences in Foucault's work, I find it useful to first reconstruct Kant's views. Kant taught geography forty-nine times, compared to the fifty-four occasions he taught logic and metaphysics and the forty-six and twenty-eight times he taught ethics and anthropology respectively. He went out of his way to gain an exemption from university regulations in order to teach geography in place of cosmology. He explicitly argued that geography and anthropology define the 'conditions of possibility' of all knowledge. He considered these knowledges a necessary preparation – a 'propaedeutic' as he termed it – for everything else (May 1970, 132–6). While, therefore, both anthropology and geography were in a 'pre-critical' or 'pre-scientific' state, their foundational role required that they be paid close attention. The question is: why did he think so?

Kant felt strongly that metaphysics and ethics should not be based upon some view of the transcendental subject given by theology or speculative cosmology. It had to be based, therefore, on a scientific understanding of the human experience. His early battle against the teaching of cosmology and its displacement by geography was the first shot in this war. If theology and cosmology could no longer provide adequate answers to the question 'what is man?' then something more scientific was needed. Where was that 'science of man' to come from if not from anthropology and geography? The distinction between geography and anthropology rested, in Kant's view, on a difference of perspective between the 'outer knowledge' given by observation of 'man's' place in nature and the 'inner knowledge' of subjectivities (which sometimes comes close to psychology in practice) (May 1970, 107–18). The fact that he began teaching the geography first (in 1756) and that much of what he there examines concerns the physical processes that affect the earth's surface and human life upon it, suggests a certain initial attraction to an underlying theory of environmental determinism as providing a potentially secure scientific basis for metaphysical reflection. Many of the examples he cites in his geography invoke environmental deterministic themes. Even more problematic, as Bernasconi shows, was Kant's initiatory attempt to found a science of racial differences (2001). Kant's later turn to anthropology (which he began teaching in 1772), and the fact that he paid far greater attention to elaborating upon it (even preparing it for publication) in his later years, suggests that he increasingly found the inner knowledge of subjectivities more relevant to his philosophical project. He seems to have given up on the idea of a scientific geography in mid-career and the students' notes on his lectures give abundant evidence of the incoherence and often prejudicial qualities of Kant's thinking on geographical matters. It is significant, furthermore, that the final passages of the anthropology address the whole question of the cosmopolitan ethic directly, indicating a certain connection between his anthropology and his ethics. There is no mention of this topic in the geography. The passage from the anthropology is worth quoting in full. Human beings, he says:

> Cannot be without peaceful coexistence, and yet they cannot avoid continuous disagreement with one another. Consequently, they feel destined by nature to develop, through mutual compulsion and laws written by them, into a cosmopolitan society which is constantly threatened by dissension but generally progressing toward a coalition. The cosmopolitan society is in itself an unreachable idea, but it is not a constitutive principle ... It is only a regulative principle demanding that we yield generously to the cosmopolitan society as the destiny of the human race; and this not without reasonable grounds for supposition that there is a natural inclination in this direction ... (W)e tend to present the human species not as evil, but as a species of rational beings, striving among obstacles to advance constantly from the evil to the good. In this respect our intention in general is good, but achievement is difficult because we cannot expect to reach our goal by the free consent of individuals, but only through progressive organization of the citizens of the earth within and toward the species as a system which is united by cosmopolitical bonds. (Kant 1974, 249)

The mission of Kant's anthropology is, therefore, to define 'the conditions of possibility' for that 'regulative principle' that can lead us from an initial condition 'of folly and childish vanity' to 'our destiny' of a cosmopolitan society. This entails an analysis of our cognitive faculties, feelings (of pleasure and displeasure), and of desire. The influence of Kant's project on Foucault's subsequent work is obvious. Kant also sought to reflect on how and why natural endowments ('temperaments') are transformed by human practices into 'character'. Kant writes: 'what nature makes of man belongs to temperament (wherein the subject is for the most part passive) and only what man makes of himself reveals whether he has character' (1974, 203). Kant derived the general proposition that 'man makes himself' from Rousseau and it carries over very strongly, of course, into the Marxist tradition. As Foucault observed in his commentary, 'man is not simply "what he is", but "what he makes of himself". And is this not precisely the field that *Anthropology* defines for its investigation?' (Foucault 2005). This formulation allows Kant to reflect more specifically in the last part of the text on differentials in national character. Human beings have made themselves differently in different places.

What Kant seems to have done here, is to move from one side to the other of what we in geography have long known as a divide between 'environmental determinism' on the one hand and 'possibilism' on the other (Tatham 1957). Kant's relative neglect of the geography later on then makes sense, for it is indeed racked with environmental determinism, racism and prejudicial commentary. But there are deeper reasons for Kant's difficulties with respect to the geography. Kant simply could not make his ideas about final causes work on the terrain of geographical knowledge. 'Strictly speaking,' he wrote (in a passage that Glacken regards as key), 'the organization of nature has nothing analogous to any causality known to us' and this problem blocked his attempt to construct geographical understandings in a style akin to Newtonian natural science (Glacken 1967, 532). His metaphysics and his ethics could not find in the geography the solid scientific foundation he considered essential. He would have to revert to the sphere of mere speculation, to the cosmology. Compared to that the turn to anthropology must have appeared attractive.

But there is yet another difficulty with Kant's formulations. Geography, he argued, organized knowledge synthetically according to the ordering of space. History is considered distinctive because it provides a narration in time. Space and time are therefore considered quite distinct from each other in the Kantian scheme of things. History and geography are quite different from each other. This positions Kant firmly in the Newtonian (rather than in, say, the Leibnizian) camp in conceptualizing space and time (May 1970, 118–31; Harvey forthcoming). Kant's geography is then defined as an empirical form of knowledge about spatial ordering and spatial structures. This may explain why Foucault formulated his questions to the geographers as solely questions about space and spatiality. His earlier unpublished essay on spatial differentiations as 'heterotopias', bears all the marks of acceptance of Kant's absolute view of space and of its separation from time (Foucault 1987). The particularity of positioning in absolute space is marked by contingency, fragmentation and uniqueness. This contrasts radically with the universality that attaches to the concept of a unidirectional time that points us teleologically towards some destiny, such as Kant's cosmopolitan governance. Foucault embraces the absolute theory of spatiality and deploys a panoply of spatial metaphors (the Panoptican being by far the most prominent). In Kant's scheme of things, spatial ordering necessarily produces regional and local truths and laws as opposed to universals. Here, too, there is a strong accord between Kant's findings and Foucault's mission. In his commentary on Kant's essay on 'What is Enlightenment' Foucault embraces the specificity of location and of difference, contingency and the micropolitics of power as he does battle with teleology and macro theory. 'The historical ontology of ourselves must turn away from all projects that claim to be global or radical,' he writes, because 'we know from experience that the claim to escape from the system of contemporary reality so as to produce the overall programs of another society, of another way of thinking, another culture, another vision of the world, has led only to the return of the most dangerous traditions' (Foucault 1984, 46).

The trouble, however, is that Kant's whole approach to geography and space rests on a pure Newtonian foundation of absolute space and time. What happens to the argument when the constraint of the absolute conception of space is lifted? There are good arguments, which I have laid out at length elsewhere, to the effect that space must be viewed dialectically as simultaneously absolute, relative and relational (Harvey 2006). In this case, the distinction between geography and history disappears. All we have is historical geography or geographical history in which the dialectics of socio-environmental change are centrally implicated. And the philosophical shift that Foucault notes from the spatial perspective of Kant to the temporal perspectives of Hegel, Bergson and Heidegger would appear as a grand diversion (Foucault 1980, 149).

Foucault's acceptance of the Kantian conception of space and time poses a whole raft of difficulties and militates against any development of geography as a 'condition of possibility' of all other forms of knowledge. Take, for example, Foucault's concept of heterotopia (like Kant's geography it was an early foray into a problem that remained unpublished for many years). I have profound objections to Foucault's conception precisely because of its basis in a purely Kantian (Newtonian) interpretation of spatiality. Foucault's conception of heterotopia is a very undialectical rendering of what space is and can be about (Harvey 2001). Lefebvre, in his characteristic off-hand and indirect way (and possibly as a direct rebuttal to Foucault's arguments of which he had almost certainly heard from his architecture friends), pointed to a far more dialectical way of discussing heterotopias as spaces of possibilities in *The Urban Revolution* (published three years after Foucault's talk) (Lefebvre 2003). And Foucault himself seems to have partially recognized the difficulty. He surely had his own formulation of heterotopia in mind, when he worried about the way 'space was treated as the dead, the fixed, the undialectical, the immobile' while 'time, on the contrary, was richness fecundity, life, dialectic' (1980, 70; this volume, 177). His further assertions that if 'space is fundamental in any form of communal life' then space must also be 'fundamental in any exercise of power' and that 'a whole history remains to be written of *spaces* – which would at the same time be a history of *powers* (both these terms in the plural)' open up possibilities that remain frustratingly undeveloped (1980, 252). While these comments may signal a certain unease with the Kantian theory of absolute space, Foucault fails to develop a viable critical theory of what space and time might be about.

So why, then, did Foucault stick so resolutely to this Kantian conception of space and time? There is here one other piece in the jigsaw of possibilities. Kant classified archaeology as a distinctively spatial science and Foucault adopted an archaeological approach to knowledge as his guiding principle. Could it be that abandonment of the Kantian sense of space would have undermined this archaeological approach to knowledge? The fixity of Kantian space sadly ends up deadening Foucault's approach to knowledge and power. His arguments, full of initial spatial insights, collapse into stasis. Foucault imprisons himself in a Kantian panoptican of his own making. The result is an unbearable rigidity in his writings.

The idea of geography not as a dead science of spatial ordering but as a live science of historical geography, as a discipline that can openly embrace the dialectic and perform its radical function alongside anthropology as a 'condition of possibility' of all other forms of knowledge, must have seemed threatening to Foucault. Much as I admire many of Foucault's achievements, I have to say that the richness and fecundity of his own contributions would have been much enhanced had he followed up on his promise to take up the critique of Kant's anthropology more directly and then, perhaps, considered a critique of geography (not as spatial science but as a dialectical examination of space, place and environment in human development). It remains for us, however, to realize the powers of both anthropology and geography as 'conditions of possibility' not only for all other forms of knowledge but also, as Kant himself envisaged, as a preparation for living in more enlightened ways.

References

Allen, Amy (2003) 'Foucault and Enlightenment: A Critical Reappraisal.' *Constellations*, 10 (2), 180–98.

Bernasconi, Robert (Ed.) (2001) *Race (Blackwell Readings in Continental Philosophy)*. Oxford: Blackwell.

Foucault, Michel (1980) *Power/Knowledge*, edited by Colin Gordon. London: Harvester.

Foucault, Michel (1982) *The Archaeology of Knowledge*. New York: Pantheon.

Foucault, Michel (1984) *The Foucault Reader*, edited by Paul Rabinow. Harmondsworth: Penguin.

Foucault, Michel (1987) 'Of Other Spaces.' *Diacritics*, 16 (1), 22–8.

Foucault, Michel (2005) 'Commentary of Kant's *Anthropology from a Pragmatic Point of View*.' Translated by Arianna Bove, http://www.generation-online.org/p/fpfoucault1.htm.

Glacken, Clarence (1967) *Traces on the Rhodian Shore: Nature and Culture in Western Thought from Ancient Times to the End of the Eighteenth Century*. Berkeley: University of California Press.

Harvey, David (2001) 'Cosmopolitanism and the Banality of Geographical Evils.' In J. Comaroff and J. Comaroff (Eds) *Millennial Capitalism and the Culture of Neo-Liberalism*. Durham: Duke University Press, 271–309.

Harvey, David (2004) 'Geographical Knowledges/Political Powers.' In J. Morrill, (Ed.) 'The Promotion of Knowledge.' *Proceedings of the British Academy*, 122, 96–112.

Harvey, David (2006) 'Space as a Key Word.' In N. Castree and D. Gregory (Eds) *David Harvey: A Critical Reader*. Oxford: Blackwell.

Harvey, David (forthcoming) *Cosmopolitanism and the Geographies of Freedom*. New York: Columbia University Press.

Kant, Immanuel (1964) *Anthropologie*. Translated by Michel Foucault. Paris: Vrin.

Kant, Immanuel (1974) *Anthropology from a Pragmatic Point of View*. Translated by V. Dowell. The Hague: Martinus Nijhoff.

Lefebvre, Henri (2003) *The Urban Revolution*. Minneapolis: Minnesota University Press.

May, J. A. (1970) *Kant's Concept of Geography and Its Relation to Recent Geographical Thought*. Toronto: Toronto University Press.

Tatham, George (1957) 'Environmentalism and Possibilism.' In Griffith Taylor (Ed.) *Geography in the Twentieth Century*. London: Methuen.

Zammito, John (2002) *Kant, Herder, and the Birth of Anthropology*. Chicago: University of Chicago Press.

Chapter 8
Geography, Gender and Power

Sara Mills

Q.3. It seems to me that you link the analysis of space or of spaces less to production and to 'resources' than to the exercise of power. Could you outline what you understand by power? (Through relation to the State and its apparatuses, through relation to class domination.) Or do you consider that the analysis of power, of its mechanisms, of its field of action, is still at the outset and that it is too soon to give general definitions? In particular, do you think one can reply to the question: who has power?

The last part of this question from Foucault is particularly important for feminist research. This has been most debated amongst feminist linguists and geographers who have analyzed how we can talk about gender in relation to power without assuming that there is a simple correlation between male/powerful and female/powerless (for example, Thornborrow 2002, and in the journal *Gender Place and Culture: A Journal of Feminist Geography*). Many feminist geographers have focused on the relation between gender, power and spatial relations (Blunt 1994; 2000; Blunt and Rose 1994). The relation between power and space is complex, particularly if one defines power in a productive way as Foucault has, and insists that power is a network of relations between people, which is negotiated within each encounter, and also if one defines space relationally and relatively as Foucault suggests. This complex network of shifting relations is both productive and problematic for feminist research. Productive in that it allows for theorizing of power which stresses that each institutionally-sanctioned power relation is negotiated at a local level and can therefore be challenged overtly or covertly; problematic in that it fails to adequately take account of the fact that there are differences in power which have material effects. A strictly Foucauldian analysis would militate against asking 'who has the power?' since no-one within this framework can be said to possess power. Power instead is seen as something that is negotiated within interactions. Thornborrow's (2002) working through of this Foucauldian position leads to an analysis which stresses resistance and challenges by those who are in institutionally weaker positions; however, even though that is clearly the case, it has to be acknowledged that police officers, teachers and doctors are affirmed in their statements because of their position within a hierarchy in a way in which suspects, students and patients are not. Thornborrow focuses on the complex negotiation between one's institutional status which is accorded to you and which gives you a position in a hierarchy, and the local status which you can negotiate for yourself, through your quickwittedness or verbal skill for example. Thus, for example, you may be relatively low in the hierarchy within an institution, but you may be able to locally negotiate a more powerful position for yourself because of your skills and ability. This distinction between two types of power is important in being able to assess which positions of power are negotiable and which are not. One can negotiate local status but your institutional status is not so flexible. Thus, although Foucault's questioning of the notion that one can possess or have power is a useful opposition to very fixed views of power, it nevertheless suggests that everything is up for grabs and sometimes ignores the very real institutional power that certain people do indeed work on and use, even if they do not possess it.

In relation to spatiality and power, I have been trying in my recent works to work out how we can discuss spatiality in relation to power (Mills 2005). Particularly in the colonial context in the 19th century, it is clear that there are a number of spatial frameworks interacting with each other, some of them ratified by the colonial powers and others not. What I have been interested in is the way that colonial power informs the acceptance of certain of these spatial frameworks as normal, but does not necessarily erase the existence of other spatial frameworks.

In Foucault's recently published *Collège de France* lectures *Abnormal,* there is an interesting vignette about spatial relations and power. He describes in some detail the development of confessional practice within the Roman Catholic Church and shows how there was a shift from a focus on penance, with all the practices that this involved, to a focus on confession of sins, which demanded a new set of relations between priests and the faithful. These were spatial and power relations, not in any simple sense, but they demanded a set of spatial arrangements which were very precise and which were responses to a perception of problems to be resolved. The confession of sins demanded that the sin be confessed to someone and that this person was able to give remission of sins; in that sense the role of the priest within the Church changed radically and Foucault argues that power relations therefore changed. This change was manifested in the development of a new set of spatial relations that became evident in the confessional booth, which is as Foucault puts it 'the material crystallization of all the rules' (Foucault 2003/1975, 181). The confessional booth, like other architectural features, determined spatial relations and was determined by a consciousness of the importance of spatial relations being managed 'appropriately' – people had to be arranged in space in order for the confession to be heard effectively: the penitent and the priest had not to be seen by each other, but the penitent needed to be heard well by the priest; the priest and penitent had to be separated from one another and from the rest of the congregation, but it was important, because of the priest's celibacy and because of the temptations attendant on the confession of sexual sin, they had not to be isolated entirely from the rest of the congregation. These very particular discursive demands necessitated this strange architectural configuration of grilles and half-curtains, encouraging certain revelations and discouraging other discourses. Foucault's careful analysis of the discursive pressures that gave rise to the development of the confessional booth to organize space in particular ways strikes me as a very productive way of thinking through the way that power relations develop in tandem with spatial relations, each exerting a distinct but not necessary deterministic pressure on the other.

References

Blunt, Alison (1994) 'Mapping Authorship and Authority: Reading Mary Kingsley's Landscape Descriptions.' In A. Blunt and G. Rose (Eds) *Writing Women and Space: Colonial and Postcolonial Geographies*. New York: Guilford Press. 51–73.

Blunt, Alison (2000) 'Spatial Stories Under Siege.' *Gender Place and Culture*, 7(3), 229–46.

Blunt, Alison and Gillian Rose, Eds (1994) *Writing Women and Space: Colonial and Postcolonial Geographies*. New York: Guilford Press.

Foucault, Michel (2003/1975) *Abnormal*. Edited by V. Marchetti and A. Salomina. Translated by G. Burchell. London: Verso.

Mills, Sara (2005) *Gender and Colonial Space*. Manchester: Manchester University Press.

Thornborrow, Joanna (2002) *Power Talk*. London: Routledge.

Chapter 9

Overcome by Space: Reworking Foucault

Nigel Thrift

Introduction

I have had an on-off relationship with the works of Michel Foucault since I first read his works in Cambridge in the 1970s in those glossy, black, and so authoritative-looking Tavistock editions. I have always connected Foucault with a certain austerity ever since: the man in black. That impression – of a certain rather gloomy outlook – has stayed with me, even as I have read and taught his subsequent work and, even though, as a result, it has become clear that this impression is, in a large part at least, mistaken.

I doubt that anyone would dispute that Foucault was a genius, a man of extraordinary erudition who changed how we view the world. But that obeisance made, he had blind spots and these were, I think, systematic and, in certain senses, politically disabling. They are not therefore trivial addenda but go right to the heart of his project. I want to fix on four of these blind spots and then, having identified them, suggest what they might indicate needs to be done, not in the sense of simply backfilling Foucault's project but in terms of changing the sense of the political that he worked with.

Four Blind Spots

The first of these blind spots arises out of Foucault's poststructuralist antihumanism, a position that has all but dominated intellectual life in large parts of the social sciences and humanities over the last thirty years or so. In particular, this antihumanism has deleted many of the ambiguities of phenomenology from the record. If Dosse (1997) is to be believed, much of this deletion is the result of Foucault's growing antipathy to phenomenology after the publication of his first book, *Madness and Civilization.*[1] In his archaeological studies of the early 1970s, and most notably in *The Order of Things* and *The Archaeology of Knowledge*, processes are conceived

---

1 Indeed, one of the many texts to which Foucault was explicitly responding in *The Order of Things* was Husserl's *Crisis of the European Sciences*.

without subjects, a position that rapidly becomes an orthodoxy; 'it is probably impossible to give empirical contents transcendental value, or to displace them in the direction of a constituent subjectivity, without giving rise, at least silently, to an anthropology' (Foucault 1970, 248). Notwithstanding Foucault's later work, this blindspot continued, typified by his lumping of Merleau-Ponty and Sartre together[2] in the introduction to Canguilhem's *The Normal and the Pathological* (1991), even though Merleau-Ponty was attempting to produce a 'more robustly intuitive account of knowledge, one not predicated on the prior existence of the subject, but rather productive of its very phenomenal appearance' (Carman and Hansen 2005, 20). In other words, Foucault was certainly one of the unwitting contributors to the current knee-jerk tendency to label almost any discussion of issues like perception or the ontological correlation of human beings and the world as mere subjectivist humanism, or as part of an attempt to constitute a transhistorical subject which can itself be subjected to the procedures he outlines.[3]

The second blindspot arises out of his seeming aversion to discussing affect explicitly. Nearly every practice Foucault fixes on comes charged with affect, sometimes of the most extreme kind. Thus, he is quite clearly drawn to practices that involve actual bodily violence and death or the traces of suppressed or channelled violence and death. Pain and torture often figure large. Periodically, too, Foucault discusses negative affects – like anger and flattery (cf. Butler 2005, Foucault 2005). Again, Foucault's work itself seems to me to be full of a suppressed anger which is only rarely allowed to surface and be confronted and most of the time is allowed an existence only in the forensic quality of his prose, and in the kind of polished and glittering rancour that is the critic's bite. Still, the lacuna is an odd one. The obvious explanation is Foucault's concentration on power, in contradistinction to desire, thus flagging up the essential difference between the Foucauldian and Deleuzian systems. But this is surely too glib. Another explanation may be Foucault's attachment to discourse, and yet Foucault's notion of discourse could hardly be more corporeal. One more explanation may arise out of an apparent desire to resurrect the Stoic premeditation of evils in a contemporary form and, relatedly, the need to guard against a loss of control of the passions. Whatever the exact case, it is surely a pity. One thinks only of what Foucault might have done with, say, the history of obsession, and with its contemporary forms, such as stalking.[4]

---

2   Stuart Elden has pointed out to me that Merleau-Ponty was associated with Sartre, especially around the political project of *Les Temps Modernes*, making this a no doubt understandable reaction, one that was only strengthened by the impact of Heidegger's *Letter on Humanism* which offered authors like Foucault, Althusser and Derrida an alternative reading of Heidegger.

3   Though in his later works it becomes clear that Foucault's position was more complicated than this.

4   Another possible explanation may be that Foucault, like so many social theorists, never really considers the importance of early years in laying down affective relays in the precognitive realm. Foucault on babies – an interesting thought, surely.

Another of those blind spots was space. It is certainly true that Foucault had a spatial sensibility (conventionally signalled by the interviews in which he addresses the topic), one that arose naturally from his critique of the architectonic space of the transcendental project, but it was not, I would argue, a sensibility that he did much with. Foucault could clearly produce subtle spatial analyses – most notably in *The Order of Things* (where he set himself the task of uncovering a 'table' across which terms of thought were distributed) and *Discipline and Punish* (where so much of the analysis is distributional). Still it seems to me that, even though he argues that the very possibility of thought arises out of an inherently spatial imaginary which (at least in the Deleuzian interpretation of Foucault) emerges 'from both the movements of bodies *and* the images those bodies produce of each other' (Colebrook 2005, 191), still Foucault tended to think of space in terms of orders, and I think that this tendency made him both alive to space as a medium through which change could be effected and, at the same time, blind to a good part of space's aliveness. Thus, when he wanted to signal this spatial quality he often found other not-categories for it, like heterotopia.[5] Everywhere in Foucault there are, in other words, markers of his sensitivity to spatial order – as a key to the constitution of power, as a marker of the self, as a requisite co-ingredient of numerous practices, as a key ingredient of 'eventalization', as an acceptance of the importance of the qualities of territory – but it has been left largely to other authors to construct a Foucauldian spatiality (e.g. Philo 2004). Yet this is a spatiality that still seems strangely muted to me, neutered especially by its inability to systematically think co-incidence (except as the aleatory). Perhaps part of this neutering comes from the lack of attention to energy in Foucault's account of power, an account which has to be reworked by others who were more concerned with motion (Deleuze 1998).[6] Perhaps part of it comes from a difficulty (shared by many, it has to be said) to imagine how different contents can inhabit the same space (Thrift 2006). And perhaps part of it comes from a blindness to the outcome of his own reasoning when it comes to space, and especially to the difference between the archive, understood as 'a form of spatial reasoning (a map, a ground-plan of a building, a city-view, a flow-chart) that classifies and categorizes inherited knowledge. Like a historical map, it is a repository of facts and, by the nature of its form, it lets the past be treated in terms of its organization and mode of presentation' and the diagram, understood as 'a map destined not just to sum up the past, but to shape the ways that the present and future will be understood and lived. A diagram seeks to impose control of the future by inflecting the ways that the past can be thought insofar as they will also determine current and future behaviour that its authors want to impose' (Conley 2004, 567).

Another blind spot was things. Though Foucault writes about technology, one of his most frequently quoted passages revolves around a catalogue of Chinese things, and some of his most acute pieces of writing centre around art works, still his work

---

5   Although, as Stuart Elden has pointed out to me, the wider register that the French *espace* has may have meant that space presented less of a problem.

6   Not least because of an interest in cinema.

is curiously devoid of thingness, except insofar as this involves the dividing hand of architecture. I still find this the oddest of absences in that a turn to animating imprecise and stubborn inanimate objects would clearly have been an easy enough move for Foucault to make, not least because of his acutely practical rendition of the manifestation of the world. Perhaps it is the result of Foucault's general avoidance of economy, except as an episteme or, later, as the circulation of people and goods. Or perhaps it is the result of his later emphasis on the self, even though in modern cultures it could be argued that the self comes not so much wrapped in as modelled by things: the technology of the self is precisely that. Or perhaps it is part of a more general emphasis on language and texts that pervades so much of his work, on the verbal part of the world made into a whole. Or perhaps it was just a simple omission: after all, no one can write about everything – philosophers most of all. But it is still a tragic omission. One thinks of what Foucault might have done with objects like barbed wire (now brilliantly brought to life by Netz 2004) or guns – or prescription drugs.

The reader who knows my work will hardly be surprised by these emendations. I have a vested interest in them, for they are four of the things that I have concentrated much of my theoretical career on. They explain why I have often found Foucault so relevant and, at the same time, so peculiarly irrelevant. His texts seem to exude a kind of theoretical colour blindness which combines an obvious ontological verve with a leaching out of many of the most vital ingredients of the world. It is why, for me, his texts often seem to exist in a kind of existential gloom. It is a bit like being in the witch wood that is the central actor of so many fairy tales but this is a witch wood in which the spell only bring its denizens partially (and sometimes grotesquely) to life.

What is to be Done

My world is going somewhere else. I want to follow a different associationist trajectory, one in which the insights of certain kinds of new phenomenological thought, the full range of work on affect, various kinds of thinking about space, and the profusion of recent research on things can be given a breathing space. However I also want to argue that it might be possible to think of that trajectory as at least partially Foucauldian and I want to use the final part of this very brief chapter to outline what I mean, in part by using Butler's (2005) recent thoughts on the issue of 'unknowingness', themselves heavily influenced by Foucault's writings on the constitutive 'outside' of thought, as a starting point.

To begin with, a different trajectory would need to go beyond arguing that the subject is simply the effect of a discourse (howsoever understood) and that the forms of rationality established by these discourses always come at a price. This is, by now, such a trite observation that it conceals as much as it reveals.

> ... we must recognize that ethics requires us to risk ourselves at moments of unknowingness, when what forms us diverges from what lies before us, when our willingness to become

undone in relation to others constitutes our chance of becoming human. To be undone by another is a primary necessity, an anguish, to be sure, but also a chance – to be addressed, claimed, bound to what is not me, but also to be moved, to be prompted to act, to address myself elsewhere, and so to vacate the self-sufficient 'I' as a kind of possession. If we speak and try to give an account from this place, we will not be irresponsible, or, if we are, we will surely be forgiven. (Butler 2005, 136)

Then, a different trajectory would need to go beyond a notion of the practices of reflexivity on thought as only having the three major forms it has in the Foucauldian corpus (memory, meditation and method). As I have tried to show here and elsewhere, what counts as thought needs to be understood as distributed more widely, taking in some of the actors I have directly and incidentally signalled. And, of course, these actors are often recalcitrant to reflection, which brings me to the issue of intelligibility.

For, finally, in this different trajectory, it would be important to comprehend that not everything that is intelligible is knowledge. Foucault understood this insight very well but he did not take it very far:

Care of the self … is not just a knowledge. So if, as I would like to show today, the care of the self is always strongly linked to the problem of knowledge, even in its most ascetic forms, those closest to exercise, it is not fundamentally, exclusively, and from end to end a movement and practice of knowledge. It is a complex practice, which gives rise to completely different forms of reflexivity. (Foucault 2005, 462)

One of the great tragedies of Foucault's early demise is that we will never find out what he thought many of these different forms of reflexivity might be.

It seems to me that the issues merely raised here need to be taken up in a different way if we are to ever get out of the loop of a certain kind of philosophy whose vaulting level of generality makes it of limited use in discussing issues like identity and the self, let alone other markers of association. To do that requires precisely the Foucauldian notion of the subject as constituted as a form of not just knowingness but also unknowingness. But, or so I would argue, understanding more about that unknowingness requires precisely a consideration of phenomenology (shorn of its claim of a general human horizon), of affect, of space and of things if we are ever to forge a sense of responsibility that is, well, responsible.

In turn, that recognition may make it possible to forge a politics that goes farther than Foucault's, a more nuanced politics of hope that does not allow the future to be sealed off by the rule of the worst, a politics that is active and is still worthy of the epithet 'revolutionary', not least because it no longer needs to comfort itself with, for example, the hackneyed belief that Western thought is in a terminal crisis (as Foucault told a Zen Buddhist priest in 1978) or the trite consolation that the capitalist West is necessarily 'the harshest, most savage, most dishonest, oppressive society one could possibly imagine' (cf. Afary and Anderson 2005). Instead, it would take Foucault's insight that the self has never just been about self-knowledge but has always been about care and broaden this out so that it becomes a means of constructing 'social-cum-spatial' associations that truly are different, thereby diffusing more possibilities to reflect upon ourselves (Tarde 2000, 2004).[7] These possibilities would include the ethical possibilities that Foucault held dear, a whole battery of different responsibilities to the world that 'recognize that ethics requires us to risk ourselves precisely at moments of unknowingness, when what forms us diverges from what lies before us, when our willingness to become undone in relation to others constitutes our chance of becoming human' (Butler 2005, 136).

References

Afary, J. and K.B. Anderson (2005) *Foucault and the Iranian Revolution. Gender and the Seductions of Islam*. Chicago: University of Chicago Press.

Butler, J. (2005) *Giving an Account of Oneself*. New York: Fordham University Press.

Canguilhem, G. (1991) *The Normal and the Pathological*. (With an introduction by Michel Foucault.) New York: Zone Books.

Carman, T. and M.B.N. Hansen, Eds (2005) *The Cambridge Companion to Merleau-Ponty*. Cambridge: Cambridge University Press.

Colebrook, C. (2005) 'The Space of Man: On the Specificity of Affect in Deleuze and Guattari.' In I. Buchanan and G. Lambert (Eds) *Deleuze and Space*. Edinburgh: Edinburgh University Press, 189–206.

Conley, T. (2004) 'The Historical Atlas: Archive or Diagram.' *Journal of Historical Geography*, 30, 264–8.

Deleuze, G. (1998) *Foucault*. Minneapolis: University of Minnesota Press.

Dosse, F. (1997) *History of Structuralism*. (Two volumes.) Minneapolis: University of Minnesota Press.

Foucault, M. (1970) *The Order of Things. An Archaeology of the Human Subject*. London: Tavistock.

Foucault, M. (2001) *Power. The Essential Works. Volume 3*. London: Allen Lane.

Foucault, M. (2005) *The Hermeneutics of the Subject. Lectures at the Collège de France, 1981–1982*. New York: Palgrave Macmillan.

Netz, R. (2004) *Barbed Wire. An Ecology of Modernity*. Middletown, CT: Wesleyan University Press.

Philo, C. (2004) *A Geographical History of Provision for the Insane from Medieval Times to the 1860s in England and Wales*. Lampeter: Edwin Mellen Press.

Tarde, G. (1897/2000) *Social Laws. An Outline of Sociology*. Kitchener: Batoche Books.

Tarde, G. (1896/2005) *Underground (Fragments of Future Histories)*. Brussels: Les Maîtres de Formes Contemporains.

Thrift, N.J. (2006) 'Space.' *Theory Culture and Society*, 23, 139–46.

---

7   The flagging of Tarde's work also points to the thoroughgoing revision of what constitutes 'the social' that is now being worked through by numerous authors. Foucault's writings can be interpreted as an early symptom of this ambition.

Chapter 10

Foucault Among the Geographers

Thomas Flynn

Historian Paul Veyne described Foucault as a 'kaleidoscopic' thinker. He took the circle of our received opinions and turned the cylinder ever so slightly, causing entirely new configurations to appear such that one could never recapture the previous set in its entirety. One had lost one's naivete, as it were. This visual metaphor is doubly apt because Foucault was a spatializing thinker. In fact, I would argue that his preference for spatial metaphors and arguments, more than a personal quirk, was his way of combating the dialectical thinking that had pervaded French thought since the 1930s (see Baugh 2003; Flynn 2005). Among other effects, it undermined the 'totalizing' effort of Sartre's *Critique of Dialetical Reason* and of more orthodox Marxists by insisting on the multiplicity of rationalities, which Foucault called the 'polyhedron of intelligibility', with which one could approach any historical event. But his spatialized thinking also countered the emphasis on 'vision' that attracted Husserlian phenomenologists, a major alternative to dialectical thinking in his day, but one from which Foucault insisted his generation of thinkers strove to free themselves.

That same Paul Veyne suggests that history has more in common with 'comparative geography' than with literature (1984, 284–6) and that Foucault has revolutionized history in our day. Though space and vision seem to go together and thus ally Foucault with the phenomenologists, for whom the intuitive grasp of an essence or intelligible contour is the model of knowledge, Foucault's 'archaeological' method is comparative and his vision diacritical (Flynn 2005, 93–5). 'Archaeology,' he explains, 'is in the plural.' It does not home in on an essence to be articulated in a definition. In fact, it is more 'nominalist' in its turn away from 'essences' entirely. Taking a cue from structural linguistics, Foucault, in his early works at least, seeks intelligibility in contrasts: binary oppositions between reason and unreason, normal and abnormal, genuine and pseudo knowledge. Of course, the matter is more complex than that, but it does underscore his concern with otherness (alterity) and opposition. Power, for example, presumes and elicits resistance.

Moreover, *pace* Plato and William James, there is no 'carving at the joints'. These boundaries are conventional; they have a history; they can be changed. This is the 'critical' aspect of Foucauldian 'archaeology'. As if to gloss Veyne's analogy, Foucault in his Tanner Lectures admitted: 'Experience has taught me that the history of various forms of rationality is sometimes more effective in unsettling our certitudes and dogmatism than is abstract criticism' (1997, Vol. 3, 323). This conjunction, if not synthesis, of the spatial and the temporal, of the here and there with the before and after, is characteristic of Foucauldian reasoning.

The best known examples of Foucault's 'spatialized' reasoning appear in his masterwork, *The Order of Things*, and in his genealogy of the penal system *Discipline and Punish*. In the former, he analyses the various perspectives both within and outside Velasquez's painting 'Las Meninas' in order to demonstrate both the impossibility of representing representation, which exemplified an epistemic change from the 'Classical' (ca. 1650–1800) to the 'Modern' era (ca. 1800+). Like a docent in a gallery, Foucault leads us to discover the missing viewer among this collection of on-lookers in the painting, namely, the subject outside the picture who has assumed the perspective of the implied subject of the painting, depicted in a mirror at the back of the scene. The 'argument' is in the layout of the scene and its ordering of perspectives, each traceable by linear angles of vision. The missing subject (the 'man' of modern humanism) is present yet absent from the depiction; he is as necessary as he is invisible. He will emerge at centre stage with the advent of the 'human sciences' in the modern era.

Foucault's second arresting example of spatialized reasoning appears in his analysis of Jeremy Bentham's 'panopticon' as the model for houses of surveillance such as prisons, military caserns, factories, hospitals and schools. This time the arrangement of the building is intended to maximize the visibility of the subjects while minimizing that of the overseers so that the 'inmates' become their own custodians due to their constant liability to supervision. This time the 'argument' is in the architecture. And once more Foucault extends it to the larger thesis, the 'disciplinary' society of the modern age. Henceforth, one cannot ignore the omnipresence of surveillance devices, the vulnerability of our various communication systems to external review and interpretation, and the insinuation of authorities into the most private portions of our personal lives. The kaleidoscope has turned again and we cannot ignore the new configurations that have emerged.

Foucault Among the Geographers

The foregoing is a way of contextualizing my response to Foucault's questions addressed in 1976 to theoretical (philosophical) geographers in an issue of their journal, *Hérodote*. Let me avow at the outset that I am a philosopher, not a geographer. I've been intrigued by Veyne's likening of history to comparative geography, especially when conjoined with his view that Foucault has revolutionized historiography in our day. What makes Foucault's four questions particularly interesting for me is his admission that these are questions he is asking himself, not queries to which he already holds the answers. In like manner, my responses are not definitive answers but rather reflections on the possibilities inherent in these questions themselves from the perspective of what we know of Foucault's interests and work at that time. I am trying to surmise what Foucault might have responded to his own questions.

1. The notion of strategy is essential if one wants to make an analysis of knowledge [*savoir*] and its relations with power. Does it necessarily imply that through the knowledge in question one wages war?
   Does strategy not allow the analysis of relations of power as techniques of *domination*?
   Or must we say that domination is only a continued form of war?
   Or alternatively, what scope would you give to the notion of *strategy*?

Foucault has already warned us against reducing knowledge to power and yet he has invited such reduction by exhibiting their almost universal overlap. And he has suggested that history be conceived not in terms of knowledge and meaning but in terms of 'strategy and tactics'. For his part, he has shown more interest in the social mechanisms for the use of knowledge (what he calls the 'games of truth') than in the epistemology of knowledge or truth itself. Thus, while an experienced nurse may have a better reading of the symptoms, it is the MD alone who is legally empowered to make the diagnosis and prescribe the medication. Foucault raises the larger question of who has the 'right' to utter the 'definitive' statement in a controversy; who has the right to speak the 'truth'. That is strategic knowledge; knowledge as power.

Perhaps a better example of this relation between knowledge and power is the rise of information technology and the sobering thought that it is arguably those who control the information (its retrieval, storage and communication) that exercise the most decisive power in contemporary society rather than those who control its wealth. These networks or 'webs' are spatial relations, shrinking distance and augmenting the possibilities of communication while creating a strategic relation of dependency-control on the part of the public and the technician respectively. As Foucault admitted, in the original *Hérodote* interview:

> Metamorphizing the transformations of discourse in a vocabulary of time necessarily leads to the utilization of the model of individual consciousnesses with its intrinsic temporality. Endeavouring on the other hand to decipher discourse through the use of spatial, strategic metaphors enables one to grasp precisely the points at which discourses are transformed in, through and on the basis of relations of power. (1980, 69–70; this volume, 177)

It appears the old adage 'knowledge is power' gains plausibility if knowledge is conceived strategically, one might say 'pragmatically', and power is understood as the ability to control. But then the extension of the term 'strategy' would be as broad as that of 'knowledge' itself.

Does this support the optimistic conclusion that, with the advent of globalization, the likelihood of total war is implausible because too many around the world have too much at stake in a community of interconnected goods and services? Foucault does not ignore the force of enlightened self-interest within the international community and he would seem more sympathetic toward neo-Stoic cosmopolitanism than with nationalism. But we should not forget his (later chastened) enthusiasm for the 'spirit' and 'spirituality' of the Iranian revolution in the days of the Ayatollah Khomeini, if we doubt his respect for 'regionalism' or his distrust of homogeneity.

And his well-known warning that, though not everything is bad, 'everything is dangerous' counsels a studied caution about any utopian schemes of world peace or universal democracy. Foucault seems to share Camus's Sisyphian notion that the only hope is to know there is no ultimate hope. In other words, hopes should be limited; utopianism is counterproductive.

Foucault has already distinguished power from force in the sense that power presumes freedom and entails resistance whereas force requires neither. Can one equate 'power' with 'domination'? Yes, if one interprets 'domination' as 'efficacity', otherwise no. The more plausible interpretation would be to equate 'power' with 'control' including the self-control of Nietzsche's 'overman'. In that way its positive character that Foucault defends but which he often overlooks is more readily conveyed.

2. If I understand you correctly, you are aiming to constitute a knowledge of spaces [*un savoir des espaces*]. Is it important for you to constitute it as a science?
    Or do you find it acceptable to say that the break which marks the threshold of science is only a means of disqualifying certain knowledges [*savoirs*], or to make them evade examination.
    Is the division between science and non-scientific knowledge [*savoir*] an effect of power linked to the institutionalization of knowledges [*connaissances*] in the University, research centres, etc.?

This is a form of the previous question. Certainly 'scientific' is an honorific in our society and it is accorded by institutions of various sorts, the most prominent being universities and professional societies as well as grant- and award-conferring organizations like the Royal Academy or the Nobel Prize selection Committee. Though Foucault has a keen sense of the historically constructed nature of such institutions, this did not keep him from campaigning for election to the Collège de France. Yet, in a society where all values are socially constructed, one can scarcely be faulted for 'playing the game' to the best of one's abilities.

But Foucault, doubtless with a certain irony and in direct response to a remark made by Sylvie le Bon de Beauvoir in *Les Temps modernes*, has referred to himself as a 'happy positivist'. He has never been critical of the scientific character of the so-called 'hard' sciences, but, like his mentor Georges Canguilhem, he was more interested in the marginalized sciences and in those like phrenology or alchemy, that lost their scientific status. For Foucault, the interesting consideration is what allowed or excluded some disciplines' entrance into the realm of science.

If one contrasts the modern as temporal (dialectical) reasoning and the postmodern as spatial (diacritical) reasoning, then comparative geography would be a properly 'postmodern' discipline as would Foucauldian 'archaeology'. But whether either would merit, or even seek, the honorific 'science' is a matter of preference. As a matter of strategy, the goods of social prestige, of financial support and the rest offer incentives for geography's gaining this honorific. And the 'geopolitical' nature of the discipline, as Foucault remarks, places geographical 'science' squarely in the field of strategies and tactics as the very word 'gerrymander' attests.

3. In particular, do you think one can reply to the question: who has power?

This query seems aimed at the Marxist bent of the journal *Hérodote*, though it extends to so-called 'structuralists' as well and hence to the followers of structuralist Marxist Louis Athusser. As professor at the prestigious École normale supérieure, Althusser might well have instructed some of the editors of the publication as he did Foucault himself. Foucault was infamous for his quasi-structuralist pronouncement that the 'man' of modern humanism has a history and might disappear as quickly as it appeared in the first place. That, of course, earned him the enmity of humanists of all stripes. And we have just noted his quasi structuralist concern to determine the social roles of certain individuals empowered to speak authoritatively.

But in his popular polemics, Foucault sounds more 'existentialist' in his search for those individuals whose 'little acts of cowardice' left them mute about a particular injustice (specifically, the firing of a prison psychiatrist who complained about the mistreatment of prisoners by their guards) and deaf to the 'voice that says "I"' in accusation (1994, Vol. 2, 238). In other words, the meanness is not entirely in the system. Consider the following, which could have been written by Sartre:

> Situations can always give rise to strategies. I don't believe we are locked into a history; on the contrary, all my work consists in showing that history is traversed by strategic relations that are necessarily unstable and subject to change. Provided, of course, that the agents of those processes have the political courage to change things. (Foucault 1997, Vol. 3, 397)

Here and on other occasions when Foucault speaks of 'political courage' we encounter the thorny issues of 'responsibility', ethics, and morality in general, that Foucault was beginning to address in his books and lectures toward the end of his life.

4. Do you think it is possible to undertake a geography – or a range of geographies – of medicine (not of illnesses, but of medical establishments along with their zone of intervention and their modality of action)?

One can view this as a natural outgrowth of Foucault's interest in the government of bodies and populations (biopower) at about this time in his career. Again, we witness the turning of the kaleidoscope to consider alternatives to our present manner of conceiving the practice of medicine and public health.

First we should employ Foucault's distinction between power *situations*, a certain distribution or economy of power in a given moment, and power *institutions* such as the army, the police, the government (1996, 260). His question here begins with the institutions such as hospitals, boards of certification, the medical 'profession' itself and numerous related service institutions like the pharmaceutical industry, retail pharmacies, and various governmental agencies for the dispensing and surveillance of health care and the like. In other words, the health care industry whose relations can be 'charted' in terms of the power relations that trickle down and bubble up from the government and its citizens and noncitizens.

Examples of power situations would be the introduction of conditions for the implementation of medical services to the indigent or the concrete exchange between a welfare recipient and his or her case worker.

From another angle and along another axis of power relations, one thinks of various technologies for communication of information around the globe, where the medical centres would be simultaneously centres of medical technology and information technology. And this creates the possibility of a multiplication of such medical 'service centres', which, in turn, raises the question of the role of the face-to-face and of compassion in the dispensing of such 'services' to the general population. The problem of science with a human face.

Additionally, the matter of decentralization of services versus centralization of sources suggests the possibility of the dispersal of individuals 'authorized' by virtue of their training to provide 'borderline' services between humanistic treatment and 'scientific' information and techniques.

All of these considerations point not to one underlying cause such as economic or technological determinism but to a 'polyhedron of intelligibility'. Foucault, who declined to play the prophet, was willing to open the historical present to the possibilities that it harboured within, if only one would turn the kaleidoscope once again and have the courage to peer in.

References

Baugh, Bruce (2003) *The French Hegel*. New York: Routledge.

Foucault, Michel (1980) 'Questions on Geography.' In Colin Gordon (Ed.) *Power/Knowledge. Selected Interviews and Other Writings, 1972–1977*. New York: Pantheon. 63–77.

Foucault, Michel (1994) *Dits et écrits*. Edited by Daniel Defert and François Ewald. Four Volumes. Paris: Gallimard.

Foucault, Michel (1996) 'Clarifications on the Question of Power.' In Sylvère Lotringer (Ed.) *Foucault Live, Collected Interviews, 1961—1984*. New York: Semiotext(e), 2nd edition.

Foucault, Michel (1997) *Essential Works*. Edited by Paul Rabinow. Three Volumes. New York: New Press.

Flynn, Thomas (2005) *Sartre, Foucault and Historical Reason Volume 2: A Poststructuralist Mapping of History*. Chicago: University of Chicago Press.

Veyne, Paul (1984) *Writing History*. Translated by Mina Moore-Rinvolucri. Middletown, CT: Wesleyan University Press.

PART 4
Contexts

Chapter 11

Strategy, Medicine and Habitat: Foucault in 1976

Stuart Elden

I study things like a psychiatric asylum, the forms of constraint, exclusion, elimination, disqualification, of which reason is always precisely incarnate, in the body of the doctor, in medical knowledge, medical institutions, etc., exercised over madness, illness, unreason, etc. What I study, is an architecture, a spatial organisation; disciplinary techniques, modalities of dressage, forms of surveillance; actually what I study is what I have called governmentality (a word which is undoubtedly much too grand): what are the practices which are put to work to govern men, that is to enable a certain manner of conducting them, government as the conduct of conduct, how to conduct the conduct of men. (Foucault 1978: 9)

The mid 1970s were productive years for Foucault. *Discipline and Punish* – his first full book for six years – appeared in February 1975, and the first volume of *History of Sexuality* in December 1976 (1976a). In 1976 he also gave the lectures which comprise *Society Must be Defended* (1997), and re-edited a French translation of Bentham's *Panopticon* (Bentham 1977), to which he contributed an interview as an introduction (reprinted in 1994, III, 190–207). A collection that he facilitated, and to which he contributed an important chapter, 'The Politics of Health in the Eighteenth Century' appeared (Foucault et al. 1976), and the researches continued under his direction for another almost unknown book, *Politiques de l'habitat (1800–1850)* (1977). A busy year certainly, but not perhaps especially unusual.

And yet this is merely the visible tip of his research. As in most years, he lectured widely across the world. These lectures included a largely unpublished one on alternatives to the prison in Montreal (1976b), and a very important study given in Brazil in November which has been available in French only since 1994 and is translated for the first time in this volume: 'The Meshes of Power' (1994, IV: 182–201; 2007a). In addition, as is well known, the first volume of the *History of Sexuality, The Will to Knowledge*, was only intended to serve as an introduction to a six volume series.

1. *La volonté de savoir* [*The Will to Knowledge*]
2. *La chair et le corps* [*The Flesh and the Body*]
3. *La croisade des enfants* [*The Children's Crusade*]

4. *La femme, la mère et l'hystérique* [*Woman, Mother, and Hysteric*]
5. *Les pervers* [*The Perverse*]
6. *Population et races* [*Population and Races*]

The logic of these volumes is that Foucault sees Christian practices of confession as central to understanding the birth of psychoanalysis and the discourse of sexuality; and that the four constituent subjects of sexuality were the masturbating child, the hysterical woman, the perverse adult and the Malthusian couple.

The projected volumes traded upon themes rehearsed in the *Collège de France* lectures of the previous few years. These courses provide valuable insight into these concerns: confession, perverts and childhood masturbation receive treatment in the *The Abnormals* course; hysterics are treated briefly in *Le pouvoir psychiatrique*; and the volume on *Populations and Races* would undoubtedly have traded upon the research presented in *Society Must be Defended*. But Foucault did not merely outline these ideas in lectures. At least two of the volumes were actually drafted: *Les pervers* and *Le chair et le corps*.[1]

Foucault's research in his first five courses at the *Collège* had therefore contained some remarkable analyses. Early concerns with systems of knowledge had been partnered by some important studies of questions of power, initially in relation to punishment and the penal society in *Théories et institutions pénales* and *La société punitive*, and then in *Le pouvoir psychiatrique* through a rereading of the last parts of *Histoire de la folie* in the light of the new conceptual tools of power and genealogy. The course immediately preceding *Society Must be Defended*, *The Abnormals*, had broadened the institutional analysis of prisons and asylums to concentrate on themes already in Foucault's work, but now brought to the fore. Society as a whole becomes an issue, with the techniques of normalization, categorization and control broadening beyond the institutions; mechanisms of confession in the production of truth are brought to bear on the individual and collective subject; and truth is conceptualized as a political force, opening to a later concern with modalities of government (see 1999, 45). Despite his very critical self-assessment in *Society Must be Defended*, about how little he felt he had achieved (1997, 5–6/3–4) Foucault had, in nascent form, outlined many of the concerns that would occupy him for the next few years.

The year 1976 was, then, a hinge point for Foucault. On the one hand it marks the culmination of several years of research, finding outlet in the programmatic first volume of the *History of Sexuality*; on the other it opens up many of the themes he would treat from 1977 until the end of his life. But few could have predicted how the next eight years would have turned out. Instead of the projected volumes at the proposed rate of one a year, he published the second and third volumes, on entirely different subject matter, in 1984.

Foucault delivered *Society Must be Defended* in the first three months of 1976. He completed *The Will to Knowledge* in August. Given his seeming discontent with what he had done, as evidenced by remarks in the course, it is perhaps surprising

---

1   For more detailed discussions, see Elden 2001; 2002; 2005; 2006b.

that he spent the next five months perfecting the plan of a project he had already decided was flawed. As his partner Daniel Defert suggests in his chronology that accompanies the French collection of his shorter works, even at this point he did not intend to write them (Foucault 1994, I, 49). Instead, he worked on a related yet different set of concerns. In a letter he wrote in August 1974, at the time of completing *Discipline and Punish*, cited by Defert in his chronology, he confessed he was bored with the subjects he had been working on and that 'political economy, strategy, politics' would be his new concerns (1994, I, 45). We get an initial glimpse of this in *Society Must be Defended*, when he suggests that if the previous five years had been given over to the disciplines, in 'the next five years, it will be war, struggle, the army' (1997, 21/23). Had this been followed through, Foucault would have been treating these subjects until 1981, but of course we know that this was not the case. Equally in April 1976 he notes that 'his next book will treat military institutions' (1994, III, 89).[2]

Sécurité, Territoire, Population was the first course Foucault gave after he returned from his 1977 sabbatical year. In this course, given in early 1978, Foucault begin with three important case studies of town planning, famines and smallpox, in order to illustrate the themes of the spaces of security, the aleatory, normalization and the birth of the modern conception of population, before giving the famous 'Governmentality' lecture. It is clear now how this lecture, for so long seen out of context, is the opening up of a problematic that Foucault then treats in sustained detail for the rest of that year's course and the next, *Naissance de la biopolitique* (2004b). Foucault is concerned with rereading the history of the state from the perspective of practices of government, and he suggests three key models for the West. These are the Christian pastoral, with its themes of the flock, confession and the government of souls; the diplomatic-military technology that emerges following the Peace of Westphalia; and the notion of the police. Population, police and governance: all themes that had been in his work before, but now given new pre-eminence and a much more explicitly political twist.[3]

The 1980 course was entitled *Du gouvernement des vivants*, and returned explicitly to Christian practices, and offers what is effectively a draft of sections of the unpublished fourth volume of the *History of Sexuality* series, *Les aveux de la chair*. As *Naissance de la biopolitique* (2004b, 3) makes clear, government can be understood in a range of ways, as government over children, families, a household, of the soul, of communities. Such issues re-emerge in Foucault's final two courses, both under the title of *Le gouvernement de soi et des autres* – but in *Naissance de la biopolitique* it is exercised in political sovereignty. Tracing the theme of government, which emerges out of the ashes of the abandoned original plan of the *History of Sexuality*, and then leads into the version of the *History of Sexuality* that Foucault left incomplete at his death is a long and complicated story, of which I have offered a fuller account elsewhere (Elden 2005; 2007a). These few points of indication are,

---

2   On this theme see also Vol. III, 123—4, 268, 515.
3   The best analysis of these lectures to date, making use of the tapes, is Lemke 1997.

I hope, useful and important. They are not merely important because they show that the lecture courses provide many of the clues necessary for the reconstruction of these paths, but because they help to illuminate the questions Foucault asked the geographers in 1976.

Some broad themes thus emerge as contexts for these questions: notably strategy and war, and medicine and habitat. These will be the topics of this chapter, which then re-examines the questions in their light, and in so doing raises issues concerning the question of science and knowledge. It concludes by raising some general issues around Foucault's work and how it might be appropriated in the future.

Strategy and War

For a perspective on Foucault's view of strategy in 1976 the obvious place to turn is *Society Must be Defended*. This course, as is becoming well known, was not entirely accurately portrayed in the summary Foucault published at the time. Rather than a study of war in itself, the course is directed at the war of races that constituted modern states, something that is only briefly mentioned in the summary (1997, 244/271–2). Issues of security, violence, revolution, class struggle and the themes of the final chapter of *The Will to Knowledge* all make an appearance.[4] Coming out of the researches of both the *History of Sexuality* and *Discipline and Punish* this was Foucault concentrating on the most spectacularly brutal, on power in its naked, violent, repressive form. Despite his later analyses to the contrary, such as the empowering passages of *The Will to Knowledge*, the initial formulations and analyses of power in Foucault are almost invariably overwhelming and encompassing.

Strategy had been a concern of Foucault's for some time, and had been proposed explicitly as a mode of analysis in *Le pouvoir psychiatrique*. Foucault notes that he is attempting to avoid psycho-sociological vocabulary, and rather decides on 'a pseudo-military vocabulary' (2003, 18). The manuscript of the course notes that examples of the exercise of power should be understood not as stories or 'theatrical episodes' but as 'a ritual, a strategy, a battle' (cited in Lagrange 2001, 139).

Foucault's fascination with the military was equally not new, indeed it had been a major topic in *Le pouvoir psychiatrique*, and *Discipline and Punish*, even before the extended treatment of war in *Society Must be Defended*. Indeed, in the interview with the geographers at the *Hérodote* journal, Foucault notes that in terms of spatial analyses Marx's work on 'the army and its role in the development of political power' had been unjustly neglected (1994, III, 39; this volume, 182). It is also notable that Foucault is interested in the second volume of *Capital*, both because of its analysis of the genealogy of capital (1978, 1), but also because of the material on circulation (see 2004a). Excepting Marx himself, Foucault had equally long claimed that struggle was a neglected issue in political theory, and that even Marxists had largely ignored it in their analysis of class struggle (1994, III, 206, 268, 310–11). In the *Hérodote*

---

4   For fuller accounts, see Stoler 1995; Zarka (Ed.) 2000; Zancarini (Ed.) 2001; Elden 2002; Dillon and Neal (Eds) 2007.

interview he suggests that the metaphors they claim are geographical are actually political, juridical, administrative and military. His interlocutors respond that some of these terms are both geographical and strategic, which is not surprising, given that geography 'grew up in the shadow of the military' (1994, III, 32–4; this volume, 176–7). This interest can be seen as continuing in the discussions of the Peace of Westphalia and the invention of standing armies in the late 17th century. Later, in *Sécurité, Territoire, Population*, Foucault discusses the permanent military apparatus, the advent of professional soldiers, the infrastructure of fortresses and transport, and the sustained tactical reflection that dates from this time (2004a, 308–13).

Strategy and war were thus at the heart of Foucault's concerns when he addressed his questions to *Hérodote*. Strategy was, he contended, essential to understanding the relation of power and knowledge, but the relation of strategy to domination and war was more complicated. How, he asks, do we need to rethink strategy? A similar set of questions are asked in the course summary to *Society Must be Defended*, written at the conclusion of the course and thus at the same time as the questions to *Hérodote*. They are worth reproducing at length:

> Must war be regarded as a primal and basic state of affairs, and must all phenomena of social domination, differentiation, and hierarchization be regarded as its derivatives?
>
> Do processes of antagonism, confrontations, and struggles among individuals, groups, or classes derive in the last instance from general processes of war?
>
> Can a set of notions derived from strategy and tactics constitute a valid and adequate instrument for the analysis of power relations? (1997, 239–40/266)

We can see, in these questions, not what is perhaps most obvious. Rather than Foucault suggesting that these are programmes for future work, the way that he turns seems to indicate that his answers would largely have been in the negative. Indeed a sense that is tangible from the lectures is that of Foucault running ideas to ground, working them through to their logical conclusions and exhausting their possibilities before turning to other avenues. These are lectures after all, not polished works. Instead, as I have argued elsewhere, Foucault is more concerned with strategies for waging peace, how mechanisms employed in what he calls governmentality are only indirectly from mechanisms of war (2005; 2007a).

Medicine and Habitat

One of the best instances of seeing how Foucault was more concerned with peacetime strategies and tactics comes in his work on medicine and habitat. Traditionally for a reading of Foucault's work on these topics, the sources had been *Birth of the Clinic* and the essay 'The Politics of Health in the Eighteenth Century' (1963; 1994, 273–89). The former is a remarkable book, but largely concentrates on the form of medicine that emerges in the clinical hospital itself, in other words clinical medicine. There Foucault discusses three forms of spatialization: the location of a disease in a family, the taxonomies of disease; the location of disease in the body; and the way diseases are located in society as a whole, in political struggles, economic constraints, and social confrontations. Foucault claims that it is in the last of these that the changes that led to a reformulation of medical knowledge occurred (1963, 14–15). This is the space of imaginary classifications, space of corporal reality, and space of social order.

Histoire de la folie and other writings on madness and mental illness broaden this work, but it is really in the 1970s that Foucault returned to these earlier themes with a more explicitly political and social twist. Central sources include his lecture courses *The Abnormals* and the recently published *Le pouvoir psychiatrique* (1999; 2003; see Elden 2001; 2006). In the former course he notes that eighteenth and nineteenth century psychiatry was a branch of public hygiene rather than of general medicine (1999, 109). A series of important lectures in Rio in 1974 (1994, III, 40–58, 207–28, 508–21; 2000, 134–56; 2004c; 2007b; see Elden 2003) and seminar research on 'the history of the hospital institution and hospital architecture in the eighteenth century' (Foucault 2003, 352) developed these themes.

Foucault's work in this seminar opens up an important and neglected area of study. As well as the books and occasional pieces, which give the impression of a lone scholar, Foucault also collaborated on a number of projects. From the early 1970s he worked with Gilles Deleuze, Félix Guattari and François Fourquet on the issue of urban infrastructure. This led to the book *Les équipements du pouvoir* by François Fourquet and Lion Murard (1976), which included contributions in discussion by Foucault, Deleuze and Guattari. This was funded by the Ministere de l'ámenagement du territoire, de l'équipment, du logement et du tourisme, and undertaken under the auspices of the Centre d'Études, de Recherche et de Formation Institutionnelle (CERFI), a group founded in 1967 by Guattari. Work undertaken in these projects and others led to the collective work *Les machines à guérir (aux origines de l'hôpital moderne)*, and eventually to the book Foucault edited *Politiques de l'habitat (1800–1850)*. Foucault's project on 'equipments of normalization', one of the parts of the CERFI research, had looked beyond the institutions of hospitals and schools to wider concerns with sanitary norms and 'the power of the state in the determination of sanitary mechanisms'.[5]

The collaborative work raises an important issue. Despite the title of this collection, Foucault should not be treated as a privileged subject, but as part of a network of researchers and relations. The Collège de France had acted as an inspiration to this work. Before 1976 the seminars accompanying his lecture courses had led to the publication of the texts around the case of Pierre Rivière (1973), and in various ways contributed to the 'Dangerous Individual' project (1994, III, 443–64), the publication of the Herculine Barbin memoir and dossier (1980), and *Le désordre des familles*, a collection of 'lettres de cachet' with commentary by Arlette Farge and Foucault (1982). The archives of the Institut Mémoires de l'Édition Contemporaine

---

5   The papers for this and other projects are archived at the Institut Mémoires d'Édition Contemporaine, now based in Caen, Normandy. Subsequent codes refer to their cataloguing system. See http://www.imec-archives.com/. This quote is from D.2.3/FCL2.A04-04.

in Paris carry details of a range of other projects, mostly abortive, but fascinating in terms of an understanding of Foucault's concerns.[6]

Foucault's claim in the interview that preceded the re-edition of Bentham's Panopticon text is that the organization of space, especially architecture in the eighteenth century, became explicitly tied to the 'problems of population, health and the town planning ... it became a question of using the management of space for economic-political ends' (1994, III, 192). In an oft-quoted phrase from the same interview, Foucault declares that:

> ... a whole history of spaces – which would be at the same time a history of powers – remains to be written, from the grand strategies of geopolitics to the little tactics of the habitat, institutional architecture from the classroom to the design of hospitals, passing via economic and political institutions ... anchorage in space is an economic-political form which needs to be studied in detail. (Foucault 1994, III, 192–3)

If the danger here is that Foucault is still privileging time over space – it is a history of space that is to be written – there is certainly a recognition of the need to transform the doing of history through attention to its spatial context.

A further indication of the direction Foucault is taking at this time is another piece from 1976, entitled 'Bio-histoire et bio-politique'. This is Foucault's brief review of Jacques Ruffié's *De la biologie à la culture* (Ruffié 1976). Ruffié is, for Foucault, 'one of the most eminent representatives of the new physical anthropology' (1994, III, 96). One of the conclusions that Foucault draws from the research here is that 'although the species cannot be defined by a prototype but by a collection of variations, race, for the biologist, is statistical notion – a "population"' (1994, III, 96). Overall, as well as contributing to the bio-politics that Foucault has been advancing since 1974, Ruffié also helps with a project of bio-history, 'that would no longer be the unitary and mythological history of the human species through time' (1994, III, 97). Several of these themes are explored in the lectures on the history of governmentality (2004a; 2004b).

Green Spaces

Before I move to looking at the questions Foucault asked, I want here to provide a brief reading of another collaborative project, which was unpublished and from the extant records, appears to have only got to proposal stage, which aimed to analyze the 'green spaces', *les espaces verts*, of Paris.[7] This project was intended to provide a mapping [*un repérage*] and analysis of the material and cultural underpinnings of 'green spaces'. The attempt was to trace the relation between administrative strategies

---

6　For a fuller discussion of these see Elden 2007b. A project on the green spaces of Paris is discussed below.

7　See D.2.1/FCL2.A04-05.

concerning public hygiene more generally with the development of green spaces.[8] The budget for the 18 month project was over a quarter of a million francs, and as well as the core team of Foucault, Blandine Barret-Kriegel, Jean-Marie Alliaumé and Anne Thalamy, who had been involved in other collaborative work, others, including the architect Henri Bonemazon and the geographer Alain Demangeon were included.[9] Demangeon would later collaborate with another colleague of Foucault's, Bruno Fortier, on the book *Les vaisseaux et les villes*, on the arsenal at Cherbourg, which appeared in the same series as *Les machines à guérir* (Demangeon and Fortier 1978).

One of the issues that seems remarkable to the team is the increasing scarcity of green spaces in Paris, but also in other major towns, and the way in which the discourse of salubrity 'seems to have systematically excluded green space from its preoccupations'.[10] Why this should have come to pass is a crucial issue, and needs to be situated in relation to the control of dangerous populations. The fairly detailed proposal notes the Ministry of Construction's realization in 1958 that Paris was very lacking in both parks and gardens, and that the key was to preserve what did exist from development and autoroutes.[11] While distancing themselves from the *paysagistes* or *Robinsonnades* with their wish for a return to nature, the research recognized their importance, and particularly their use by the old, women, and children.[12]

The key green spaces – the Bois de Vincennes and the Bois de Boulogne to the east and west – need to be related to the planning of Baron Haussmann in the Third Republic, and the establishment of the Parc de Buttes Chaumont and Parc Montsourris can also be traced to this time. This is all related to the transition from Royal gardens to public parks and squares, but questions need to be asked about the 'amputation' of parks such as the Luxembourg gardens just south of the Latin Quarter, and the Parc Monceau, and why public places such as the Champs Elysses and Les Invalides did not become parks. These are perhaps the most crucial issues, but what is also notable is the lack of individual or communal private gardens on the English model.[13] While the research looked to this specific example, it is clear that it is an opportunity to approach familiar questions from a new angle. The stated aims of the research to be undertaken in this project cover the recurrent themes of public hygiene; urban surveillance; and industrialization and the urban.[14] In sum though, this projected research was concerned with the problem of 'verdure', of 'greenery', what they state to be 'a recent problem, more exactly environmental' of green belts, barriers.[15] Of marginal interest perhaps, but notable for the geographical

---

8   D.2.1/FCL2.A04-05, 1.
9   D.2.1/FCL2.A04-05, 3, 7.
10 D.2.1/FCL2.A04-05, 1–2.
11 D.2.1/FCL2.A04-05, 4.
12 D.2.1/FCL2.A04-05, 5.
13 D.2.1/FCL2.A04-05, 16–17
14 D.2.1/FCL2.A04-05, 17–22.
15 D.2.1/FCL2.A04-05, 24.

and environmental angle on a continuing set of concerns, and for the interest in this shown by Foucault, not noted for his love of the non-built environment.

Consequences for Questions

In the light of all this work, we can return to the four questions Foucault posed. As he says, 'they are inquiries that I am asking myself', rather than work to which he already feels he has the answers. We can clearly see how the first question relates to the work of *Society Must be Defended*, concerning questions of strategy, war and domination, and how this continues into his work on governmentality. As Arnold Davidson has noted, a 'full study of the emergence of this strategic model in Foucault's work would have to begin with texts written no later than 1971' (2003, xviii–xix), namely the first lecture course at the Collège de France *La volonté de savoir*. As Davidson outlines, the model of strategy can be seen as the bridge between works such as *The Order of Things* and *The Archaeology of Knowledge* (1966; 1969), where he looks at how discourses are structured, to the later work where discourses are relations of force, of battle in, for example, both the lecture course and the book from 1976 (1976a; 1997).[16]

The second question is less obviously related to the work undertaken at this time, but returns to the questions posed in works such as *The Order of Things* and *The Archaeology of Knowledge* about taxonomies of scientific knowledge. One of the interesting issues that arises from reading the lecture courses is how frequently Foucault did refer back to those earlier works, to rework their analyses and engage in self-critique. For example, *Le pouvoir psychiatrique* returns to *Histoire de la folie*, and both *Society Must be Defended* and *Sécurité, Territoire, Population* explicitly politicize questions raised in *The Order of Things* about mathesis and the birth of man respectively, in the latter case seeing the development of biology, political economy and linguistics as the emergence of the 'living being, the working individual, the speaking subject', which 'can be understood from the emergence of population as a correlative of power and as an object of knowledge [savoir]'. While this brings the application of the power-knowledge couplet to the analysis, Foucault also broadens and politicizes the notion of man itself: 'man … is nothing other, in the last analysis, than the figure of population' (2004a, 81).

Right at the beginning of *Society Must be Defended* Foucault does something similar, where he discusses what he calls 'subjugated knowledges' both as 'historical contents that have been buried or masked in functional coherences or formal systemizations', and 'a whole series of knowledges that have been disqualified as nonconceptual knowledges, as insufficiently elaborated knowledges: naïve knowledges, hierarchically inferior knowledges, knowledges that are below the required level of erudition or scientificity' (1997, 8–9/7). This is important because Foucault wants to see his own work of genealogies as 'antisciences':

---

16 On the 1971 course see Defert 2001.

They are about the insurrection of knowledges. Not so much against the contents, methods or concepts of a science; this is above all, primarily, an insurrection against the centralizing power-effects that are bound up with the institutionalization and workings of any scientific discourse organized in a society such as ours. That this institutionalization of scientific discourse is embodied in a university or, in general terms, a pedagogical apparatus, that this institutionalization of scientific discourse is embodied in a theoretico-commercial network such as psychoanalysis, or in a political apparatus – with everything that implies, such as in the case of Marxism – is largely irrelevant. It is thus against the power-effects characteristic of any discourse that is regarded as scientific that genealogy has to engage in combat. (Foucault 1997, 10/9)

The question that needs to be asked, Foucault suggests, is 'what types of knowledge are you trying to disqualify when you say that you are a science? What speaking subject, what discursive subject, what subject of experience and knowledge are you trying to make minor [minoriser] when you begin to say: "I speak this discourse, I am speaking a scientific discourse, and I am a scientist"' (1997, 11/10). These reflections, Foucault contends, allow us to understand the relationship between archaeology and genealogy:

Archaeology is the method specific to the analysis of local discursivities, and genealogy is the tactic which, once it has described these local discursivities, brings into play the desubjugated knowledges that have been released from them. That just about sums up the overall project. (Foucault 1997, 11–12/10–11)[17]

The third question might be asked back to Foucault, and this is certainly one of the key things that he is known for today. Despite offering a clear analytic in *The Will of Knowledge*, especially in the chapter 'Method' (1976a, 121–35), Foucault often returned to this question with clarifications and fine-tunings. This is perhaps particularly the case in two important interviews: 'The Subject and Power' and 'The Ethic of Care for Self as a Practice of Freedom' (1994, 222–43; 708–29). Once again, *Society Must be Defended* offers an important perspective on this question:

What is power? Or rather – given that the question 'What is power?' is obviously a theoretical question that would provide an answer to everything, which is just what I don't want to do – the issue is to determine what are, in their mechanisms, effects, their relations, the various power-apparatuses that operate at various levels of society, in such very different domains and with so many different extensions? (Foucault 1997, 13–14/13)

Question four seems to be a question directed at Foucault's own ongoing work with the collaborative projects, and retrospectively to the Rio lectures and even back as far as *Birth of the Clinic*. While there is a developed spatial sense to these investigations, and though they undoubtedly provide rich insights and tools for geographers today, it is questionable if they would have been recognized as such in either the Anglophone

---

17 For a fuller discussion see Chris Philo's chapter in this volume.

or Francophone academy at the time. Gerry Kearns's chapter in this volume provides a number of helpful pointers.

Consequences for Foucault Studies

It is also important to consider the more general context of 1976. The editors of the course note that the events of 1968 in Czechoslovakia and France were hardly a distant memory, and that recent history included the transitions from dictatorship to democracy in Spain, Portugal and Greece; terrorism in Ireland, Italy and Germany; war in Indochina and the Middle East; and civil war in Africa and South America (Fontana and Bertrani 1997: 257). The first September 11th, in Chile when Salvador Allende was overthrown by a CIA-backed Augusto Pinochet, had taken place in 1973. Industrial unrest had recently led to the defeat of Ted Heath's government in England. This led to his replacement as Conservative leader by Margaret Thatcher, and the beginning of the neo-liberal dominance. The French left was struggling with the ideas of Eurocommunism and the unrepentant Stalinism of Georges Marchais, and it took five more years before a semi-united left facilitated the election of François Mitterand.

In the academy, numerous changes were taking place, particularly in the wake of the 1968 protests and those against the Vietnam war. Anglophone political theory was reenergized by debates around John Rawl's *A Theory of Justice* (1971), and in geography, David Harvey's *Social Justice and the City* (1973) was having an important catalyzing effect, particularly in the light of its move from positivism through liberalism to an explicitly Marxist agenda. The journals *Antipode* and *Hérodote* can be seen as having a similar role in the Anglophone and Francophone geographical circles, both attempting to shake up established patterns in research. Matthew Hannah and Juliet J. Fall's chapters in this volume provide much useful context and comparison here.

As Anglophone studies of Foucault's works continue to develop, both within geography and in wider disciplinary contexts, a number of general points can be drawn. First, the material that is now available. The 1994 four volume *Dits et écrits* not only collected together texts published in a range of hard to find places – it was the first collection as such in French, despite several collections in, for example, English, German and Italian – but also made a number of shorter texts available in French for the first time. Disparate texts originally published in Japanese, Italian and Portuguese were resituated in Foucault's work, often providing new insights and analyses. In terms of the themes explored here I would particularly highlight the lectures on medicine given in Rio in 1974. Much of this material is yet to appear in English, as the *Essential Works* collection is merely a selection from the much richer French edition. In addition to *Dits et écrits* the lecture courses also provide much new and contextual material. The impact that *Society Must be Defended* has had, in the short time since its publication, especially in the field of International Relations, is testament to what lies in store as further volumes are translated. In those volumes already published in France there are rich pickings for geographers – the analysis of the family and domestic space, asylum architecture, racial politics, colonial ordering and Foucault's relation to Marx, to name only some of the most obvious.

Second, Foucault as collaborator. The collaborative works were published during his lifetime, but with the exception of *I, Pierre Rivière* have not appeared in English. This is regrettable, because here the researches of Foucault and his colleagues are often much more concrete. *The Foucault Effect* was of course a collaborative work of sorts. It appeared after Foucault's death, and included work by many of his colleagues and friends, and has initiated a whole series of important studies utilizing the notion of 'governmentality' (Burchell, Gordon and Miller (Eds) 1991). Perhaps here we can see the model for the sort of work that Foucault hoped would follow him, work that was going on in his lifetime but remains largely unknown even in his native France. In this respect we should also look at some of the work done by colleagues since that date, including studies by Jacques Donzelot (1979), François Ewald (1986), Blandine Kriegel (1989) and Michel Senellart (1995).

Third, implicit in the above is the relation between this work and the later Foucault, shaped by two complementary concerns into the conduct of conduct and the question of the self. The question of the government of others depends centrally, as Foucault recognized, on the question of the government of the self. This is a theme that can be seen in the collaborative work mentioned, the detours through liberal modes of governance, Christian practices of confession and pastoral power, and the work of his last years on antiquity.

Finally, these questions demonstrate Foucault's openness to debate and productive dialogue. The rest of this book continues that discussion.

References

Bentham, Jeremy (1977) *Le panoptique*. Paris: Pierre Belfont.

Burchell, Graham, Colin Gordon and Peter Miller, Eds (1991) *The Foucault Effect: Studies in Governmentality*. Chicago: University of Chicago Press.

Davidson, Arnold I. (2003) 'Introduction.' In Michel Foucault, *Society Must be Defended*. Translated by David Macey. London: Allen Lane, xv–xxiii.

Defert, Daniel (2001) 'Le «dispositif de guerre» comme analyser des rapports de pouvoir.' In J-C. Zancarini (Ed.) *Lectures de Michel Foucault Volume 1: A propos de «il faut défendre la société»*. Lyon: ENS Éditions, 59–65.

Demangeon Alain and Bruno Fortier (1978) *Les vaisseaux et les villes*. Bruxelles: Pierre Mardaga.

Dillon, Michael and Andrew Neal, Eds (2007) *Foucault: Politics, Society, and War*. London: Palgrave.

Donzelot, Jacques (1979) *The Policing of Families: Welfare versus the State*. Translated by Robert Hurley. London: Hutchinson.

Elden, Stuart (2001) 'The Constitution of the Normal: Monsters and Masturbation at the *Collège de France*.' *boundary 2*, 28(1), 91–105.

Elden, Stuart (2002) 'The War of Races and the Constitution of the State: Foucault's *«Il faut défendre la société»* and the Politics of Calculation.' *boundary 2*, 29(1), 125–51.

Elden, Stuart (2003) 'Plague, Panopticon, Police.' *Surveillance and Society*, 1(3), 240–53.

Elden, Stuart (2005) 'The Problem of Confession: The Productive Failure of Foucault's *History of Sexuality.' Journal for Cultural Research*, 9(1), 23–41.

Elden, Stuart (2006) 'Discipline, Health and Madness: Foucault's *Le pouvoir psychiatrique.' History of the Human Sciences*, 19(1), 39–66.

Elden, Stuart (2007a) 'Governmentality, Calculation and Territory.' *Environment and Planning D: Society and Space*, 25(4).

Elden, Stuart (2007b) 'Strategies for Waging Peace: Foucault as *collaborateur.'* In Michael Dillon and Andrew Neal (Eds) *Foucault: Politics, Society, and War*. London: Palgrave.

Ewald, François (1986) *Histoire de l'état providence: Les origins de la solidarité*. Paris: Éditions Grasset et Fasquelle.

Farge, Arlette and Michel Foucault (1982) *Le désordre des familles: Lettres de cachet des Archives de la Bastille*. Paris: Julliard.

Fontana, Alessandro and Mauro Bertrani (1997) 'Situation du cours.' In Mauro Bertani and Alessandro Fontana (Eds) *Michel Foucault, «Il faut défendre la société»: Cours au Collège de France (1975-1976)*. Paris: Seuil/Gallimard.

Foucault, Michel (1963) *Naissance de la clinique: Une archéologie du regard médical*. Paris: PUF.

Foucault, Michel (1966) *Les mots et les choses – Une archéologie des sciences humaines*. Paris: Gallimard. Translated by Alan Sheridan as *The Order of Things – An Archaeology of the Human Sciences*. London: Routledge, 1970.

Foucault, Michel (1969) *L'Archéologie du savoir*. Paris: Gallimard. Translated by Alan Sheridan as *The Archaeology of Knowledge*. New York: Barnes & Noble, 1972.

Foucault, Michel, Ed. (1973) *Moi, Pierre Rivière, ayant égorgé ma mère, ma sœur et mon frère: Un cas de parricide au XIXᵉ siècle*. Paris: Gallimard.

Foucault, Michel (1975) *Surveiller et punir – Naissance de la prison*. Paris: Gallimard. Translated by Alan Sheridan as *Discipline and Punish – The Birth of the Prison*. Harmondsworth: Penguin, 1976.

Foucault, Michel (1976a) *Histoire de la sexualité I: La volonté de savoir*. Paris: Gallimard. Translated by Robert Hurley as *The History of Sexuality Volume I: The Will to Knowledge*. Harmondsworth: Penguin, 1978.

Foucault, Michel (1976b) 'Conference de Michel Foucault, Montréal, le 29 mars 1976.' Unpublished typescript. Some extracts appeared as 'Points de vue.' *Photo*, 24–25, Summer-Autumn 1976, 94, in 1994 III, 93–4.

Foucault, Michel, Ed. (1977) *Politiques de l'habitat (1800–1850)*, CORDA.

Foucault, Michel (1978) 'Interview of 3rd April 1978 with Paul Patton and Colin Gordon.' Unpublished typescript.

Foucault, Michel, Ed. (1980) *Herculine Barbin: Being the Recently Discovered Memoirs of a Nineteenth Century Hermaphrodite*. Translated by Richard McDougall. New York: Pantheon.

Foucault, Michel (1994) *Dits et écrits 1954–1988*. Edited by Daniel Defert and François Ewald. Paris: Gallimard, Four Volumes.

Foucault, Michel (1997) *«Il faut défendre la société»: Cours au Collège de France (1975–1976)*. Edited by Mauro Bertani and Alessandro Fontana. Paris: Seuil/Gallimard. Translated by David Macey as *Society Must be Defended*. London: Allen Lane, 2003. English pagination is provided after /, and though I have utilized Macey's translations I have modified them in a few places.

Foucault, Michel (1999) *Les Anormaux: Cours au Collège de France (1974–1975)*. Edited by Valerio Marchetti and Antonella Salomani. Paris: Seuil/Gallimard. Translated by Graham Burchell as *Abnormal*. London: Verso, 2003.

Foucault, Michel (2000) *Power: The Essential Works of Michel Foucault 1954–1984 Volume Two*. Edited by James Faubion. Translated by Robert Hurley and others. London: Allen Lane.

Foucault, Michel (2003) *Le pouvoir psychiatrique: Cours au Collège de France (1973–1974)*. Edited by Jacques Lagrange. Paris: Seuil/Gallimard.

Foucault, Michel (2004a) *Sécurité, Territoire, Population: Cours au Collège de France (1977–1978)*. Edited by Michel Senellart. Paris: Seuil/Gallimard.

Foucault, Michel (2004b) *Naissance de la biopolitique: Cours au Collège de France (1978–1979)*. Edited by Michel Senellart. Paris: Seuil/Gallimard.

Foucault, Michel (2004c) 'Crisis of Medicine or Anti-Medicine?' Translated by Edgar C. Knowlton Jr, William J. King and Clare O'Farrell. *Foucault Studies*, 1, 5–19.

Foucault, Michel (2007a) 'The Meshes of Power.' Translated by Gerald Moore (chapter 16 in this volume).

Foucault, Michel (2007b) 'The Incorporation of the Hospital into Modern Technology.' Translated by Edgar C. Knowlton Jr, William J. King and Stuart Elden (chapter 15 in this volume).

Foucault, Michel, Blandine Barrett Kriegel, Anne Thalamy, François Beguin and Bruno Fortier (1979 [1976]) *Les machines à guérir: Aux origines de l'hôpital moderne; dossiers et documents*. Paris: Institut de l'environnement. Reprinted as *Les machines à guérir (aux origines de l'hôpital moderne)*. Bruxelles: Pierre Mardaga.

Fourquet, François and Lion Murard (1976) *Les équipements du pouvoir: Villes, territories et équipments collectifs*. Paris: Union Générale d'Éditions (originally *Recherches*, 13, December 1973).

Harvey, David (1973) *Social Justice and the City*. Oxford: Basil Blackwell.

Kriegel, Blandine (1989) *L'état et les esclaves: Réflexions pour l'histoire des états*. Paris: Éditions Payot.

Lagrange, Jacques (2001) 'Versions de la psychiatrie dans les travaux de Michel Foucault.' In Philippe Artières and Emmanuel da Silva (Eds), *Michel Foucault et la médecine: Lectures et usages*. Paris: Éditions Kimé, 119–42.

Lemke, Thomas (1997) *Eine Kritik der politischen Vernunft: Foucaults Analyse der modernen Gouvernementalität*. Berlin: Argument.

Rawls, John (1971) *A Theory of Justice*. Oxford: Clarendon.

Ruffié, Jacques (1976) *De la biologie à la culture*, Paris: Flammarion, Two Volumes.

Senellart, Michel (1995) *Les arts de gouverner: Du* regimen *medieval au concept de gouvernement*. Paris: Seuil.

Stoler, Ann Laura (1995) *Race and the Education of Desire: Foucault's* History of Sexuality *and the Colonial Order of Things*. Durham: Duke University Press.

Zancarini, J.-C., Ed. (2001) *Lectures de Michel Foucault Volume 1: A propos de «il faut défendre la société»*. Lyon: ENS Éditions.

Zarka, Yves Charles, Ed. (2000) 'Michel Foucault: De la guerre des races au biopouvoir.' *Cités: Philosophie, Politique, Histoire*, 2.

Chapter 12

Formations of 'Foucault' in Anglo-American Geography: An Archaeological Sketch

Matthew Hannah

I do not believe that a search for *the* elusive, original 'meaning' of Foucault's various writings will necessarily be productive. (Driver 1985, 443)

I do not think that things have 'significance'. Things happen. That's all. What will have happened to Foucault remains (as always) undecidable. (Doel questionnaire 2004, see this volume, 87 note 3)

David Macey's biography of Michel Foucault is distinguished by an uncommon sensitivity to the difficulty of settling upon a unitary representation of his life and work (Macey 1993). This difficulty is fully consistent not only with Foucault's rich corpus of writings on the many dimensions of subjective (self-) formation, but also with his practice of periodically returning to and re-interpreting his own work in the light of new intellectual directions or discoveries. Foucault himself: 'insisted that his texts were a toolkit to be used or discarded by anyone and not a catalogue of theoretical ideas implying some conceptual unity' (Macey 1993, xx). Although he recognized that projections of unity would be inevitable, he nevertheless hoped his work would serve to open up new questions and unexpected directions, not to tether his inheritors to debates over the bounds of the properly 'Foucauldian' (Driver 1994, 116–17; Philo 2004, 122). This explains his experimentation with authorial anonymity, most notably in a 1980 *Le Monde* interview published in English as 'The Masked Philosopher' (Foucault 1989). Paradoxical though it may seem, then, 'Foucault' is most 'faithfully' understood as a multiple and dispersed construction, a construction which it does not make sense to measure against some originary, orthodox 'real Foucault'. This chapter is an attempt to trace some important aspects of the discursive construction of 'Foucault' in Anglo-American academic geography since roughly the early 1980s. Because I draw on Foucault as a theorist to understand 'Foucault' as a construction, the danger arises of confusing these two modes in the course of the narrative. Accordingly, quote marks will bracket the name 'Foucault' whenever he is discussed as a construction rather than as the source of guidelines on how to construct him.[1]

Method

In keeping with what Chris Philo quite rightly terms the 'spirit' of Foucault's approach to discursive materials (Laurier and Philo 2004, 424), the focus here is on systematic description of the different discursive things that have been done with and to 'Foucault' in the pages of geographical books and journals, at the podia of academic conferences and around university seminar tables. The specific descriptive method most appropriate to the task of turning Foucault's approach back upon his own academic discursive construction is what he termed 'archaeology' (Foucault 1972). Many accounts of Foucault's writings concur in the view that his archaeological method had run into insurmountable difficulties by the early 1970s. Chief among these were a blindness to extradiscursive power relations and an inadequate theorization of temporality in his histories of scientific discourse (Gutting 1989; Dreyfus and Rabinow 1983). Accordingly, the 'genealogical' method he subsequently developed, while perhaps not a decisive solution to the lingering questions about temporality, has generally been hailed as a clear improvement because of its explicit incorporation of issues of power.[2] Both archaeology and genealogy were additionally condemned in some quarters for an apparent neglect of individual human agency (Megill 1985; Frank 1989; Fraser 1989). However, as explained below, these dismissals were based on a misrecognition of 'Foucault' as a grand theorist of society, and on a correspondingly selective reading of his work. And in any case, his later research on 'technologies of the self' (Foucault 1988) and on governmentality (Foucault 1991) finally demonstrated with unmistakable clarity what he had been insisting upon all along, namely, that the lack of attention paid to individual human freedom or autonomy was a matter of methodological emphasis or bracketing, not of ontological principle. Be that as it may, because of multiple problems, the archaeological method appears to have been consigned by many commentators to the dustbin. Alan Megill,

---

1   Because Foucault's bodily life, as well as the texts he wrote, pre-existed his academic geographical construction, some ambiguity will be inevitable in the use of this device. A gray zone surrounds especially the liminal process of 'incoporation': at what point in the 'reception' of Foucault's work is his subjective authorship of discourse absorbed and transformed into an attribute of 'Foucault' the discursive object? Throughout the chapter, this ambiguity will express itself also in an oscillation between the language of construction and that of reception.

2   There are important exceptions to this pattern among interpretations in the geographical literature. Chris Philo, for example, argues that in fact 'for Foucault the conceptual distance between genealogy and archaeology is not as great as is sometimes thought' (Philo 1992, 143, note 6). Stuart Elden recommends seeing archaeology and genealogy as 'two halves of a complementary approach' (Elden 2001, 104).

though offering a more detailed reading than most, nevertheless ultimately dismisses it as a 'parody' of method (Megill 1985, 227).

If there is something to the 'parodic' reading of Foucault's *intent* in writing the *Archaeology*, an archaeological perspective should (and even the 'post-archaeological' Foucault would) caution us against giving this issue too much weight. Whether or not intended as a parody, the judgment that the *Archaeology* fails by 'serious' standards is too hasty, for a number of reasons. First, Foucault did in fact address extradiscursive power relations, both in his theoretical treatise on the archaeological method (Foucault 1972, 51–2; 67–8; 162–4) and in his classic, if underappreciated, empirical study *The Birth of the Clinic* (Foucault 1973b). Second, adequate attention to the issue of temporality is in no sense theoretically *forbidden* by the strictures of archaeology; indeed, various forms of temporality are given an explicit, if underemphasized place in the archaeological system (Foucault 1972, 56–7; 166–77; 186–9). This lack of emphasis on temporality is an artifact of Foucault's positioning of archaeology as an alternative to traditional history of ideas and history of science, which have always granted time a privileged causal status. In short, it actually presents no great difficulties to update or adapt the archaeological method to address these long-standing criticisms, while remaining basically faithful to Foucault's historiographical sensibility. But even if this is the case, it might be asked, why bother? Why not opt for the genealogical method Foucault and many of his interpreters consider superior? An important part of the answer is that the genealogical method has never been spelled out in anything like the rich detail in which Foucault laid out the archaeology. The scattering of essays (Foucault 1977b) and of 'methodological precautions' (Foucault 1980a, 96–103) found here and there throughout Foucault's writings from the 1970s do not actually move beyond a fairly rudimentary set of principles regarding the study of power-relations. Most of what Foucault 'meant' by genealogy must be pieced together from his empirical studies of discipline, biopower and governmentality (Foucault 1977a; 1980b; 1991; 2003). The piecemeal methodological discourse found in these studies is, to be sure, tremendously subtle, but it is not systematic. The great advantage of systematicity in an account like the one undertaken here is that the explicit ordering of the archaeological method constitutes a much more efficient window on its animating sensibility than would a painstaking search through the vast and complicated landscape of Foucault's various empirical studies.

To clarify what I mean here, it is helpful to engage with one final critical approach to the use of archaeology as a method. Chris Philo, the foremost interpreter of Foucault's archeological work in Geography, follows Lemert and Gillan (1982) in insisting that the *Archaeology* does not lay out a method in any straightforward sense but is instead a 'topical' study of the ontology of discourse (Philo 1989, 220; Laurier and Philo 2004, 424). Indeed, he suggests that archaeology is more of a 'style', and thus that 'one can easily be a competent and sensitive archaeologist without ladening one's analyses with the bewildering profusion of notions and terms that constitute this precarious edifice' (Laurier and Philo 2004, 428). I agree fully that archaeology is a style, and also that there is no imperative to use all the details of Foucault's system in any given archaeological study. But why, then, did Foucault himself bother to elaborate this system in such exhaustive and exhausting detail? In my view, he did so for two reasons. First, he did so *precisely to demonstrate a style of relating to discursive materials*. The crux of his demonstration is the mimetic relation between the style of exposition in the *Archaeology* and the descriptive sensibility it seeks to impart to those who might wish to use it to study other discursive materials. In other words, it is the painstaking and patient systematicity of Foucault's presentation in the *Archaeology of Knowledge*, more than the specific categorial structure he introduces, that constitutes its fundamental lesson. The 'bewildering profusion' characteristic of the *Archaeology* is only bewildering as long as the reader clings to an ideal of the possibility of re-emerging from the thicket of terms to establish a synthetic overview. Once that ambition is abandoned, it is possible to enter into a less reductive relation to textual traces, and to be satisfied kneeling alongside Foucault and absorbing the ethos of patience required by 'the slow removal of soil with delicate hand-tools, the troweling, picking, and brushing away of dirt, to reveal the shape of the object normally surrounded by soil' (Laurier and Philo 2004, 424). The second reason Foucault bothered to write the *Archaeology* as he did is that its categories do in fact constitute relevant descriptive dimensions within which to place organized discourses. Although Philo is right to point out that any particular archaeological study need not make use of all of these categories, as I hope to show below, they can in fact be extremely useful in the task of outlining discursive 'spaces of dispersion'. For both these reasons, I believe the *Archaeology* provides appropriate conceptual equipment with which to illustrate the specificity of the ways in which 'Foucault' has existed in Anglophone academic geographic discourse. Illustration is indeed the chief aim of the account offered here.

The details of the archaeological method, and adjustments introduced into it, will be spelled out as the narrative proceeds. One of these adjustments will be a heightened level of attention to temporalities. The other major issue raised by critiques of archaeology, the issue of non-discursive power relations, is not actually so pressing in the context of an account of changes within a professional academic discourse, but will be addressed insofar as it arises. Individual human agency will be given a more prominent place here than in Foucault's archaeological studies, but the emphasis will be laid less on personal uniqueness than on institutional or discursive positioning of the people under discussion. I will also not shy away, to the extent that Foucault might have done, from drawing limited generalizations out of the discursive material at hand.

A good deal of the empirical material on which this chapter is based comes from the familiar published literature (articles, book chapters, etc.). This material will be supplemented by responses to an informal survey sent out to geographers (as well as a few non-geographers who have participated in the geographic discourse) in 2003 and 2004 regarding the details of their engagements with Foucault's writings and geographic work based upon them.[3] The responses do not in any sense provide a complete overview of what Foucault's work has meant to geographers. First, questionnaires were sent only to scholars who have had some sort of sustained intellectual engagement with Foucault. This means that the constructions of 'Foucault' documented below are largely 'positive'; the many geographical encounters with his work that have left colleagues indifferent, dismissive or downright hostile find no expression here. This is, in one sense, not such a great loss, as it is safe to say that negative constructions of 'Foucault' tend to be far less nuanced and far lower-dimensional than the sustained constructions likely to accompany positive encounters. Secondly, the survey was not sent to *every* scholar who has made detailed use of Foucault, but first only to contributors to this volume, followed by a few others selected in an unsystematic way. Even within the relatively small category of Anglophone geographers who have delved deeply into Foucault's work, the sample taken here is thus fundamentally incomplete. Thirdly, only some of those to whom the survey was sent were able to respond, further compounding the incompleteness. Nevertheless, despite the extremely partial nature of the sources, it was possible to obtain responses from geographers whose encounters with Foucault began at different times during the past three decades, and in different institutional settings. A fourth limitation of the discursive field out of which material is taken is the neglect of conversations, marginal notes and other elements of the Foucauldian archive that would be difficult or impossible to obtain in any case. Finally, and most obviously, 'Foucault' as an archival corpus continues to grow and mutate, not only through secondary works such as this book but, even more importantly, as primary works, including the translation published in this volume, and at a larger scale his lecture courses from the 1970s continue to appear in print. As Stuart Elden notes, these lecture courses will 'radically overhaul our view of Foucault, particularly around the *History of Sexuality* and recontextualizing the governmentality literature' (Elden questionnaire 2004). All of this means that the particular formations of 'Foucault' traced here can at best be provisional and indicative.

The argument will proceed according to the system laid out by Foucault in the *Archaeology*, but will cover only a limited selection of the descriptive categories elaborated there. More specifically, I will focus primarily on the various dimensions of the 'formation of objects' within a 'discursive formation', and treat 'Foucault' as a discursive object fitted out through the discourse of critical human geography with a range of attributes (including, of course, the properties associated with an 'author'). Brief attention will be paid at the end to the 'formation of enunciative

---

3   I am extremely grateful to colleagues who responded to this survey: Jeremy Crampton, Marcus Doel, Felix Driver, Stuart Elden, Margo Huxley, Gerry Kearns, Steve Legg, Sara Mills, Miles Ogborn and Chris Philo. I also completed a survey. Because these respondents have characterized their responses in many cases as provisional and off-the-cuff, and because not all of their comments were intended for publication, I have been fairly conservative in decisions about direct quotation. For these reasons, too, there are places where I simply indicate a common or general opinion expressed in the surveys, without listing specific names of respondents. A copy of the (blank) survey questionnaire is available upon request.

modalities', in order to indicate how possible subject positions in the academic geographical discourse about Foucault have changed in recent years. For reasons of space, it is necessary simply to leave aside the other major categories through which an archaeology would complete the description of 'Foucault' as a discursive phenomenon. However, despite its unfinished character, the account offered here should suffice to trace important features of his geographical construction, and to illustrate the archaeological sensibility or style.

The Formation of 'Foucault' as an Object

Foucault's concern in elaborating the archaeological method was to get away from traditional histories of ideas in which what is said and written is understood to take its meaning from, or to express, *something else* (for example, the march of Reason, Progress, Enlightenment, the Spirit of the Age, or even just the intentions or personality of an author). Simply put, archaeology is an attempt to stop asking what lies outside of discourse that it 'expresses', an attempt to confine the study of discourse to the specificity of what is actually said and written, to the irreducible complexity of relations between actual discursive events: to study discourse as 'monument' rather than as 'document', as *existence* rather than as *expression* (Foucault 1972, 7). However, Foucault does not wish simply to record the multiplicity of discursive events in a nominalistic fashion, without reference to any kind of system (Philo 1992, 143). His aim is to locate the complex relations that tie units of discourse together *within, not outside* the level of discursive events. The relevant units of discourse for him are thus not the familiar ones of 'genre', 'work', 'author', 'discipline', 'oeuvre', 'school', etc. Each of these points to something beyond discourse as the seat of its 'real' meaning. In place of such terms, Foucault proposes to study 'discursive formations', 'statements' and 'archives'. In order to describe 'Foucault', it will be most useful to concentrate on the first of these concepts. A discursive formation is a provisional unity of discourse whose coherence is attributed to a delimitable range of common objects, subject positions, concepts and argumentative strategies (Foucault 1972, 31–9). The descriptive task of archaeology is to show how a given set of objects, subjects, concepts and strategies have been *formed* and perhaps altered through actual discursive events. The result is an account of what Foucault calls 'spaces of dispersion' (Foucault 1972, 10). Because of this focus on modes of discursive existence, such an approach will not have as its end result an account of 'what Foucault means' but rather an account of how 'Foucault' has existed as a discursive phenomenon. What follows is thus not an interpretation of Foucault's thought, nor of geographers' thoughts about Foucault. Nevertheless, as I hope to show here, an archaeology of 'Foucault' can in fact offer interpretive insights unavailable through more traditional lenses.

For the purposes of the argument here, it suffices to take Anglo-American critical human geography of the past 25 years as the relevant discursive formation, and to ask how 'Foucault' appeared, and has been formed and re-formed as an object, within its spaces of dispersion. 'Foucault's' specificity as an object also involves his constitution as a subject, as an 'author' or 'thinker', a very distinct type of discursive phenomenon. Thus, before considering other, generic issues associated with the delimitation of a discursive object, it is necessary to dwell on this matter of authorship. In his essay 'What is an author?', which originated as a lecture given in 1969 (Macey 1993, 209) and is thus roughly contemporaneous with the *Archaeology*, Foucault indicates a number of peculiarities attaching to discourses (among them, academic discourses) in which the 'author function' is important (Foucault 1984, 107). In such discourses:

> [t]he author's name serves to characterize a certain mode of being of discourse: the fact that the discourse has an author's name, that one can say 'this was written by so-and-so' or 'so-and-so is the author', shows that this discourse is not ordinary everyday speech that merely comes and goes, not something that is immediately consumable. On the contrary, it is a speech that must be received in a certain mode and that, in a given culture, must receive a certain status. (Foucault 1984, 107)

More particularly, 'Foucault' has been designated by geographers as an author whose work intersects with discourses of history, philosophy, social, cultural or political theory, but who cannot be set squarely in any of these boxes (Driver 1985, 426). A survey of the bewildering variety of shelving categories imprinted by publishers on the covers of Foucault's books gives a strong sense of this ambiguity and multiplicity. According to whether his writings are labelled as history, philosophy or social theory, they will be approached from within somewhat different conceptual economies, standards of argumentation, rules of evidence, baseline assumptions, and so on. The fact that, taken in the aggregate, Foucault's works appear to exceed or criss-cross all such boundaries only heightens the importance of his name as a unifying device for all of his work. What Foucault writes of the role of the author in literary criticism is thus certainly applicable to the geographic construction of 'Foucault':

> ... the author provides the basis for explaining not only the presence of certain events in a work, but also their transformations, distortions, and diverse modifications (through his biography, the determination of his individual perspective, the analysis of his social position, and the revelation of his basic design). The author is also the principle of a certain unity of writing – all differences having to be resolved, at least in part, by the principles of evolution, maturation or influence. The author also serves to neutralize the contradictions that may emerge in a series of texts: there must be – at a certain level of his thought or desire, of his consciousness or unconscious – a point where contradictions are resolved, where incompatible elements are at last tied together or organized around a fundamental or originating contradiction. Finally, the author is a particular source of expression that, in more or less completed forms, is manifested equally well, and with similar validity, in works, sketches, letters, fragments, and so on. (Foucault 1984, 111)

As will be shown in more detail below, the interdisciplinary span of Foucault's authorship made it very tempting in early geographic encounters to line him up alongside what Chris Philo terms 'other grand theorists of society and space' (Philo

1992, 137). And having done so, the unity ascribed to his writings was easy to see as an overarching vision of modern society *tout court*. To assemble a more supple context for exploring this and other features of Geography's 'Foucault', we return to the wider systematics of archaeology.

Authorities of Delimitation

In the *Archaeology of Knowledge*, Foucault distinguishes three dimensions along which the constitution of objects in a discursive formation can be described: 'authorities of delimitation', 'surfaces of emergence', and 'grids of specification'. Here I take the liberty of adding another dimension between the second and third of these: what might be called 'temporalities of emergence'. As noted above, Foucault's downplaying of temporality in his exposition of archaeology was a matter of strategic emphasis. To balance out this reticence, it will be helpful to highlight temporality as a distinct dimension, but in a way consistent with Foucault's overarching commitment to dispersion and specificity.[4]

By 'authorities of delimitation', Foucault means subject positions from which individuals have or acquire the right to define and delineate the objects of a discursive formation. In his work of the 1960s, Foucault was interested chiefly in the initial emergence of the human sciences, so the issue of who gained the authority to delineate objects was an important one. Unlike Foucault, I am concerned here with features of a discursive formation that has its home within an already well-established and institutionalized academic discourse, a discourse firmly anchored in departments, journals, professional associations, and the like. Thus the question of

---

4   The nature of these dimensions is immediately suspect. Despite Foucault's professed desire to restrict archaeology to the level of discursive events themselves, his descriptive framework is clearly a coordinate system imposed from outside. The discursive formations he studies of course make no mention of 'authorities of delimitation', 'surfaces of emergence' or 'grids of specification'; such terms must be read into the material under investigation. Yet the way these dimensions function in archaeological description is very different from the role played by general interpretive frameworks in traditional modes of historical analysis. They only make up a 'coordinate system' in a very loose sense, and do not in fact allow comparison of different discursive phenomena in terms of 'degree', 'development', etc. Every discursive formation involves 'authorities of delimitation', 'surfaces of emergence' and 'grids of specification', and in this sense the archaeological framework is universalistic. But it cannot be used, for example, to argue that one discursive formation is more 'advanced' than another. Like traditional interpretive frameworks, archaeological dimensions take the first step of differentiating discursive elements into categories. However, these categories are intentionally designed not to support the second step in traditional intellectual history, the reduction of the resulting diversity by arrangement of specific details along a common 'scale'. Thus, while the archaeological dimensions are indeed imposed from outside, once they have been used to describe a discursive formation they do not allow a return to the outside, a reductive abstraction. This is what Foucault means when he writes that archaeology is conceived as 'general history' (Foucault 1972, 9).

who it is that first identified 'Foucault' is not a story of major shifts in disciplinary structures. Two points of entry proved important in the 1980s in giving his work a profile within Anglophone geography (though Gerry Kearns, for example, recalls seeing Foucault's work in bookstores in the mid-1970s (Kearns questionnaire 2005), and indicates that some geographers were already aware of Foucault before this period (personal conversation March 2004)). The first discursively productive node was clearly the Cambridge Department of Geography and in particular Derek Gregory, who, though he did not himself initially publish on 'Foucault', encouraged his then-doctoral students Felix Driver and Chris Philo in the early 1980s to read different works by Foucault in connection with their dissertation projects on the historical geography of institutions for the poor and the 'mad', respectively (Driver 1993; Driver questionnaire 2004; Philo 2004; Philo questionnaire 2004; Ogborn questionnaire 2004). Philo also recalls that Allen Pred mentioned Foucault while visiting Cambridge in 1982–1983 (Philo questionnaire 2004). Later, while visiting Penn State University as a distinguished lecturer in 1987, Gregory likewise put me on to *Discipline and Punish* upon learning that I was interested in power relations and censuses (Hannah questionnaire 2004). The second influential node was Ed Soja, whose 1989 *Postmodern geographies* gave prominent place to Foucault as a social and cultural theorist helping to correct the long-standing marginalization of space in critical social theory. Writing in 1992, Chris Philo singled out Soja's book as the only exception, up to that point, to the general absence of any 'sustained theoretical engagement with Foucault on the part of theoretically minded geographers' (Philo 1992, 138). Many of the respondents to the questionnaire on which this chapter is based concur in the judgment that Soja's approach to Foucault was very influential in placing 'Foucault' for geographers.

As noted above, Derek Gregory and Ed Soja were already ensconced in institutional frameworks of considerable stability, and their engagements with Foucault's writings did not challenge their basic professional positionalities. Nevertheless, it is not necessarily true that anyone within the field of human geography could have launched a sustained engagement with Foucault's work. This is one place where the archaeological sensibility should encourage us to pause and consider the specific characteristics of those who first delineated 'Foucault'. An intriguing comment by Gregory in the preface to his 1994 book *Geographical Imaginations* points to a field of questions that can only be briefly explored here. Gregory recalls the experience of reading David Harvey's *Social Justice and the City* in the following terms: 'This was the first book in geography I knew I didn't understand. Trained in spatial science, I was accustomed to technical difficulty, but conceptual difficulty was an altogether different order of things, particularly when it required an engagement with social theory and a recognition of ethical and political responsibility' (Gregory 1994, xi). By the time Foucault began to be read in geography, Gregory, like Soja, belonged to a cadre of scholars (also including David Harvey, Doreen Massey, Allen Pred, Gunnar Olsson, Michael Dear, Nigel Thrift, Trevor Barnes, Susan Hanson, and others) who had successfully managed the transition from 'technical' to 'conceptual' difficulty. All had made their reputations in established areas of human geography but had subsequently distinguished themselves as *importers of difficult but useful ideas from other fields*. By the early 1980s, both Gregory and Soja were well-versed in the arcana of Marxist theory, the first great critical theoretical tradition to become a regular part of the education of human geographers after the 1960s. And their credentials as practitioners of spatial analysis may have helped to underwrite their authority as importers of abstract social-theoretic constructs.[5] Gregory's 1978 *Ideology, Science and Human Geography* was important in that it expanded the scope of the Marxist tradition available to geographers beyond the initial focus on economic and urban dynamics (Gregory 1978). In so doing, it inaugurated the more wide-ranging exploration of European social theory that would form the general context for the initial engagement with Foucault. The point of an archaeological account, however, is not to single out individuals for their ground-breaking achievements (though Gregory and Soja can and should, of course, be given credit for their work). It is rather to describe those features of their subject-positions that may have lent a certain degree of initial 'credibility capital' to 'Foucault' as he was introduced into geography.

Surfaces and Temporalities of Emergence

The second major dimension in the formation of objects Foucault terms 'surfaces of emergence'. Here he is concerned to attend to the specific discursive and institutional sites in which objects first emerge or are re-configured. In the case of Anglophone critical human geography, important surfaces of emergence included the university bookstores in which English translations of Foucault's works became available (often initially because of interest generated in disciplines other than geography), interdisciplinary seminars (Gerry Kearns and Felix Driver both recall an important Foucault discussion group in the mid-1980s involving scholars from around the UK: Nikolas Rose, Colin Gordon and others (Driver questionnaire 2004 and personal communication 2005; Kearns personal conversation 2004)), and conversations with fellow academics. As geographers started to process Foucault's writings, academic conferences became important points of dissemination, as did books and the pages of journals, particularly *Society and Space*, where in 1985 Felix Driver published the first full-length research paper primarily concerned with Foucault's work (Driver 1985). *Society and Space* is of particular significance from an archaeological perspective concerned with the *mode of existence* of discourses: in providing the most amenable outlet for early publications on 'Foucault', its range of articles in the mid- to late-

---

5   This claim is admittedly speculative, but it points to an interesting and as yet relatively under-explored issue in studies of the recent history of human geography. The turn to critical social and cultural theory among some human geographers was at one level clearly a turn away from positivism, scientism, etc. as philosophical positions. But at the level of practice, that is, of forming and communicating concrete arguments, and at the level of discursive authority, the continuities between spatial analysis and critical social theory may be more numerous than the 'paradigm shift' approach would suggest.

1980s constituted a matrix of topics, genres and problematics that could not help but lay down some of the main coordinates along which 'Foucault' would be understood. For example, a survey of articles appearing around Driver's in the 1985 issues of *Society and Space* reveals emphases on industrial and urban geographies, social-theoretically informed takes on regional economies and planning, and engagements with fundamental issues in social theory such as social reproduction, naturalism, and the spatial and temporal constitution of social action. As will become clearer below, this material, printed co-presence alongside articles (or within an article, alongside references to other thinkers) placed 'Foucault' in intellectual company that made certain interpretations of his significance for geographers more difficult than others.

Again, in comparison with the kinds of nascent disciplinary situations Foucault chronicled in his studies, the surfaces of emergence of new objects within the well-established and institutionally anchored field of human geography was perhaps not so remarkable. But in keeping with his insistence on attention to specificity, it is possible to give the description additional detail by asking about the *temporalities* of 'Foucault's' emergence as an object. In the *Archaeology*, again, temporality is a complex question. Here we can discern a number of different levels of temporality relevant to the ways in which 'Foucault' emerged: (1) the order of appearance of Foucault's writings in English (both the sequence of major works and the translation of papers, collections of essays, lectures, etc.); (2) the different sequences in which geographers have read Foucault, and thus formed and elaborated usable understandings of his work; and (3) the longevity, duration, or 'shelf-life' of 'Foucault' as a discursive object remaining in circulation over a span of time in the traditional sense. In general terms, the availability of Foucault's major works in English was substantial, though by no means complete, by the mid-1980s, when the last volume of his *History of Sexuality* appeared. Collections of interviews, papers and the secondary literature lagged the major works somewhat, and, as noted above, Foucault's lecture courses continue to appear in the 2000s. However, most of the work on which the early geographical encounter with Foucault was based was already accessible in the early 1980s. Thus the second type of temporality, the sequence in which geographers have read Foucault, has been quite important. The questionnaire responses I was able to gather show one clear pattern: the writings on power-relations, particularly *Discipline and Punish* and/or *Power/Knowledge*, have remained the most common points of entry into Foucault's corpus since the early 1980s. This accords well with the general perception among respondents that Foucault has been constructed by geographers up to now primarily as a 'geometer of power' (Laurier and Philo 2004, 422 note 3), and that this construction has led to what Elden terms an 'incessant emphasis on the Panopticon' (Elden 2001, 3). At the level of substantive argument, not all work focusing on the relations of power idealized in Bentham's plans for the Panopticon can be said to support this construction of 'Foucault' (Felix Driver, especially, has been very careful to spell out the scope of 'transferability' of panoptic strategies from one setting to another; Driver 1994). But at the archaeological level of material discourse, it is undeniable that Driver's (1994) and Hannah's (1992, 1997) somewhat different approaches to the question of how disciplinary techniques of power can operate outside situations of bodily confinement have strengthened the association 'Foucault – power – Panopticon'.

It is not unreasonable to assume, further, that the process of entering the sequence of Foucault's writings in what most commentators consider his 'middle period', and reading both backward (into the archaeological work) and forward (into the work on subjectivity, sexuality and self-constitution) from there, has indelibly coloured the sense many of us have of how these other themes are to be understood. For example, having myself followed the well-trodden path into 'Foucault' that begins at *Discipline and Punish*, I can attest to the difficulty, in particular, of subsequently reading his archaeological work without seeing it as indelibly marked in part by the 'presence of an absence', namely, the absence of an account of non-discursive power-relations (Hannah questionnaire 2004). It is probably not an accident that Chris Philo, the only geographer responding to the survey whose course of reading began with works from Foucault's earlier period, has remained almost alone in representing 'Foucault' primarily as an archaeologist (though see Hannah 1993) (Philo questionnaire 2004).[6]

Interestingly, the tendency among geographers to start with writings on power relations is not followed by any consistent pattern regarding what comes next in the chain of subsequent readings. To adopt Marcus Doel's succinct formula for his own reading of Foucault, 'fragments at random' (Doel questionnaire 2004), would be to exaggerate the unsystematic character of most scholars' readings, which have clearly been structured in part by the pursuit of specific questions. But once engaged, geographers appear to have moved through Foucault's archive in every imaginable sequence, some roughly following the order of publication of major works, some tackling each of the three major 'phases' but not in their original order of appearance, some skipping around. The array of temporalities evident here is a clear illustration of a 'space of dispersion'. However, when the occasion has arisen to re-read Foucault's works systematically, whether as a result of the acquisition of new language competencies, as part of the completion of a doctoral degree, or because of a specific publishing project, the tendency has been to start at the beginning and proceed chronologically (Elden questionnaire 2004; Hannah questionnaire 2004). This means that traditional temporality, in the sense of the chronological sequence of publication of Foucault's works, is still tacitly understood to be centrally important to the strengthening of one's grasp of his corpus. Comparing accounts of re-reading Foucault with responses to questions about more and less valuable aspects of the geographical reception of 'Foucault' suggests that re-reading Foucault's work, whether chronologically or otherwise, may be experienced as a form of emancipation from the effects of having originally started with *Discipline and Punish*.

---

6   Sara Mills, approaching Foucault's work initially from Linguistics rather than Geography, was the only one of the questionnaire respondents who actually began her reading of Foucault with the *Archaeology* (Mills questionnaire 2004). At the other end of Foucault's publishing career, only Miles Ogborn reports beginning with *The History of Sexuality* (Ogborn questionnaire 2004).

To all of these senses of the temporality of emergence must be added the effects of discursive *duration*, or, less formally, of *shelf-life*. Especially in cases where a thinker or concept is constructed as potentially dangerous to pre-existing intellectual or political traditions, the establishment of a certain durability by means of repeated discursive citation and iteration can eventually alter the perception of 'danger'. 'Foucault', like 'Derrida' and others often lumped together under the rubric of postmodernism or poststructuralism, was sometimes portrayed in the 1980s as potentially a dire threat to human geographic discourse. Yet, at the archaeological level of discursive monuments or events, their persistence as objects and anchors for clusters of concepts is decisive evidence that they were after all no more difficult for this discursive formation to incorporate than were 'Marx' or 'Habermas'. A long shelf-life not only tends to defuse perceptions of danger but in doing so also tends to lend objects like 'Foucault' an increasingly general availability for use in a wide range of scholarly projects.

There are a number of aspects to 'Foucault's' 'shelf-life' that could be approached archaeologically. For illustrative purposes, I will focus on just one issue: the shifting discursive placement of 'Foucault' in relation to other 'authors' considered relevant to geographical thought. Margo Huxley evokes what is probably a quite common sense that 'Foucault' is no longer counted among the sources of 'innovative' thought at the cutting edge of geographical readings. In one department she is familiar with, Huxley reports that the staff consider Foucault 'a bit old hat'. 'We're all Deleuzian, ANT, Nature/Culture, hybridity, post-colonial now' (Huxley questionnaire 2004). An archaeological study would seek to link such perceptions to concrete changes in the existence and co-existence of discursive phenomena. An interesting opportunity to do so is afforded by the re-publication of Chris Philo's 1992 paper 'Foucault's Geography' in the edited collection *Thinking Space* (Crang and Thrift 2000). A full sense of how Foucault would have approached this instance of re-publication archaeologically would require a far more exhaustive discussion of his concept of the 'statement' (*énoncé*) than there is room for here (Foucault 1972, 79–134). Nevertheless, something of the basic flavour of an archaeological analysis can be hinted at by a selection of passages on the problem of a statement's repeatability:

> The same person may repeat the same sentence several times: this will produce the same number of enunciations distinct in time. The enunciation is an unrepeatable event; it has a situated and dated uniqueness that is irreducible. Yet this uniqueness allows of a number of constants – grammatical, semantic, logical – by which one can, by neutralizing the moment of enunciation and the coordinates that individualize it, recognize the general form of a sentence, a meaning, a proposition. [...] But the statement itself cannot be reduced to this pure event of enunciation, for despite its materiality, it cannot be repeated: it would not be difficult to say that the same sentence spoken by two people in slightly different circumstances constitute only one statement. And yet the statement cannot be reduced to a grammatical or logical form because, to a greater degree than that form, and in a different way, it is susceptible to differences of material, substance, time and place. What, then, is that materiality proper to the statement, and which permits certain

special types of repetition? [....] What, then, is this rule of *repeatable materiality* that characterizes the statement? (Foucault 1972, 101–2)

The affirmation that the Earth is round or that species evolve does not constitute the same statement before and after Copernicus, before and after Darwin; it is not, for such simple formulations, that the meaning of the words has changed; what changed was the relation of these affirmations to other propositions, their conditions of use and reinvestment, the field of experience, of possible verifications, of problems to be resolved, to which they can be referred. (Foucault 1972, 103)

Committed as he was to viewing discourse as *monument* rather than as *document*, to tracing its conditions of existence in a non-reductive way, it should not be surprising that Foucault was unsatisfied with the traditional units of discourse (sentences, meanings, propositions). These units can all be detached from the material context of any specific enunciation and are therefore well suited to the traditional task of reducing the complex web of the actually said or published to the more manageable set of key ideas, advances, lines of thought. Since such reduction is what Foucault strives consciously to avoid, his 'units' must not be so easily detachable. Thus Foucault's concept of the statement may be characterized as the basic building block of discursive formations, the basic unit of discursive *existence*, in a manner parallel to the sense in which a proposition or sentence forms the basic building block of discursive *meaning* (Foucault 1972, 79–80). But because it is a unit of existence rather than meaning, the statement is trapped to or chained within its concrete conditions of production: it cannot be a 'floating signifier'. Despite all the space he devotes to explaining and defining statements in the *Archaeology*, Foucault is not actually interested in identifying instances of statements. The term 'statement' serves him practically speaking less as a category of identification than as a sort of discursive 'torch' he carries on his exploration of the catacombs of the history of ideas, in order to throw into sharp relief all the different ways such concepts as 'sentence' and 'proposition' obfuscate the manner in which discourses exist.

Nevertheless, it is possible, with all due caution and a silent entourage of caveats, to understand Chris Philo's paper in its initial appearance as a statement. According to the general approach spelled out in the above passages, it is doubtful whether its reappearance in 2000 constitutes the same statement. Just to illustrate where such a problematic might lead us, we can begin by examining the published work that forms the immediate context of the two iterations of Philo's paper. It initially appeared in 1992 in volume 10 of *Society and Space*, which was very strongly oriented (across most of the numbers that year) around the modernism-postmodernism debate. 'Foucault' was certainly linked to that debate, as suggested in part by the fact that Philo was one of four geographers appearing that year in *Society and Space* who were associated with Foucault's work (Felix Driver, David Matless, and Miles Ogborn being the other three). Driver's and Ogborn's papers were in no sense primarily about postmodernism, but since an archaeological approach is sensitive above all to how discursive phenomena exist, it would encourage us to assign some significance to the repeated co-existence in print of Foucault scholars with papers on postmodernism.

Philo's 1992 paper would have been taken by many readers as an intervention in the debate on postmodernism, which not only structured the context but also played a prominent role in Philo's abstract. There he argued that Foucault's 'sensitivity to spaces of dispersion' as well as his substantive geographies emerge 'directly from his own suspicion of the certainties (the order, coherence, truth, reason) supposed by most historians and social scientists to lie at the heart of social life, and as such I think that it can be adjudged a 'truly' postmodern human geography in a manner that, say, Edward Soja's postmodern geographies cannot' (Philo 1992, 137).

In its 2000 iteration, Philo's paper inhabits a somewhat different discursive matrix, one populated by some of the same figures who would have been familiar in 1992 (Benjamin, Lefebvre, Bourdieu, Wittgenstein, Lacan, Deleuze), but also by a range of thinkers who, even if known, were not particularly central to the postmodernism debate in the early 1990s (Bakhtin, de Certeau, Cixous, Fanon, Latour, Serres, Said, Minh-ha, Virilio). The editors conceive the theme of the book (*Thinking Space*) in a specific way: 'We have asked contributors who have engaged with particular writers and thinkers to unpack the way their approaches utilize spaces rather than appropriate them unproblematically' (Crang and Thrift 2000, xii). In other words, unlike in the postmodernism debate, here there is no particular approach or school whose definition or reputation is at stake. In this different context, the same paper by Chris Philo will undoubtedly be read with different emphases. The issue of whether Foucault's geography is more fully postmodern than Soja's is likely to recede into the background and be replaced in the foreground by Philo's rich, two-layered exposition of Foucault's spatial sensibilities and substantive geographies. Of course, one need not adopt an archaeological perspective to argue that the same work means different things in different contexts. But archaeology does encourage us to dwell on the specifics, to pay close attention to exactly how (in material detail) the discursive contexts differ, and to hold open the possibility that such mundane changes in *how discursive phenomena exist* play a significant role as the 'substratum' of phenomena such as 'paradigm shifts', 'advances' and so on which are typically portrayed in a more ideational or semantic register.

Grids of Specification

In the larger project of describing the formation of discursive objects, authorities of delimitation, surfaces of emergence and temporalities of emergence may all influence and be influenced, finally, by what Foucault terms 'grids of specification', that is, classificatory dimensions along which an object is located within a discursive formation. 'Foucault' has been formed and re-formed in critical human geography in important ways according to where he has been located in terms of *genre* (is he an historian, a philosopher, a social theorist?), as noted above, and further, in terms of a series of key distinctions or spectra. Where he has been placed in terms of modernist vs. postmodernist approaches, in terms of the distinction between theorist and empirical researcher, or along the political continuum stretching from radical through progressive and liberal to conservative, has been intimately bound up with basic judgments about the meaning of his writings. In the mid- to late-1980s, as Foucault's discursive presence in Geography deepened, one of the key grids of specification was constituted by geographers' intensive engagement with critical social theory. This engagement had a number of dimensions, but the central problematic was clearly the agency-structure issue associated especially with the work of Anthony Giddens. In a manner that combined (and often confused) analytical and political interests, debates raged around the degree to which socio-spatial determination left room for, or could perhaps even provide the medium for, relatively autonomous individual human agency (Giddens 1979, 1981, 1984). Emerging into this milieu, the reception of 'Foucault' could scarcely have avoided the vortex of the agency-structure debate. Chris Philo's experience is probably not untypical for the time: 'Initially, though, I tended to read Foucault – indeed everyone! – through the lens of the Giddensian 'agency-structure' debate, and so missed much of what was important about his take on power until later on' (Philo questionnaire 2004). Felix Driver lists Giddens, along with John Urry, Nicos Poulantzas, critical realism and early feminist geography as important components of this context (Driver questionnaire 2004).[7] In the later 1980s, when I was delving into Foucault, the agency-structure problematic remained the single most important axis organizing the intellectual field (Hannah questionnaire 2004).

It is not surprising that Foucault was generally understood by critics to privilege determination over agency (Matless 1992, 46), especially since many geographical studies have focused on his analysis of power relations (Driver 1985; 1994; Soja 1989; Hannah 1992; 1993; 1997; Doel and Clarke 1999). A similar pattern is undoubtedly evident in other social sciences. But to relate 'Foucault' to the agency-structure nexus in this way was to include him in the category of 'grand theorists of society and space' (Philo 1992, 137). Doing so meant saddling him with specific responsibilities both theoretical and practical/political in nature. As a grand theorist, for example, he was held responsible by non-geographers as well as geographers for *comprehensive* representations of 'society'; as a *critical theorist*, he was further taken to have assumed the burden of *political implications* (the readings of Foucault offered by Habermas (1987) and Fraser (1989) are symptomatic of this pattern). I do not want to suggest that it is only for critical theorists 'proper' that the issue of politics should be of interest. However, the label does tend to carry with it a stronger expectation that an author's theoretical contributions will at least be consistent with, if not serve as a foundation for, an acceptably progressive political programme. Once arrayed alongside such figures as Giddens or Bhaskar, Foucault's claims about how power worked, or about the role of humans in social scientific explanation, could be understood both as general claims about modern society *tout court* and as guidelines

---

7   Driver also reports (personal communication 2005) that his own reception of Foucault was influenced as well by French interpretations from Paul Claval and Claude Raffestin, and it is worth stressing here, too, that my portrayal of the context of reception already necessarily reduces the complexity of the spaces of dispersion marked out in every survey response.

for politics. It was against this complex of assumptions and categorizations that Driver cautioned as early as 1985, when he remarked that 'Foucault, above all, would reject the label of 'abstract theoretician' with which he has sometimes been saddled' (Driver 1985, 426). In a similar vein, Philo urged geographers 'to pause for a moment in our projects of combining Foucault with Giddens, Lefebvre, Mann, or whomever – the projects of turning Foucault into the 'same' – and instead we should recognize (and perhaps marvel at) the 'otherness' of his perspective on geography and postmodernism as something really quite 'alien' to all manner of current ways for proceeding as geographers' (Philo 1992, 140).

In this passage, Philo hints at something I would like to make more explicit: that it was this placement of Foucault in the ranks of grand social theorists that lent whatever urgency there seemed to be to the question of where to place 'Foucault' on the modernist-postmodernist grid. More generally, as I have argued elsewhere, much of the angst animating discussions of postmodernism can be traced to the survival of expectations of comprehensiveness, originally associated with grand social theory, in the assessment of the claims of writers such as Derrida or Foucault (Hannah 1999). The somewhat paradoxical figure of the 'grand theorist of postmodernity' appearing at the intersection of these two grids of specification is an inherently unstable subject-position for Foucault (or anyone else) to inhabit. Although he would undoubtedly (and perhaps quite rightly) object to the simplifications involved in holding him responsible, Ed Soja is singled out in published accounts (Philo 1992, 138–9; Matless 1992, 43; Elden 2001, 3, 157–8, notes 12 and 14) as well as in a number of the questionnaires as the geographer who has done the most to cast Foucault in the guise of an implicitly universalist socio-spatial theorist.

More recently, as the geographers have slowly extricated 'Foucault' from the field of social theory, other grids of specification have become important. Race studies, colonial studies and postcolonial theory have provided an increasingly rich grid of specification for Foucault, through the writings of Derek Gregory (Gregory 1994), Phil Howell (2000), Dan Clayton (2000), and others. Through this scholarship, 'Foucault' has been more clearly identified as a *white European* author, but as one who cannot simply be dismissed for his Eurocentric sins. These sins notwithstanding, scholars such as Edward Said (1978) and Gyan Prakash (1999) have found his subtle empirical accounts of the techniques for constructing sameness and otherness in European society relevant to accounts of colonialism and postcolonial movements in the Global South. This strand of scholarship has been given a boost by the 2003 publication of Foucault's lecture course from 1976, which reveals a more extended and detailed discussion of the political process of racial 'othering' than is to be found in his major works (Foucault 2003). This lecture course, and the anticipated publication of his courses from the subsequent two years (1977 and 1978) will also continue to expand the archive of easily available material on 'biopower' and 'governmentality'. The latter concept has made it possible for scholars to begin to connect Foucault's work with the larger critical discourse on neo-liberalism (Dean 1999; Rose 1999), and may lead to a new, more positive, assessment of the relation between Foucauldian and Marxian accounts of the capitalist state.

Formation of Enunciative Modalities

In the method outlined by the *Archaeology* for describing a discursive formation, the formation of objects, which I have been concerned to illustrate here, is only the first of four dimensions. It is followed in Foucault's narrative by investigation of the formation of 'enunciative modalities', that is, the positions and practices attaching to subjects qualified to speak about the object, investigation of the formation of concepts, and lastly, of the formation of argumentative strategies. Each of these sections is quite detailed and complicated, and each would certainly lend additional depth to the archaeological excavation of 'Foucault's' place in the discursive formation of critical human geography. Although there is no space to cover these dimensions here, it is still worth making a brief foray into the second of the four, the formation of enunciative modalities. Concluding comments will follow.

Foucault understands enunciative modalities as a different issue from the identification of 'authorities of delimitation' reviewed above. This distinction is traceable in large part to his interest in the early stages of emergence of the human sciences, when those who first identify an object ('disease', 'society', or what have you) are not necessarily those who will come to be seen later as the experts qualified to unfold a detailed scientific account of it. As in the case of authorities of delimitation, Foucault would want to ask: (1) who can speak?; and (2) from what institutional positions? But in addition, the description of enunciative modalities involves asking, (3) with what specific relations to the object? (Foucault 1972, 50–55.) In the discursive formation of Anglophone critical human geography, the first two of these questions would have the same answers as noted in the discussion of authorities of delimitation above. The third, relation to the object, becomes a more complicated issue as the engagement with an object is elaborated over time. Whereas authorities of delimitation were essentially those people who first identified 'Foucault' as a relevant object, scholars who participated in the subsequent geographical engagement could take a wider array of positions: popularizer, exegete, re-interpreter, critic, defender, expander or contractor of applicability or relevance, primarily empirical user, gesturer, invoker.

As it happened, the roles typically assumed by geographers interested in Foucault were closely related to the problem noted above in the section on grids of specification: once Foucault was identified more or less explicitly as a grand theorist of society with a suspect politics, it became more difficult simply to *use* his work. What I mean by this is that some geographers who, though not originally interested in becoming 'Foucault authorities' but simply in making use of aspects of his approach to interrogate specific empirical problems, have nevertheless felt themselves obligated to inhabit the position of 'defender of Foucault'. I can certainly recall many such moments in hallway conversations or reading groups in graduate school (Hannah questionnaire 2004). 'Defending' Foucault, however, involves buying into more assumptions about the self-consistency of 'author', the coherence of the '*oeuvre*', etc., than would merely making limited use of specific concepts. In his questionnaire response, Chris Philo notes the irony in the fact that he 'risk[s] being a Foucault 'puritan', demanding greater fidelity to his texts and getting annoyed with what I see as shallow or mis- readings: this seems rather counter to his own position' (Philo questionnaire 2004). It is perhaps not too much of an exaggeration to say that until recently, geographers interested in Foucault's ideas have also found themselves 'holding shares' in 'Foucault the author', and have had a hard time embracing the sort of relation to 'Foucault' put so inimitably by Marcus Doel: 'I have never been interested in either the "author" or the "*oeuvre*". Foucault never existed for me' (Doel questionnaire 2004).

A key text that has spurred an ongoing, fundamental re-description of 'Foucault' in geography, Stuart Elden's *Mapping the Present* (Elden 2001), was launched from a position that can perhaps best be described as exegetical. As a political philosopher by training, Elden can lay claim to a level of expertise in interpreting the thought of 'thinkers' that is unavailable to most geographers (responses to questionnaires show evidence of this textual authority, and reveal that Elden's book has quickly become a touchstone for geographers). Crucially, Elden's exegesis of Foucault as a philosophical thinker leads him to reject the inherited habit of seeing him as a theorist of social space. By placing Foucault alongside Heidegger and noting many formal and conceptual similarities between their approaches to problems of space and time, Elden locates Foucault in the grid of philosophical discourse. This is a less-confining home than is social theory, because philosophers can concern themselves with fundamental issues of epistemology and ontology without incurring the responsibility directly to account for social realities. Read in this context, Foucault appears not as a theorist or historian of the concept of space, but as a philosophically-informed historian whose studies underwrite a programme to '*spatialize history*, to inject an awareness of space into all historical studies, to critically examine the power relations at play in the ways space is effected and effects' (Elden 2001, 7). Not only *Discipline and Punish* (Foucault 1977a), but also *Madness and Civilization* (Foucault 1973a) and *Birth of the Clinic* (Foucault 1973b) are rich with empirical accounts of the ways in which spatial relations and techniques of confinement, exclusion and ordering are inextricably woven into wider social processes and transformations. Elden makes a persuasive argument that a narrow focus by geographers and others on the figure of the Panopticon in *Discipline and Punish* has unnecessarily reduced Foucault's account of power relations to a fairly simple model, and has thereby made it far easier than it should have been to treat this account as a transportable theory of spatial power relations (Elden 2001, 120–50). This critique lends more critical bite to the common discomfort noted above, and expressed both in the questionnaire responses I received and in some of the published work on 'Foucault', with the prominence of discussions of panopticism in geography (Laurier and Philo 2004, 422, note 3; Philo 1992, 139; Driver 1985, 433, incl. note 6; Doel and Clarke 1999).

Viewed archaeologically, in terms of the range of enunciative positions available to geographers with respect to 'Foucault', Elden's text appears to have made possible a new division of enunciative labour. From the mid-1980s until recently, many scholarly users of Foucault's ideas also took on some of the duties of explication and defence of 'Foucault', if not always in print, at least often in their departments and at conferences. If some scholars can establish exegesis as their primary role with respect to 'Foucault', this double-duty may no longer seem so necessary to the rest. Stuart Elden and some of the respondents to the questionnaire agree that what geographers have always done most successfully is what Foucault himself usually did: use Foucault's ideas in connection with specific empirical research projects. According to Felix Driver, for example, 'the best evidence of Foucault's influence on geographical scholarship is to be found in substantive works – on colonialism, mapping, statistics, social policy, madness, urbanism, sexuality. The philosophical engagements, such as they are, have been less interesting' (Driver questionnaire 2004). In this vein, Miles Ogborn's book on eighteenth century London is frequently singled out in the questionnaires as exemplary (Ogborn 1998).

The only potential obstacle to such a new division of labour within geography may be the difficulty of publishing 'pure' intellectual history or philosophical exegesis in geography journals whose editors still often ask authors 'where's the geography?' (Elden hints at this issue in his questionnaire response). This problem is probably of steadily decreasing relevance, as geographers now habitually cross disciplinary boundaries in gathering material for their work. The inauguration of the e-journal *Foucault Studies* (www.foucault-studies.com) will certainly boost whatever movement there already is toward such a new division of labour, and make it easier for scholars across the social sciences and cultural studies to focus on using Foucault's ideas without mis-using 'Foucault'.

Conclusions

As the epigraphs chosen for the beginning of this chapter indicate, there is little point in pursuing the 'true' Foucault for the sake of pinning him down once and for all. But this is a far cry from claiming that there is no point in studying his works or the ways in which he has been constructed in geography. The intervening argument was intended as an illustration of the contribution of a specifically archaeological sensibility to the latter task. Again, the archaeological method revolves around the question of the mode of existence of discourse. While this question does not and cannot exhaust the 'meaning' of Foucault or of any other thinker, it can offer insights not available through the lenses of a traditional history of ideas. Whether or not one sees changes and shifts in the Anglophone geographical construction of 'Foucault' as a matter of 'progress', questions of who has been able to delineate 'Foucault' from which specific positions, exactly where in discursive spaces has 'Foucault' emerged, in what sequences his writings have appeared and been read, and along which conceptual and genre-related dimensions 'Foucault' has been located, all considerably enrich our sense of why 'Foucault' has come to mean the variety of things he has.

Acknowledgements

In addition to the scholars thanked in note 3 above for the time they devoted to completing the questionnaire from which some material for this chapter was gleaned, I would like to thank Felix Driver, Stuart Elden, Steve Legg and Chris Philo for their thoughtful comments, clarifications and corrections.

References

Clayton, D. (2000) *Islands of Truth: the Imperial Fashioning of Vancouver Island.* Vancouver, BC: University of British Columbia Press.

Crang, M. and N. Thrift, Eds. (2000) *Thinking Space.* New York: Routledge.

Dean, M. (1999) *Governmentality: Power and Rule in Modern Society.* Thousand Oaks: Sage.

Doel, M. and D. Clarke (1999) 'Dark Panopticon. Or, Attack of the Killer Tomatoes.' *Environment and Planning D: Society and Space*, 17, 427–50.

Dreyfus, H. and P. Rabinow (1983) *Michel Foucault: Beyond Structuralism and Hermeneutics.* 2nd Ed. Chicago: University of Chicago Press.

Driver, F. (1985) 'Power, Space, and the Body: A Critical Assessment of Foucault's *Discipline and Punish.' Environment and Planning D: Society and Space*, 3, 425–46.

Driver, F. (1993) *Power and Pauperism: The Workhouse System 1834–1884.* Cambridge: Cambridge University Press.

Driver, F. (1994) 'Bodies in Space: Foucault's Account of Disciplinary Power.' In C. Jones and R. Porter (Eds) *Reassessing Foucault: Power, Medicine and the Body.* New York: Routledge. 113–31.

Elden, S. (2001) *Mapping the Present: Heidegger, Foucault and the Project of a Spatial History.* New York: Continuum.

Foucault, M. (1972) *The Archaeology of Knowledge.* Translated by A. Sheridan-Smith. New York: Pantheon.

Foucault, M. (1973a) *Madness and Civilization: A History of Insanity in the Age of Reason.* Translated by R. Howard. New York: Vintage.

Foucault, M. (1973b) *The Birth of the Clinic: An Archaeology of Medical Perception.* Translated by A. Sheridan-Smith. New York: Vintage.

Foucault, M. (1977a) *Discipline and Punish: The Birth of the Prison.* Translated by A. Sheridan. New York: Vintage.

Foucault, M. (1977b) 'Nietzsche, Genealogy, History.' In M. Foucault, *Language, Counter-Memory, Practice: Selected Essays and Interviews.* Translated by D. Bouchard and S. Simon. Edited by D. Bouchard. Ithaca, NY: Cornell University Press. 139–64.

Foucault, M. (1980a) 'Two Lectures.' In *Power/Knowledge: Selected Interviews and Other Writings.* Edited by C. Gordon. New York: Pantheon. 78–108.

Foucault, M. (1980b) *The History of Sexuality Volume I: An Introduction.* Translated by R. Hurley. New York: Vintage.

Foucault, M. (1984) 'What is an Author?' Translated by J. Harari. In P. Rabinow (Ed.) *The Foucault Reader*. New York: Pantheon. 101–20.

Foucault, M. (1988) *Technologies of the Self: A Seminar with Michel Foucault*. Edited by L. Martin, H. Gutman and P. Hutton. Amherst, MA: University of Massachusetts Press.

Foucault, M. (1989) 'The Masked Philosopher.' Translated by J. Johnston. In S. Lotringer (Ed.) *Foucault Live (Interviews, 1966–84)*. New York: Semiotext(e). 193–202.

Foucault, M. (1991) 'Governmentality.' In G. Burchell, C. Gordon, and P. Miller (Eds) *The Foucault Effect: Studies in Governmentality*. Chicago: University of Chicago Press. 87–104.

Foucault, M. (2003) *'Society Must be Defended': Lectures at the Collège de France, 1975–1976*. Translated by D. Macey. Edited by M. Bertani and A. Fontana. New York: Picador.

Frank, M. (1989) *What is Neo-Structuralism?* Translated by S. Wilke and R. Gray. Minneapolis, MN: University of Minnesota Press.

Fraser, N. (1989) *Unruly Practices: Power, Discourse and Gender in Contemporary Social Theory*. Minneapolis: University of Minnesota Press.

Giddens, A. (1979) *Central Problems in Social Theory: Action, Structure and Contradiction in Social Analysis*. Berkeley, CA: University of California Press.

Giddens, A. (1981) *A Contemporary Critique of Historical Materialism*. Berkeley, CA: University of California Press.

Giddens, A. (1984) *The Constitution of Society*. Berkeley, CA: University of California Press.

Gregory, D. (1978) *Ideology, Science and Human Geography*. New York: St. Martin's Press.

Gregory, D. (1994) *Geographical Imaginations*. New York: Blackwell.

Gutting, G. (1989) *Michel Foucault's Archaeology of Scientific Reason*. New York: Cambridge University Press.

Habermas, J. (1987) *The Philosophical Discourse of Modernity: Twelve Lectures*. Translated by F. Lawrence. Cambridge, MA: MIT Press.

Hannah, M. (1992) 'Foucault Deinstitutionalized: Spatial Prerequisites for Modern Social Control.' Unpublished PhD dissertation. University Park, PA: Pennsylvania State University.

Hannah, M. (1993) 'Foucault on Theorizing Specificity.' *Environment and Planning D: Society and Space*, 11, 349–63.

Hannah, M. (1997) 'Space and the Structuring of Disciplinary Power: An Interpretive Review.' *Geografiska Annaler*, 79 B, 171–80.

Hannah, M. (1999) 'Skeptical Realism: From Either/or to Both-and.' *Environment and Planning D: Society and Space*, 17, 17–34.

Howell P. (2000) 'Prostitution and Racialised Sexuality: The Regulation of Prostitution in Britain and the British Empire before the Contagious Diseases Acts.' *Environment and Planning D: Society and Space*, 18(3), 321–39.

Laurier, E. and C. Philo (2004) 'Ethnoarchaeology and Undefined Investigations.' *Environment and Planning A*, 36, 421–36.

Lemert, C. and G. Gillan (1982). *Michel Foucault: Social Theory and Transgression.* New York: Columbia University Press.

Macey, D. (1993) *The Lives of Michel Foucault: A Biography*. New York: Pantheon Books.

Matless, D. (1992) 'An Occasion for Geography: Landscape, Representation, and Foucault's Corpus.' *Environment and Planning D: Society and Space*, 10, 41–56.

Megill, A. (1985) *Prophets of Extremity: Nietzsche, Heidegger, Foucault, Derrida.* Berkeley, CA: University of California Press.

Ogborn, M. (1998) *Spaces of Modernity – London's Geographies, 1680–1780*. New York: Guilford.

Philo, C. (1989) 'Thoughts, Words, and "Creative Locational Acts".' In F. Boal and D. Livingstone (Eds) *The Behavioural Environment: Essays in Reflection, Application and Re-evaluation.* London: Routledge. 205–34.

Philo, C. (1992) 'Foucault's Geography.' *Environment and Planning D: Society and Space*, 10, 137–61.

Philo, C. (2004) 'Michel Foucault.' In P. Hubbard, R. Kitchin and G. Valentine (Eds) *Key Thinkers on Space and Place*. Thousand Oaks, CA: Sage Publications. 121–8.

Prakash, G. (1999) *Another Reason: Science and the Imagination of Modern India.* Princeton, NJ: Princeton University Press.

Rose, N. (1999) *Powers of Freedom: Reframing Political Thought.* New York: Cambridge University Press.

Said, E. (1978) *Orientalism*. New York: Vintage.

Soja, E. (1989) *Postmodern Geographies: The Reassertion of Space in Critical Social Theory*. New York: Verso.

Chapter 13

Catalysts and Converts: Sparking Interest for Foucault among Francophone Geographers

Juliet J. Fall

Divided Geographical Worlds

In a recent article in *L'Espace Géographique*, Jean-Marc Besse notes, almost with some surprise, that 'one of the important references of [Anglophone] postmodern writers is the work of Michel Foucault, in particular the articulation between knowledge and power'[1] (Besse 2004, 4). The fact that this is worth noting in an introductory article of a journal on postmodernism and geography articulates the gulf between *Anglo*[2] and Francophone geographies. This chapter on Foucault and Francophone geography explores the context of this comment and the corresponding fracture between two very different geographical traditions. It confronts, as Minca has put it, 'the persistence of a sort of "parallel" geographical tradition that in France

---

1   All translations, unless stated otherwise, are personal translations. The original text is included in these footnotes for clarity. Translation from: 'une des références importantes des postmodernes aux Etats-Unis est l'œuvre de Michel Foucault, en particulier cette articulation faite par Foucault entre savoir et pouvoir'.

2   In order to avoid arguing myself into a corner over whether to use Anglo-American or English-language, or any other term to describe the sort-of-geography-as-carried-out-in-multiple-places-where-English-is-used-to-write, I will use the French spoken term *Anglo*, a mildly slangy expression as in 'Les *Anglos* font comme ça, mais nous...'. In doing this, I am mindful of reifying this as an internally coherent, homogenous body of writing. Paasi, for instance, most recently explored the uneven geographies of knowledge production within geography, noting most convincingly that 'binary divisions, such as Anglophone versus the rest of the world, thus hide that these contexts are in themselves heterogeneous and modified by power geometries' (Paasi 2005, 770; see also Garcia-Ramon 2003; Samers and Sidaway 2000; Minca 2000; Agnew and Duncan 1981). Thus while simply acknowledging the breach between the two is at times intellectually unsatisfactory, and requires nuance, I am happy to risk this shorthand since the viewpoint adopted in this chapter is one from which the internal differences appear erased. It is harder to find a snappy equivalent to describe the French-language world since the term often used by the *Anglos* is 'French'. As an Anglo-Swiss, I find this unsatisfactory, and have chosen 'Francophone' as a rather less catchy alternative.

is still very much alive but (…) does not nurture a broad dialogue with the Anglo-American ("international"?) geographical universe, although it continues to exert significant influence on a number of European geographies' (Minca 2000, 286; see also Staszak 2001; Chivallon 2003; Besse 2004). Paasi most recently explored the uneven geographies of knowledge production within geography, noting most convincingly that 'binary divisions, such as Anglophone versus the rest of the world, thus hide that these contexts are in themselves heterogeneous and modified by power geometries' (Paasi 2005, 770; see also Garcia-Ramon 2003). Simply acknowledging the breach between the two is intellectually unsatisfactory. The suggestion that there is an *Anglo* versus a Francophone space of (political) geography requires nuance and is explored in detail elsewhere (Fall 2006). Yet as a contribution to this wider discussion, this chapter offers one opportunity for a reflexive look at the production of scientific discourses by comparing different contexts, mindful of their internal complexities.

Despite the seminal interview of 1976 that appeared to build a bridge between disciplines, Francophone geographers have rarely used the work of Michel Foucault. To some extent, this reflects differences in the way authors and references are used within the two traditions, a point I will return to. Yet more than just writing styles underpin these differences. This chapter seeks to explore why Foucault is such a marginal figure in Francophone geography, why he has in effect performed his own *exercice de disparition*. I start out by briefly noting the ironic absence of 'French Theory' within Francophone geography, a group Foucault is framed in the *Anglo* world as belonging to, subsequently exploring the institutional and historical contexts of university life in France and other Francophone countries that point towards explanations. I then move on to explore what parts of Foucault's writing have in fact permeated and been picked up, tracing how they got there, using the contrast of the *Anglo* world to highlight specificities, emphasizing in particular the recent work of Christine Chivallon, Michel Lussault and Jacques Lévy. Lastly, by examining in more detail the writings of Claude Raffestin, a Swiss geographer who relied heavily on certain aspects of Foucault's work, and by exploring why he has remained largely unknown outside of his immediate circles, I point to a number of further paths for reflection.

Setting the Scene: French Theory Everywhere but in France

The crux, of course, and the main point that is explored here, is that while Michel Foucault, Jacques Derrida, Gilles Deleuze and others were becoming unavoidable in universities across the Atlantic and in Britain, 'their names were being systematically eclipsed in France'[3] (Cusset 2003, 22). This absence of Foucault is especially striking within geography: heralded as manna in the various foci of *Anglo* geography, he shines by his absence – as we say in French – in Francophone geographical circles.

---

3   '… leurs noms connaissaient en France une éclipse systématique.'

Commenting on the relative absence of spatial analyses within the wider political sciences, Buléon noted 'it is particularly striking that a number of Anglo-Saxon debates within which space is considered central are being precisely fed by French authors, including Foucault and Lefebvre'[4] (Buléon 1991, 33). As two members of the established clique of French geographers put it simply: 'the French critical philosophy of the 60s and 70s is less popular in France and the Latin countries than in the United States – Barthes and Derrida are not quoted; the interest in Foucault is more evident' (Claval and Staszak 2004, 319, see also Söderström and Philo 2004, 304). Yet even if Foucault gets a special mention in that editorial to a special journal edition on 'Latin' geographers – in this case French, Swiss-Romand, Italian and Brazilian – the only explicit reference in the entire issue is to factual historical points put forward in *Les Mots et les Choses* (Foucault 1966). A short survey of the scant references to Foucault by Francophone geographers indicates that in addition to *Les Mots et les Choses*, only *La Volonté de Savoir* and *L'Ordre du Discours* have been used in any meaningful way and even then only scantly. In contrast to his comment quoted above, Claval, rather ambiguously, had written earlier that the absence of convergence between Francophone and Anglophone (political) geographies continued 'in spite of the obvious intellectual influence of French thinkers such as Foucault, Derrida and Baudrillard' (Claval 2000, 262). What this earlier comment failed to note, of course, is that these 'French' thinkers had very little influence on their (geographical) compatriots.

In his book on what has been called 'French Theory', François Cusset (2003) lays out some of the historical, social and institutional processes that participated in the creation of a global politico-theoretical arena fed by an amalgamation of key writers, firmly centred and grounded not in France, but in American universities. To a certain point, this present book on Foucault and geography is part of this global movement. The tale of reducing, reusing and recycling ideas in order to create 'French Theory' is nothing new and was first hinted at, albeit ambiguously, by Sylvère Lotringer and Sande Cohen two years earlier. The latter cannot make their mind up about the true nature of French Theory, describing it simultaneously as 'arguably the most intellectually stimulating series of texts produced in the postwar area' (Lotringer and Cohen 2001, 3), or 'an American invention going back to at least the eighteenth century' (Lotringer and Cohen 2001, 1) and eventually stating that 'there was never any 'unity' to such French Theory, even among those close to each other' (Lotringer and Cohen 2001, 8). Cusset is much less ambiguous, stating that the unity within French Theory is indeed no more than a juxtaposition, a proximity and promiscuity forced through systematic intertextuality, a position also adopted in this present chapter (see also Agnew 2001, 11). This in no way diminishes the individual contributions of the various authors, nor does it deny their tentative collaborations.

---

4   '... il est d'autant plus piquant que nombre de discussion anglo-saxonnes où l'espace tient une place de choix soient justement nourries d'auteurs français parmi lesquels Foucault et Lefebvre.'

Institutions, Rituals and Personalities across Francophone Geography

Foucault's absence is particularly surprising in France since geography is institutionally still largely associated with history, a fact that has been called a 'bidisciplinarité relative', dating back to the institutionalization of the disciplines in the 1880s (Garcia in Djament 2004). French historians have tackled Foucault's proposals on archaeologies of knowledge and genealogies to a certain extent and a historian, Olivier Razac is for instance credited by Michel Lussault – a geographer – with having written the best 'Foucaldian' essay on space in his *Histoire Politique du Barbelé* [barbed wire] (Razac 2000). When looking at Foucault's very different impact on geography, Raffestin wrote that 'I don't know if (...) M. Foucault revolutionized the study of history, only historians can endorse this or not, but in any case the Foucauldian method provided, together with the archaeology of knowledge, a precious method for "genealogical" researches that the human sciences are often confronted with'[5] (Raffestin 1992, 23). The link between geography and history is far from benign in France:[6] in many ways it reduces geography to the role of little sister of the more glamorous sibling, in contrast to the context of, say, French-speaking Switzerland where geography is institutionally more likely to be associated with the social sciences, the earth sciences or the natural sciences.

A quick parenthesis on the French system of universities is useful, if a little laborious, at this point, particularly as it is so alien to Anglo-Saxon ways of organizing the academy. It is also different from the much more decentralized structures prevalent in other French-speaking contexts such as Switzerland or Quebec. Understanding the intricacies of the French system and its potential for immobility helps to understand the non-emergence of Foucault within French geography. It is a cliché to say that France remains a centralized country, with official lists of required reading set on a national level by a committee of respected elders: the CNU or *Commission Nationale Universitaire*. Research is still largely directed centrally within programmes defined by the Ministry of Research and the *Centre Nationale de la Recherche Scientifique* (CNRS) (Collignon 2004: 376). It is not surprising that such a system led Foucault to come up with the term 'groupe doctrinal' (Foucault 1971, 47), describing particular *sociétés de discours* whose functions were to 'conserve or produce discourses in order to circulate them within an enclosed space, but only distributing them according to strict rules and without this bringing about any loss of control for the

---

5 '... je ne sais pas si P. Veyne a eu raison d'écrire que M. Foucault a révolutionné l'histoire, seuls les historiens peuvent ou non en témoigner mais, en tout cas, la méthode foucaldienne a fourni, avec l'archéologie du savoir, une précieuse méthode aux recherches 'généalogiques' auxquelles les sciences humaines sont souvent confrontées.'

6 Both Jean-François Staszak and Louis Dupont commented on this proximity of geography and history in France in the debate reprinted in *L'Espace géographique*, 2004 (1), 18 and 19. Dupont notes in particular that 'On est surpris de l'ancrage incroyable qui fixe et limite le savoir géographique à l'histoire'.

holders'[7] (Foucault 1971, 42). Although naturally not restricted to the academy, such a definition seems convincingly apt in France, land of the supposedly reason-led *planification nationale*. In this context, confusing to outsiders, a whole host of academic-oriented *concours* [competitions] are organized on a national level, each requiring about a year of preparation within designated schools of varying prestige, creating a highly guarded clique of people able to discuss any topic at very short notice (see Lévy 1995 for a personal description; Bourdieu 1984 for an outline).[8] Another formal step on the way to an academic career is the *Maître de conférence* exam which is more like a competitive registration: having finished a doctoral thesis, candidates have it validated by the CNU. Approximately 40 per cent of candidates get through and can then apply for lectureships at universities, paying out of their own pockets to attend interviews around the country.

This is not a system designed for rapid innovation or the rise of freethinkers – innovation for innovation's sake is frowned upon and pointed out as something uniquely *Anglo*, and therefore intrinsically suspect (Cusset 2003, 230). Instead, as one anonymous colleague put it, the system rewards cooptation through spiritual formatting from an early stage, rewarding those who are strong enough to navigate through a jungle of implicit and explicit rules, gaining substantial diplomatic and strategic skills in the process and wisely choosing well-placed mentors (Anon 2004, *personal communication*). In comparison to British or North American contexts, the French geographical world is like a small family within which – as one geographer put it – *il faut montrer patte blanche* (Chivallon, 2005, *personal communication)*, that is to say that individual acceptance is obtained by demonstrating one's worth, as in many exclusive peer groups, as well as by conforming and not sticking out too much. Furthermore, a form of intellectual immobility is maintained by hierarchy: academics only get to supervise doctoral theses towards the end of their careers,

---

7   '... conserver ou de produire des discours, mais pour les faire circuler dans un espace fermé, ne les distribuer que selon des règles strictes et sans que les détenteurs soient depossédés par cette distribution même.'

8   To start with, candidates must pass the *CAPES* in order to qualify as *histoire-géographie* high-school teachers, with the further option of the *Agrégation* to be better-paid teachers or academics. The latter in particular, while not formally an academic degree, is run by academics and rewards candidates' ability to produce a *leçon magistrale*, an academic lecture, at short notice on any topic. Its highly selective nature and formal ranking of individuals confers substantial prestige on its holders, making them better placed when applying for academic jobs and reinforcing a dominant clique firmly centred on Paris, grounded in a particular body of thought. While not formally required when applying for academic jobs, successful candidates are more likely than not to be *Agrégés*, having passed both geography and history sections. For geographers, this would mean three written parts in geography and one in history, for both written and oral exams. In recent years, to give an idea of the scale of these competitions, roughly 10,000 people sign up for the geography/history *Capes*, of which 800–900 are selected. At the next level, the 'Capétiens', assuming they have four years of university-level studies, can apply for the *Agrégation*. Each year, about 1000 people apply of which 35 are finally selected. Their individual rank is published in a formal *classement*.

once they have attained the status of full professors, after passing another hurdle, similar to the German *Habilitation,* by writing what amounts to a second thesis. Paradoxically, however, or maybe in consequence of this hierarchical system, 'belonging' to a particular school of thinking is not highly regarded in France – in contrast, I would suggest, to the *Anglo* world – and instead being 'outside' and 'unclassifiable' is valued (Lévy and Debarbieux 2004, *personal communication;* Chivallon 2005, *personal communication*).[9]

Anything identified as jargon is savagely frowned upon. This could be seen as a rejection of clear *doctrines* (Foucault 1971, 45), although to suggest there are none within geography would be to misunderstand Foucault's point. Likewise, labels ('postmodern', 'poststructuralist', 'constructivist', 'feminist' and so on), are seen to enclose and are largely rejected in France (Chivallon in Antheaume et al. 2004, 13) and sometimes feared. Indeed, in another piece, Chivallon writes that 'it is scarcely possible to speak of 'postmodern geography' in France without suspicion of scientific heresy' (Chivallon 2003, 406). This, however, does not mean there is no cult of particular individuals on a national level, each engaged in very actively promoting themselves within the media, often at the cost of actual debates about ideas[10] (Lévy 2004, *personal communication*). As Bourdieu (1984) has noted, this need to position oneself within the academia has an important effect on how ideas are spread and appropriated, relating to the varying visibility of different thinkers. Another substantial difference in France is the rarity of public debates, partly due to the absence of recent paradigmatic change, due mainly to reduced generational renewal. This institutional fixity has largely contributed to a certain climate of comfortable conformity and the corresponding strategy of remaining within the accepted *pré carré,* the designated field assigned to the discipline, rather than seeking inspiration from the outside – such as from social theorists like Foucault. This may well be simply a current trend linked to individual waves of recruitment, as the current pattern is in contrast to more vivid debates in the 1970s and 1980s pitching the *Nouvelle Géographie* against established conservative paradigms (Chivallon 2005, *personal communication*).

In consequence of this highly codified French system, the smaller, marginal or peripheral schools in Switzerland and Quebec have sometimes acted as catalysts and innovators, largely by simply staying outside of partisan politics in France. In the case of Quebec – where historically, and rather ironically, geography departments have been part of an American regional division of the Association of American Geographers, not a Canadian one – more intense exchanges with the US has also ensured the circulation of alternative ideas. In the past thirty years, many French academics have moved to Switzerland, for instance, not only lured by the substantially

---

9   Ironically, a counter-trait is also common: describing oneself, even late in one's career, as 'a student of' a particular recognized professor. See for instance several of the short biographies of authors in EspacesTemps 43/44 (1990).

10   Cusset makes a similar comment in his scathing attacks on Luc Ferry, Bernard-Henri Levy, Pierre Nora and Alain Renaut (Cusset 2003, 323–30).

higher salaries and better material conditions, but also for the perceived intellectual freedom, rejoining what Söderström rather prettily described as an 'archipelago of thinkers'[11] (Söderström 2004, *personal communication*), very different from the centralized French system of large centrally-funded *laboratoires*. Individuals such as Jacques Lévy, most recently, found it easier to obtain a full professorship in the neighbouring country, having upset French cliques since the 1970s and having never really had a formal mentor. Others, such as Bernard Debarbieux or Jean-François Staszak, made a strategic choice to move outside the established French system. In fact, it almost seems as if some French geographers have idealized Switzerland as an innovative periphery, as Guy Di Meo romantically stated (Di Meo 2004, *personal communication*) that Swiss-Romand geographers have historically had an impact on geography far beyond the objective size of the academy,[12] in particular through the work of Claude Raffestin, Jean-Bernard Racine and Antoine Bailly, a point also made by Claval (1998, 439). The first of these will be discussed more at length towards the end of this chapter. French-speaking Canada on the other hand has also played a different role of catalyst, a point I will also return to subsequently, by translating and bringing into French much of the trends and literature prevalent within the *Anglo* world (Racine; Lévy 2004, *personal communication*), although not as much as might have been expected.

Foucault Enters the Scene

It is in these particular academic contexts that Foucault's writing appeared on the scene in the 1970s. At the time, academic geography in France was undergoing violent and highly personalized fistfights and struggles (Orain 2003, 267) in which official national geographical institutions such as the Comité National de Géographie were seen as nothing less than the 'hateful emanation of an over-hierarchical system of mandarins that systematically marginalized progressive groups, specifically financially'[13] (Orain 2003, 264). The time was one of volatile rejection of the orthodox Vidalian *Géographie Classique* and the corresponding renegotiation of a theoretical grounding within the quantitative and positivist *Nouvelle Géographie*. At this time, the publication of an interview of Foucault constituted a first, indicating a welcome change in the nature of academic debates in a country where these have often centred on individuals, not ideas (Lévy 2004, *personal communication*). The return of Foucault, to the extent that there has been one in Francophone geography, took place most recently via those *Anglo* interpretations within 'French Theory', particularly via Quebec. Guy Di Meo, for instance, recalled hearing about the enthusiasm for

---

11 '... un archipel de penseurs.'

12 Within the four French-speaking geography departments in Switzerland, there are only 10 full professors in all: equivalent to one large department in many universities in Great Britain, for instance.

13 '... l'émanation haïssable d'un système mandarinal par trop hiérarchique, marginalisant (d'abord et avant tout financièrement) les équipes "progressistes".'

Foucault in geography when colleagues such as Vincent Berdoulay and Olivier Soubeyran moved back to France, bringing Foucault with them, so to speak (Di Meo 2004, *personal communication*, see also Buléon 1990, 14). This provided a second impetus to explore his work, after the first wave provoked by Raffestin in the 1980s. Dupont makes a similar comment about the influence on location or context in discovering authors when he recalls first reading Foucault in the United States: 'I read Foucault in the English text. I thought he was brilliant, and then when I got to France I said to myself "he's not that brilliant, he just managed to express the often frozen structures of knowledge that exist in France". He simply critiqued that, and in the United States this was taken to be a revolution, when instead he was just asking the question of the limits and structures of knowledge in France'[14] (Dupont in Antheaume et al. 2004, 19). While one may of course disagree with this very narrow interpretation of Foucault's scope, his point about the context of the reception of Foucault's ideas is instructive.

Opinions differ as to whether geographers would have really read Foucault in the 1970s and 1980s, notwithstanding his public visibility: Foucault was cited in a 1981 *Lire* survey of opinion leaders as the third most important contemporary *Maître à penser* in France (Bourdieu 1984, 281). Lévy, for instance, suggests that geographers were not particularly well read at the time and that innovators were more likely either to be involved in the quantitative surge or else were reading Karl Marx instead (Lévy 2004, *personal communication*). Hepple makes a similar comment, suggesting that while 'Anglophone human geography was becoming excited by the ideas of Althusser and French structural Marxism, [French geographers] were moving to a post-Marxist analysis' (Hepple 2000, 272). More convincing, I believe, is the opinion that Foucault was read, but that the academic and political contexts were not conducive to his absorption and adaptation in any meaningful way. As Collignon suggests, 'we did not digest the authors to which they [*Anglo* geographers] refer in the same way, especially because we read them in the original versions within a different historical context – that of the 1960s and 1970s, and not the 1980s as our Anglophone colleagues did – and because these were integrated into the common grounding of the social sciences before the arrival of the postmodern society which they helped explain and describe across the Atlantic'[15] (Collignon in Antheaume et al. 2004, 22). Söderström, similarly, suggested Foucault in particular was 'strategically forgotten'

---

14 '... j'ai lu Foucault dans le texte anglais. Je le trouvais génial, puis arrivé en France je me suis dit: "il n'est pas si génial que cela, il a simplement exprimé les structures souvent figées du savoir dans la structure française". Il a simplement critiqué cela, et aux Etats-Unis on a pris cela comme une révolution, alors qu'il posait la question des limites et des structures du savoir en France.'

15 '... nous n'avons pas digéré de la même façon les auteurs auxquels ils se réfèrent, notamment parce que nous les avons lus dans leur version originale, dans un autre contexte historique – celui des années 1960 et 1970 et non pas celui des années 1980 comme nos collègues Anglophones – et parce qu'ils ont été intégrés au fond commun des sciences sociales avant l'avènement de la société postmoderne qu'ils ont servis, outre-Atlantique, à appréhender.'

(Söderström 2004, *personal communication*), something that is different from being outright ignored. Foucault, questioning universalizing knowledge – a French obsession – was also strategically avoided.

In contrast, in his review of what he considers to be French radical geography, Hepple (2000) suggests that Foucault's interview and subsequent questions to geographers did in fact have an impact on Francophone geography, specifically within Lacoste's *Hérodote* group. He notes that the interview in 1976 highlights 'the convergence between Foucauldian thought and the geopolitical perspectives of the *Hérodote* group well before Foucault's impact on the construction of Anglophone critical geopolitics by Dalby, O Tuathail and other [sic] in later years' (Hepple 2000, 292). In saying this, he suggests that Lacoste, in particular, emphasized one dimension of Foucault's power/knowledge, 'that of the pervading role of state power (including class power) and its influence on intellectual, academic and political structures' (Hepple 2000, 292). This seems like a tenuous link at best, and is largely contradicted by Hepple's later comments about Lacoste's aversion to anything that approached theory.[16] Hepple explains this by suggesting that Lacoste was scarred in his experience of Marxism and New (quantitative) Geography, as well as by Roger Brunet's *Chorèmes* (Hepple 2000, 293). This is very plausible, as is the simple fact that Lacoste's determination to create a real 'school' for his revamped *géopolitique*, considered by him to be both necessary and sufficient to replace all other geographical forms of inquiry, implied rejecting any other such attempts. Lacoste's savage review (Lacoste 1981) of Claude Raffestin's own attempt at engaging with Foucault, discussed below, shows just how unlikely this supposed convergence was.

Different Foucaults in Different Places

Translation and transposition, as well as the different way quotes and literature reviews are used in both traditions are important factors in explaining the different reception of Foucault. As hinted earlier, the need to ground an argument by referring to key authors within an initial literature review is less prevalent in the Francophone world, lessening the amplifying effects of authors invoked *de rigueur* but barely appropriated, reduced to magical incantations (Debarbieux and Lévy 2004, *personal*

---

16 The impression that Lacoste had no time for Michel Foucault's writings is reinforced on reading his review of Claude Raffestin's book *Pour une Géographie du Pouvoir* (1980) when he writes, voluntarily quoting bits out of context that 'Raffestin peut bien au début de son livre se rallier à la thèse paradoxale de Michel Foucault pour qui "le pouvoir vient d'en bas" (mais Foucault dit cela à propos de la sexualité!), il n'en reste pas moins que le pouvoir d'Etat s'exerce de haut en bas et qu'il est territorialement hiérarchisé; c'est certainement fâcheux, déplorable, injuste, mais c'est ainsi et ce n'est pas un vice de la géopolitique que de dire vrai' (Lacoste 1981, 157), adding in a footnote that 'je ne discuterais pas ici des sophismes qui reposent sur la confusion de très différentes sortes de "pouvoir" (pouvoir sexuel et pouvoir d'Etat) et sur la confusion des niveaux d'analyse (rapports entre deux personnes et rôle des appareils d'Etat sur des milliers ou des millions d'individus' (Lacoste 1981, 157).

communication). Likewise, the use of literature references is different on a very practical level: the near-exclusive use of footnotes and endnotes, rather than the Harvard system of quoting authors within the text in brackets, lessens the impact and reduces the need for what is often little more than name-dropping. Methodologically, this also means that it is harder to identify Francophone geographers who have drawn from Foucault since much of the influence is implicit – as, for instance in the work of Bernard Debarbieux or Ola Söderström – remaining in the background, rather than referred to explicitly and referenced. Additionally, translations play a role, simply 'because they are in themselves transfers and repeated appropriation, translations participate on their own level, and perhaps more powerfully than other processes, in the means of production of theoretical discourses'[17] (Cusset 2003, 101). As Dupont's quote indicated earlier, translation does not mean simply copying out a text in another language[18] but instead adapting it to a given context, be it linguistic or academic. Lussault, writing in French, states that 'in reading him [Foucault], the potential richness of his writing appears to those interested in space. A potential richness, however, because the work of critical 'translation' of Foucault into geography needs to be done almost entirely'[19] (Lussault 2003: 377), a comment applied of course exclusively to Francophone geography. This need to adapt an author to a discipline, an act of conceptual translation, may be paradoxically easier when the author is writing in another language. Cusset had further suggested that because English is a more playful language it desecrates words more eagerly than French (Cusset 2003), making it easier for *Anglos* to reinvent Foucault to suit a new paradigm.

Foucault, of course, could have predicted the *disparition* of his original texts and would no doubt have been amused by it, as he playfully recognized the lives they lead after their creation: 'many major texts are scrambled and disappear, and commentaries at times come to replace them. But even if their area of concern may well change, their function remains; and the idea of a shift is constantly replayed'[20] (Foucault 1971, 25). Foucault's comment is subtle, hinting at the Borgesian appeal of 'the playful existence of a critique that would endlessly discuss a work that does not exist'[21] (Foucault 1971, 25), paradoxically saying something for the first time and yet endlessly repeating that which was never said. Chivallon is much less amused by this desecration of Foucault and others, and notes with some irritation

---

17 '… parce qu'elle est elle-même transfert et réappropriation, la traduction participe à son tour – et peut-être plus puissamment que les autres procédés – de ces modes de production du discours théorique.'

18 I have borrowed this expression from Sophie Rey, a translator and friend, who laughs at herself in saying that 'je recopie simplement des textes dans une autre langue'.

19 '… a le lire, on s'aperçoit en effet de la richesse potentielle de ses écrits pour qui s'intéressent à l'espace. Richesse potentielle, car il faut entreprendre presque entièrement le travail de "traduction" critique de Foucault à destination de la géographie.'

20 '… bien des textes majeurs se brouillent et disparaissent, et des commentaires parfois viennent prendre la place première. Mais ses points d'applications ont beau changer, la fonction demeure; et le principe d'un décalage se trouve sans cesse remis en jeu.'

21 '… jeu … d'une critique qui parlerait à l'infini d'une oeuvre qui n'existe pas.'

that commentaries on commentaries have tended to accumulate in the *Anglo* world (Chivallon 1999, 302; see also Cusset 2003, 235). This is not as chauvinistic as it might sound, since her main point is that the marginal position of Foucault's thoughts on space in line with postmodern deconstructivist paradigms does not really justify his enthusiastic embracing by *Anglo* geographers. She suggests instead that the link between them and Foucault is tenuous and that 'the name of the famous philosopher is but a smokescreen'[22] (Chivallon 1999, 310). Instead, she suggests, the bulk of his writing on space is more largely in tune with existing more classical positions that consider space as constitutive of the social, including attempts to explore the semantics of space such as carried out by Claude Lévi-Strauss, Roland Barthes and, later on, Marc Augé, Augustin Berque and Claude Raffestin. Since 'the most explicit references to space made by Foucault are tightly linked to projects that we have not been used to calling postmodern'[23] (Chivallon 1999, 310), his lauded project for thinking *autrement* about space and his 'conception of a new way of thinking that mobilizes spatial resources is barely formulated'[24] (Chivallon 1999, 309). This comment also draws attention to a uniquely Francophone obsession with modernity, partly explaining why the term 'postmodern' is in scant use. 'In France, the limits of reason and modernity are questioned as though nothing could exist beyond them; this explains for instance why thinkers such as Foucault or Barthes are considered here, in France, to be modern, within a philosophical tradition stemming from social philosophy, questioning the limits of reason and the limits of applying reason to the organization of society by the State. Whereas in the United States, their writing is taken as a demonstration of a break from this position, at least on a theoretical level'[25] (Dupont 2004, 11). Similarly, and in contrast to what Harvey (1989) and Soja (1989) have suggested, Di Meo has argued that Foucault did not really contest the permanence of modernity for two reasons: firstly because socio-spatial segmentation and segregation as modern technologies of domination are not in decline in western countries, and secondly because reason always acts through the exclusion of *un*reason *[déraison]* or that considered as such (Di Meo 1991, 14).

Noting that Foucault pretty much ignored geographers, notwithstanding his interview with *Hérodote*, Michel Lussault admits that indeed 'symmetrically,

---

22 '... le nom du célèbre philosophe ne sert que de couverture.'

23 '... les références les plus explicites de Foucault sur l'espace entretiennent donc un étroit rapport avec des projets que nous n'avons pas eu jusqu'ici l'habitude de designer comme postmodernes.'

24 '... cette conception d'une pensée nouvelle mobilisant la ressource spatiale est à peine formulée.'

25 '... en France, on s'interroge sur les limites de la raison et de la modernité, comme s'il ne pouvait y avoir rien au-delà; c'est ce qui explique par exemple que des penseurs comme Foucault ou Barthes sont ici, en France, des modernes qui, dans une tradition philosophique issue d'une philosophie sociale, questionnent les limites de la raison, les limites de l'organisation d'une société par la raison, par l'Etat. Alors qu'aux Etats-Unis, leurs écrits sont pris comme une démonstration de la rupture, du moins théorique.'

geographers have engaged too little with the work of Michel Foucault'[26] (Lussault 2003, 377). Agreeing with Chivallon's earlier comments, Lussault suggests that space does form an integral part of Foucault's work: 'he took it abundantly into account in his work, without reducing it to an inert produced form or to a neutral substrate. It is possible to enrich our thinking about space by drawing upon Foucault'[27] (Lussault 2003, 379). Söderström has suggested that Francophone geography specifically missed out on Foucault on three levels: theoretically, in failing to understand his use of discursive formations and relational approaches; thematically in ignoring his notions of heterotopia and governmentality; and methodologically, by failing to build on his approach to the control of space (Söderström 2004, *personal communication*). Taking this suggestion seriously, I will briefly examine each of these, aiming for a brief panorama of what has actually been done.

Theory: Relations, Power and Discourse

Foucault famously stated that space is fundamental in any form of communal life; space is fundamental in any exercise of power (see for instance Elden 2003, 119). Claude Raffestin's *Pour une géographie du pouvoir*, published in 1980, implicitly built on this statement, constituting a form of response to Michel Foucault's questions to geographers. Raffestin was a driving force of what has been called the post-vidalian critique,[28] endorsing the role of senior theoretician in the linguistic and constructivist turn the discipline took in the Francophone world at the end of the Seventies. Much of his inspiration came from the work of Michel Foucault, Henri Lefebvre, Martin Heidegger and Luis Prieto, bringing a much-needed breadth of references to a discipline pitted by intellectual incest. He is one rare example of a Francophone geographer active within and not outside the wider *sciences humaines*. Söderström and Philo wrote for instance that 'the most substantial theoretical contribution to non-Anglophone social geography in the 1970s and 1980s was (...) to be found in the work of the Swiss geographer Claude Raffestin. Being rather idiosyncratic, his social geography was difficult to categorize in the neat boxes traditionally used to describe English-speaking geography (terms such as spatial analysis, humanistic geography, and radical geography)' (Söderström and Philo 2004, 304–5).

---

26 '... symétriquement, les géographes ont trop peu abordés l'œuvre de Michel Foucault.'

27 '... il l'a abondamment pris en compte dans son oeuvre, sans le réduire à une forme produite inerte ou à un support neutre. On peut donc nourrir une pensée de l'espace via le détour par Foucault.'

28 Further details on the epistemological history of French geography can be found in Olivier Orain's excellent thesis (2003) *Le plein-pied du monde: postures épistémologiques et pratiques d'écriture dans la géographie française au XXe siècle*, Thèse de doctorat, Université de Paris I Panthéon-Sorbonne. (The title is mildly misleading since many of the authors invoked belong to the French-speaking world in the wider sense, in particular from universities in the Suisse Romande.)

It is in fact not always easy to read Raffestin, as his grand theory of territory and territoriality, as well as his wider writings on the geographical intelligibility of reality, are often put forward more as proposals than polished theories. Orain notes for instance that 'his production has the character of a slowly built up mosaic in which each text takes its place as a piece, both a device and a process. It is a device in that each piece of writing refers to other contemporary ones, edging them on and adding elements through partial repetitions that can be easily pieced together'[29] (Orain 2003, 315). Raffestin's *Pour une Géographie du Pouvoir* did constitute a clear formalization of a theory of territory and territoriality within a clearly Foucaldian framework of power relations strongly influenced by *La Volonté de Savoir* published in 1976; yet even this is a far from finished theory, reflecting his rejection of finished, closed systems and his personal attachment to a *pensée en procès*. Raffestin writes beautifully, making use of a breadth of references and myths. Foucault's notion of power is a central inspiration, and he subtly gives it a more spatial dimension and rootedness:

Power, a common noun, hides behind Power, a proper noun. It hides so efficiently specifically because it is present everywhere. It is present in every relation, within every action: it insidiously uses every social fracture to infiltrate into the heart of people. It is ambiguous because there is Power and there is power. But the former is easier to grasp because it manifests itself through complex apparatuses that surround and grasp each territory, control the population and dominate the resources. It is visible, massive, identifiable power. In consequence it is dangerous and unsettling, but it inspires wariness through the very threat that it represents. But the most dangerous is that which is unseen or that which one no longer sees because it is assumed to be discarded through house arrest. It would be too simple if Power were the Minotaur locked into its labyrinth that Theseus could kill once and for all. But power is reborn worse than it was, when Theseus meets the Minotaur: Power is dead, long live power. From then on, power is assured to live forever as it is no longer visible; instead it is consubstantial to all relations.[30] (Raffestin 1980, 45)

---

29 '... sa production a le caractère d'une mosaïque lentement échafaudée, dans laquelle chaque texte prend place comme *pièce*, d'un dispositif et d'un processus. Dispositif, car chaque écrit renvoie à d'autres, contemporains, qu'il relaie et qu'il complète, avec des redites partielles qui permettent un *empiècement* assez aisé.'

30 '... le pouvoir, nom commun, se cache derrière le Pouvoir, nom propre. Il se cache d'autant mieux qu'il est présent partout. Présent dans chaque relation, au détour de chaque action: insidieux, il profite de toutes les fissures sociales pour s'infiltrer jusqu'au cœur de l'homme. Ambiguïté donc puisqu'il y a le 'Pouvoir' et le 'pouvoir'. Mais le premier est plus facile à cerner car il se manifeste à travers des appareils complexes qui enserrent le territoire, contrôlent la population et dominent les ressources. C'est le pouvoir visible, massif, identifiable. Il est dangereux et inquiétant, par conséquent, mais il inspire la méfiance par la menace même qu'il représente. Mais le plus dangereux c'est celui qu'on ne voit pas ou qu'on voit plus parce qu'on a cru s'en débarrasser en l'assignant à résidence surveillée. Ce serait trop simple que le Pouvoir soit le Minotaure enfermé dans son labyrinthe qu'un Thésée pourrait aller tuer une fois pour toutes. Le pouvoir renaît, plus terrible encore, dans la rencontre de

Space, Knowledge and Power

As this short extract illustrates, Foucault's definitions of power developed in *La Volonté de Savoir* underpins Raffestin's approach. Each relation is the place *[le lieu]* within which power manifests itself, as energy and information get manipulated: formed, accumulated, combined, and circulated (Raffestin 1980, 46). Knowledge and power are linked as insolubly as energy and information, within any relation, a point Raffestin reinforces by quoting Foucault and Deleuze's comment that any point in which power is exercised is simultaneously a place of knowledge formation.[31] Raffestin's concept of territory also draws upon Lefebvre's idea of the production of space, further spatializing Foucault. Territory, in his perspective, is a space within which work *[travail]*, that is to say energy and information, has been projected and that in consequence is constructed through and reveals power relations (Raffestin 1980, 129). His distinction between space (pre-existent to any action) and territory (produced relationally) is fundamental, enriched by an analysis of representations and the semiotics of territory that draw on sources as diverse as Ludwig Wittgenstein, Edward Soja and Umberto Eco. In an interview in 1997, Raffestin noted that 'I have been very heavily criticized for this use of Foucault and the only consolation I have is that Americans, and in particular Californian geographers, are discovering or are rediscovering Foucault today'[32] (interview carried out by Elissade 1997, quoted in Orain 2003, 306), presumably referring in this case principally to Edward Soja.

Jacques Lévy commented on *Pour une Géographie du Pouvoir* by linking it to Paul Claval's rather different (and far from Foucaldian) *Espace et Pouvoir*, noting cautiously that:

> ... despite the great interest of these books, they were scarcely taken up, perhaps because they cumulated two opposite handicaps. On one hand, they were too advanced for their readers, handling concepts perceived to be too abstract, too far from usual research fields; on the other they continued to approach politics indirectly, a topic that remains the real blind spot of the geographical *Weltanshauung*. In that, they gave up creating a political geography based on a clear epistemological and theoretical basis. This is true for France and for other Latin countries, because within the Anglophone world throughout the 1980s political geography has softly conquered a significant place within the discipline.[33] (Lévy 2003, 738)

Thésée et du Minotaure: le Pouvoir est mort, vive le pouvoir. Dès lors, le pouvoir est assure de pérennité car il n'est plus visible, il est consubstantiel de toutes les relations.'

31 '... tout point d'exercice du pouvoir est en meme temps un lieu de formation du savoir.' (Quoted by Raffestin 1980, 48, but not referenced.)

32 '... j'ai été très critiqué pour cette utilisation de Foucault et la seule consolation que j'ai, c'est que les Américains et notamment les géographes californiens découvrent ou redécouvrent Foucault aujourd'hui.'

33 '... malgré leur grand intérêt, ces deux ouvrages ont peu fait école, peut-être parce qu'ils cumulaient deux handicaps pourtant opposés. D'un côté, ils étaient trop avancés pour leurs lecteurs, maniant des concepts perçus comme trop abstraits, trop lointain des champs de recherche habituels; de l'autre, ils continuaient d'aborder de biais le politique, véritable point aveugle de la *Weltanshauung* des géographes, renonçant à fonder une géographie politique sur des bases épistémologiques et théoriques claires. Du moins en France et dans les pays latins,

This is a more guarded critique than the angry one that Lévy wrote in a volume of *EspaceTemps* in the 1980s, emphasizing the lack of definition of the 'political' that is replaced with the much wider and more global theory of *pouvoir*. He is not convinced by Raffestin's uses of Foucault, noting that *pouvoir* is neither a category nor a social science concept, but instead is only a linguistic category, upstream epistemologically from the *politique*, a notion he has personally favoured (Lévy 1994). As a notion, he believes *pouvoir* to be too general to be operational, noting simultaneously that the *politique* is really a dark spot in the social sciences, linked to psychological issues which are intrinsically taken to be suspect and difficult to cope with within existing frameworks (Lévy 2004, *personal communication*). A similar point was made by Villeneuve who wrote that Raffestin 'could be accused of practising political determinism when he argues that power is consubstantial to all relations'[34] (Villeneuve 1982, 266). Yet these, I think, are unfair to Raffestin and mainly reflect both commentators' lack of familiarity with Foucault as a theoretical grounding. They also stem from Raffestin's conscious choice not to exploit and apply his proposals on power empirically, preferring instead to assume this would be done subsequently by someone else.

It has seemed at times, to those around Raffestin, that he may have been waiting for a disciple to take on this role of polishing his proposals. Perhaps Raffestin's greatest sadness has been increasing disillusionment with other geographers, coupled with a personal frustration at not being recognized for his contributions. This is of course where the peripheral nature of Swiss geography shows its limitations: Raffestin's lack of insertion into certain guarded circles of French geography, as well as his long-standing personal feuds with people like Yves Lacoste, certainly didn't help to get his ideas spread about. However, having said that, a number of links did exist and continue, in particular inserting geographers working in Geneva into networks centred on universities and *laboratoires* at Grenoble and Pau, in France. Raffestin also has a large following in Italy where he currently spends most of his time. Nevertheless, being seen as *raffestinien(ne)* has sometimes been a dangerous card to play in certain circles, dividing individuals between loyal followers[35] and enemies.

In an article dealing with regulation and self-regulation, offering a theoretical grounding for understanding the production of scientific knowledge, Raffestin noted that:

---

car dans le monde anglophone, la géographie politique a, au cours des années 1980, conquis en douceur une place significative dans la discipline.'

34 '... pourrait être taxé de pratiquer un certain déterminisme politique quand il affirme que le pouvoir est consubstantiel de toute relation.'

35 See for instance the contributors to the *Colloque* on 'Territorialité, une théorie à construire' organized on the occasion of Raffestin's retirement from the University of Geneva. Available online http://www.unige.ch/ses/geo/recherche/colloqueRaffestin/Textes_CollCR. pdf.

... it is because there are networks of practices that there is a need for norms, both statutory and legal, and not the other way round. Likewise, it is because of the historic nature of the world [*parce qu'il y a de l'historicité*] that there is a similar need within the human sciences since their construction is always confronted with networks of practices. It is probably the great lesson left behind by Michel Foucault, and put into perspective by Paul Veyne first for historians but also for all researchers working within the human sciences, even if few within geography have claimed it. But that is another story...'[36] (Raffestin 1996, 124)

Other Catalysts and Converts

Another author to draw upon Foucault, partly via Raffestin's work, is Guy Di Meo, one of the main proponents of innovative social and Marxist geography in France. He also draws upon both Foucault and Lefebvre in his project of arming social geography and similarly also has a fondness for Heidegger. His work has included introducing the tools of historical materialism to geography, including dialectic thinking, a non-linear and evolving conception of time, and an awareness of spatial or territorial contradictions that partly give meaning to and explain social life (Di Meo 1991, 15). Di Meo also notes pessimistically, like many others, that despite certain theoretical contributions to geography such as Raffestin's 'it is nevertheless clear that up to now it is mostly sociologists and anthropologists who have theorized about spatial practises and territoriality'[37] (Di Meo 1999, 79).

Other authors have referred to Foucault mostly peripherally, using elements from his work as building blocks within a larger theoretical body based on other sources. Ideas of 'discourse' and 'discursive formation' gleaned from *l'Archéologie du Savoir* (1969) have been used successfully by Söderström (1997, 31), for example, as have the links between knowledge and power. Thematically, the idea of a 'security society' from *Naissance de la biopolitique* (2004) and notions of heterotopia have likewise also been picked up by several authors. Di Meo, for instance, used the concept in passing, noting that 'in the heterotopia that Foucault defines, all the frontiers of space whether real or imagined, only take on a very limited meaning, like an anecdote. It is the global space that has meaning. (...) In reading Foucault, it clearly appears in what way territoriality can spring out of geographical space,

---

36 '... c'est parce qu'il y a des réseaux de pratiques qu'il y a de la nécessité normative, réglementaire et légale et non pas l'inverse. De la même manière, c'est parce qu'il y a de l'historicité qu'il y a de la nécessité dans les sciences de l'homme dont la construction est toujours confrontée avec les réseaux de pratiques. C'est probablement la grande leçon, léguée par Michel Foucault, que Paul Veyne a su mettre en perspective, d'abord pour les historiens mais aussi pour tous les chercheurs en sciences de l'homme quand bien même peu s'en sont réclamés en géographie. Mais c'est une autre histoire...'

37 '... il n'empêche que ce sont à ce jour les sociologues et les anthropologues qui ont sans doute le plus théorisé sur les rapports des pratiques spatiales et de la territorialité.' Di Meo also listed Frémont et al. (1984) as another example of a foucaldian approach to space but, on reading it, I was pushed to find any hint of Foucault.

moulded by repeated use'[38] (Di Meo 1999, 85). Lussault also mentions heterotopia, stating that despite being 'an announcement of ulterior developments that manifested a series of intuitions that Foucault regrettably did not develop'[39] (Lussault 2003, 379), the opportunity Foucault left has not been taken up by geographers, neither theoretically nor methodologically. In fact, in stark contrast to Raffestin, none of these authors have drawn upon Foucault in any fundamental way. At best, he has provided theoretical fodder for thinking about power, discourse and space as part of the required backbone of requisite readings in the social sciences gleaned during individual studies, integrated but not explicitly cited (as would be expected within the *Anglo* tradition), at worst he has been used to suggest little more than research themes such as surveillance or heterotopia. Methodologically, of course, searching out for a latent, underlying Foucaldian flavour within a discipline is much more difficult than skimming lists of explicit references – a point that may be kept in mind as a nuance on some of the comments above that suggest that Foucault has had little visible impact on Francophone geography.

Chivallon, in an excellent article on British postmodern geography decoded for French-language readers, gives further compelling arguments for why Foucault has not been picked up in the same way by Francophone geographers. In particular, she notes the near-absence of any interest in France for traditionally postmodern categories such as race, gender and sexuality. This is reflected, for instance, in the near-total absence of any original Francophone feminist geography. Chivallon is in fact critical of the way Foucault has been used, in parallel with this Anglo obsession with categories. She first notes that Foucault's warning that power is everywhere and stems from everywhere is paradoxically in danger of being forgotten in the surge of enthusiasm for 'other' voices: 'at a time when the marginalized and dominated voice is considered to be the only container of truth, it is in many people's interest to demonstrate and conserve a position from which it is taken to be legitimate to speak'[40] (Chivallon 1999, 305). Chivallon directs this virulent comment particularly at certain feminist geographers, noting that 'there must also be something related to power [quelque chose de l'ordre du pouvoir] in the process of construction of

---

38 ... dans l'hétérotopie que définit Foucault, chacune des frontières de l'espace, réelle ou fictive, ne revêt qu'une signification fort limitée, anecdotique. C'est l'espace global qui fait sens. ... En lisant Foucault, l'on mesure bien de quelle façon la territorialité peut jaillir d'un espace géographique forgé par des cheminements répétitifs.' Note that the term 'territorialité' in French has a much denser meaning than 'territoriality' as used commonally within the Anglophone literature. For an analysis of 'territoire' and 'territorialité' within the two traditions, see Debarbieux 1999.

39 '... une annonce de développements ultérieurs et manifestait une série d'intuitions qu'il est dommage que Foucault n'ait pas plus développés.'

40 '... à un moment où la voix marginalisée et dominée est tenue en quelque sorte tenue pour être seule détentrice de vérité, il y a tout intérêt à démontrer et à conserver une position d'où il est censé être légitime de parler.'

women's knowledge'[41] (Chivallon 1999, 305). Such a comment goes a long way in indicating the chasm between what is considered orthodox within the two traditions, and indeed she has gone so far as to say that the total adhesion to postmodern discourses within the Anglo world is almost alienating to those on the outside, in total contrast to the proffered attempts to question hegemonic discourses (Chivallon 2005, personal communication).

Conclusion

Foucault once defined philosophy as the critical process of thought carried out on itself, that rather than legitimizing what is already known, consists of attempting to know how and to what extent it would be possible to think otherwise (Foucault 1994). To a modest extent, this chapter has sought to contrast two traditions in order to explore precisely how one author has been used to think in very different ways. By exploring what scant parts of Foucault's writing have in fact permeated and been picked up within Francophone geography and by tracing why there is so little to write about in contrast to the plethora within the *Anglo* world, I have attempted to highlight specificities and point out a number of further paths for reflection. If anything, this chapter has highlighted the near-total absence of Foucault within Francophone geography at a time when, even in France, he is slowly undergoing a renaissance. A gathering in January 2005 organized by Science Po (the prestigious political science department in Paris) and the *Centre Interdisciplinaire de Recherche en Sciences Sociales* on Foucault's work, for instance, tellingly includes a wide range of social scientists – but not a single geographer. This would not be cause for undue concern if Francophone geography were otherwise healthy and vibrant. Indeed, diversity in the face of increasing *Anglo* hegemony would be more than welcome. The sad thing is that part of the explanation lies in the immobility of the French academy. However, the strong indication of a renaissance of a critical strand of fresh thinkers within Francophone geography is cause for celebration, as authors are increasingly open to other literatures yet convincingly critical of contemporary fads.

Acknowledgements

This article could not have been written without the endless kind help of a number of people who took the time to listen to my questions, sit down and spend substantial time discussing the themes developed in this chapter and reading various drafts, both while working at the University of Geneva, in Switzerland, and later at the University of British Columbia in Vancouver, Canada. In particular, I should like to thank John Agnew, Christine Chivallon, Bernard Debarbieux, Stuart Elden, Merje

---

41 '... il doit bien y avoir aussi dans la constitution des savoirs féminins quelque chose qui est de l'ordre du pouvoir.'

Kuus, Jacques Lévy, Laurent Matthey, Guy Di Meo, Jean-Bernard Racine, Claude Raffestin, Ola Söderstrom, and the kind survivor of the French *concours* who wishes to remain anonymous.

References

Agnew, John and James Duncan (1981) 'The Transfer of Ideas into Anglo-American Human Geography.' *Progress in Human Geography*, 5(1), 42–57.

Agnew, John (2001) 'Disputing the Nature of the International in Political Geography – the Hettner-Lecture in Human Geography.' *Geographische Zeitschrift*, 89(1), 1–16.

Antheaume, Benoît et al. (2004) 'Le Postmodernisme en Géographie.' *L'Espace Géographique*, 1, 6–37.

Besse, Jean-Marc (2004) 'Le Postmodernisme et la Géographie: Eléments pour un Débat.' *L'Espace Géographique*, 1, 1–5.

Bourdieu, Pierre (1984) *Homo Academicus*. Paris: Les Editions de Minuit.

Buléon, Pascal (1990) 'Réintégrer l'Espace dans une Approche Globale des Comportements Politiques.' In Jacques Lévy (Ed.) *Géographies du Politique*. Paris: Presses de la Fondation Nationale des Sciences Politiques. 33–9. (First published in *EspacesTemps* 43/44, 14–18.)

Chivallon, Christine (1999) 'Les Pensées Postmodernes Britanniques ou la Quête d'une Pensée Meilleure.' *Cahiers de Géographie du Québec*, 43(119), 293–322.

Chivallon, Christine (2003) 'Country Reports: A Vision of Social and Cultural Geography in France.' *Social and Cultural Geography*, 4(3), 401–8.

Claval, Paul (1998) *Historie de la géographie française de 1870 à nos jours*. Paris: Nathan Université.

Claval, Paul (2000) '*Hérodote* and the French Left.' In Klaus Dodds and David Atkinson (Eds) *Geopolitical Traditions: A Century of Geopolitical Thought*. London and New York: Routledge, 239–67.

Claval, Paul and Jean-François Staszak (2004) 'Confronting Geographic Complexity: Contributions from Some Latin Countries. Presentation.' *GeoJournal*, 60, 319–20.

Collignon, Béatrice (2004) 'It's a Long Way to Other Geographers and Geographic Knowledges.' *GeoJournal*, 60, 375–9.

Cusset, François (2003) *French Theory: Foucault, Derrida, Deleuze & Cie et les Mutations de la Vie Intellectuelle aux États-Unis*. Paris: La Découverte.

Debarbieux, Bernard (1999) 'Le Territoire: Histoires an deux Langues – a Bilingual (His-)Story of Territory.' In Bernard Debarbieux (Ed.) *Discours Scientifiques et Contextes Culturels: Géographies Françaises et Britanniques à l'Epreuve Postmoderne*. Bordeaux: MSHA. 33–46.

Derrida, Jacques (2001) 'Deconstructions: The Im-Possible.' In Sylvère Lotringer and Sande Cohen (Eds) *French Theory in America*. New York and London: Routledge. 13–32.

Di Meo, Guy (1991) *L'Homme, la société, l'Espace.* Paris: Anthropos-Economica.

Di Meo, Guy (1999) 'Géographies Tranquilles du Quotidien: une Analyse de la Contribution des Sciences Sociales et de la Géographie à l'Etude des Pratiques Spatiales.' *Cahiers de Géographie du Québec*, 43(118), 75–93.

Djament, Géraldine (2004) 'Régimes d'Historicité et Régimes de Géographicité.' *EspacesTemps.net.*

Dupont, Louis (2004) 'Le Postmodernisme en Geographie; Débat Du 17 Janvier 2003.' *L'Espace Géographique*, 1, 11–12.

Elden, Stuart (2001) *Mapping the Present: Heidegger, Foucault and the Project of a Spatial History.* London: Continuum.

Fall, Juliet (2006) 'Lost Geographers: Lost Geographers: Power Games and the Circulation of Ideas within Francophone Political Geographies.' *Progress in Human Geography.*

Farinelli, Franco (2000) 'Friedrich Ratzel and the Nature of (Political) Geography.' *Political Geography*, 19(8), 943–55.

Foucault, Michel (1966) *Les Mots et les Choses: Une Archéologie des Sciences Humaines.* Paris: Gallimard.

Foucault, Michel (1967) 'Of Other Spaces, Translated by Jay Miskoviec.' *Diacritics*, 16(1), 22–7.

Foucault, Michel (1969) *L'archéologie du savoir.* Paris: Gallimard.

Foucault, Michel (1971) *L'ordre du discours.* Paris: Gallimard.

Foucault, Michel (1976) *Histoire de la sexualité, vol. 1: La volonté de savoir.* Paris: Gallimard.

Foucault, Michel (1994) 'Usage des plaisirs et techniques de soi', in Daniel Defert and François Ewald (Eds) *Dits et écrits*, Four Volumes. Paris: Gallimard. Vol IV, 539–61.

Foucault, Michel (2004) *Naissance de la biopolitique. Cours au Collège de France 1978–1979.* Paris: Gallimard-Seuil.

Foucher, Michel (1991) *Fronts et Frontières: Un Tour Du Monde Géopolitique.* Paris: Fayard.

Frémont, Armand, Jacques Chevalier, Robert Hérin and Jean Renard (1984) *Géographie Sociale.* Paris: Masson.

Garcia-Ramon, M.-D. (2003) 'Globalization and International Geography: The Questions of Languages and Scholarly Traditions.' *Progress in Human Geography*, 271, 1–5.

Giblin, Beatrice (1985) '*Hérodote*, une Géographie Géopolitique.' *Cahiers de Géographie du Québec*, 29(77), 283–94.

Harvey, David (1989)*The Condition of Postmodernity.* Oxford: Blackwell.

Hepple, Leslie W. (2000) 'Géopolitiques de Gauche: Yves Lacoste, *Hérodote*, and French Radical Geopolitics.' In Klaus Dodds and David Atkinson (Eds) *Geopolitical Traditions: A Century of Geopolitical Thought.* London and New York: Routledge. 268–301.

Lacoste, Yves (1981) 'Hérodote a lu: Paul Claval, *Espace et Pouvoir*; Claude Raffestin, *Pour une Géographie du Pouvoir*.' *Hérodote*, 22, 154–7.

Lévy, Jacques (1994) *L'espace Légitime: Sur la Dimension Géographique de la Fonction Politique*. Paris: Presses de la Fondation Nationale des Sciences Politiques.

Lévy, Jacques (1995) 'Pourquoi la Géographie?' *Cahiers de Géographie du Québec*, 39(108), 517–26.

Lévy, Jacques (2003) 'Michel Foucault.' In Jacques Lévy and Michel Lussault (Eds) *Dictionnaire de la Géographie et de l'Espace des Sociétés*. Paris: Berlin.

Lotringer, Sylvère and Sande Cohen (2001) *French Theory in America*. New York and London: Routledge.

Lussault, Michel (2003) 'Michel Foucault.' In Jacques Lévy and Michel Lussault (Eds) *Dictionnaire de la Géographie et de l'Espace des Sociétés*. Paris: Berlin. 377–9.

Mamadouh, Virgine (1998) 'Geopolitics in the Nineties: One Flag, Many Meanings.' *GeoJournal*, 46, 237–53.

Minca, Claudio (2000) 'Guest Editorial.' *Environment and Planning D: Society and Space*, 18, 285–9.

Orain, Olivier (2003) *Le plein-pied du monde: postures épistémologies et pratiques d'écriture dans la géographie française au XXe siècle*. Doctoral thesis: Université de Paris I Panthéon-Sorbonne.

Paasi, Ansi (2005) 'Globalisation, Academic Capitalism, and the Uneven Geographies of International Journal Publishing Spaces.' *Environment and Planning A*, 37, 769–89.

Parker, Geoffrey (2000) 'Ratzel, the French School and the Birth of Alternative Geopolitics.' *Political Geography*, 19(8), 957–69.

Raffestin, Claude (1974) 'Éléments pour une Théorie de la Frontière.' *Diogène*, 134(Avril-Juin), 3–21.

Raffestin, Claude (1980) *Pour une Géographie du Pouvoir*. Paris: Litec.

Raffestin, Claude (1985) 'Marxisme et Géographie Politique.' *Cahiers de Géographie du Québec*, 26(77), 271–81.

Raffestin, Claude (1986) 'Territorialité: Concept ou Paradigme de la Géographie Sociale?' *Geographica Helvetica*, 2, 91–6.

Raffestin, Claude (1991) *Géopolitique et Histoire*. Lausanne: Payot.

Raffestin, Claude (1992) 'Géographie et Ecologie Humaine.' In Antoine Bailly, Robert Ferras and Denise Pumain (Eds) *Encyclopédie de la Géographie*. Paris: Economica. 23–36.

Raffestin, Claude (1996) 'Le Labyrinthe du Monde.' *Revue Européenne des Sciences Sociales*, XXXIV(104), 111–24.

Raffestin, Claude (1997a) 'Le Rôle des Sciences et des Techniques dans les Processus de Territorialisation.' *Revue Européenne des Sciences Sociales*, XXXV(108), 93–106.

Raffestin, Claude (1997b) 'Foucault Aurait-il Pu Révolutionner la Géographie?' *Au Risque de Foucault*, Supplémentaires. Paris: Centre Georges Pompidou, Centre Michel Foucault. 141–9.

Razac, Olivier (2000) *Histoire Politique du Barbelé*. Paris: La Fabrique-Editions.

Samers, Michael and James D. Sidaway (2000) 'Exclusions, Inclusions, and Occlusions in "Anglo-American Geography": Reflections on Minca's Venetian Geographical Praxis.' *Environment and Planning D: Society and Space*, 18(6), 663–5.

Sanguin, André-Louis (2005) *Géographie Politique, Géopolitique, Géostratégie, Domaines, Pratiques, Friches*. Institut de Stratégie Comparée – Ecole Pratique des Hautes Etudes / Sciences Historiques et Philologiques. 00/03/05 2005.

Söderström, Ola (1997) *l'Industriel, l'Architecte et le Phalanstère: Discours, Formes et Pratiques du Logement Ouvrier à Ugine*. Paris: L'Harmattan.

Söderström, Ola and Chris Philo (2004) 'Social Geography: Looking for Geography in Its Spaces.' In Georges Benko and Ulf Stohmayer (Eds) *Human Geography: A History for the Twenty-First Century*. Oxford: Blackwell.

Soja, Edward (1989) *Postmodern Geographies*. London and New York: Verso.

Staszak, Jean-François (2001) 'Les enjeux de la géographie anglo-saxonne.' In J.-F. Staszak, B. Collignon and C. Chivallon (Eds) *Géographies anglo-saxonnes: tendances contemporaines*. Paris: Belin. 7–21.

Veyne, Paul (1978) *Comment on Ecrit l'Histoire et Foucault Révolutionne L'histoire*. Paris: Seuil.

Villeneuve, Paul (1982) 'Commentaire sur Pour une Géographie du Pouvoir.' *Cahiers de Géographie du Québec*, 26(68), 266–7.

Chapter 14

Could Foucault have Revolutionized Geography?

Claude Raffestin, Translated by Gerald Moore

'Foucault aurait-il pu révolutionner la géographie?' In *Au risque de Foucault*. Paris: Éditions de Centre Pompidou, 1997, 141–9.

The question – for it is a genuine question – takes its inspiration from the obviously modified title of a work Paul Veyne dedicated to Foucault, not in order to hide behind a prestigious name, but to show that a question mark can express the whole distance that there is between two disciplines, even when a certain tradition would have them as neighbours. What Foucault offered to historians, he offered just as much to geographers, although the latter have, in a certain way, refused the gift that he made them. A gift which, today, has become an inheritance on which demands are placed a bit more pressingly, albeit still timidly.

When I discovered the work of Foucault more than twenty years ago, I was, of course, dazzled – it is banal to say it – by the richness of Foucault's arsenal or, if one prefers, of his workshop. The method set out in *The Will to Knowledge* enabled me fundamentally to rethink the question of power (Foucault 1976, 121–35; 1978, 92–102).[1] The welcome set aside by geographers for this endeavour has been cold, at the least (Raffestin 1980). By contrast, it has drawn attention from other social sciences. Must we deduce from this that the intellectual distance between the thought of Foucault and geographical thinking is much more considerable than one would *a priori* have expected? Without any doubt, in any case for a certain practice of geography, which continues to think of its object more in morphological than in relational terms. This appears to me retrospectively, clearly, since at the time, I had been led, in collaboration with Luis Prieto, to reformulate the object of geography as the explication of the knowledge of practices, and of the knowledges that men have of this material reality that is the Earth. This explication as a consequence naturally erects a new geographical object, one that does not define itself through a system of forms, but rather through a set of relations to a system of forms. Social sciences do not have to study material or ideal objects, but relations to material or ideal objects, in Geography in any case.

---

1   Where translations of Raffestin's references are available we have provided them.

There are no knowledges without practices: 'Practice is not some mysterious agency, some substratum of history, some hidden engine: it is what people do (the word says just what it means)' (Veyne 1978, 211; 1997, 153). Further on, Paul Veyne adds: 'the relation determines the object, and only what is determined exists' (1978, 212; 1997, 155). Foucault, in sum, substitutes a philosophy of relation for a philosophy of the object. This positioning of the problem, by successive touches, is certainly elliptical, but for the time being it suffices to understand what Foucault essentially contributed. Armed with his philosophy of the relation, Foucault set himself to work at unravelling the most complex relations: thus, in relation to the social sciences, he played the role of Daedelus, to whom were endlessly posed the problems of *unravelling* brought about by an original *pathological* relation. This explains the continual invention of instruments and tools *for healing.*

Had his method been clearly understood and assimilated, Foucault would probably have revolutionized human geography, but for him to go this far it would have been necessary for geographers themselves to pose the essential question of the birth of the geographical gaze, just as the question of the birth of the clinical gaze was posed. Is it not in fact strange that we find nothing equivalent in human geography to the birth of the clinic? (See Foucault 1988; 1973.) It is less astonishing than one might at first think, in that the geographical gaze is not, originally, a gaze identified with any domain of experience: it is a sort of blind gaze, which is to say one that allows itself to be invaded by the unexpected but ultimately wondrous visible world. The geographical gaze's reason for being, or what one can call as such, exhausts itself in a description that has no recipient, in other words it concerns a gaze that is not extended by any practice, at least in appearance. I say: at least in appearance, for geography does not try, like medicine does with the human body, to observe the body of the earth, in order to see where the Earth 'could be ill'. The geographical gaze is not, firstly, a clinical gaze, it is the gaze of the voyeur mobilized by curiosity, which seeks on every occasion to be original, primary and thereby foundational.

The only one who, to some extent, linked the geographical gaze and the clinical gaze, who to some degree established a correspondence between the body of man and the body of the earth, is Hippocrates, who grounds the parallelism between the two gazes in the following way: 'The doctor will be informed on the majority of these points, on all of them if possible. Arriving in a town that is unknown to him, he will not be unaware of local diseases, nor of the nature of general diseases, to the effect that he will not hesitate in treatment, nor will he commit the errors into which one who had not beforehand studied the essential information would fall' (Hippocrates 1994). That said, the gaze of Hippocrates is embryonic, even prevented from exercising itself by the system of correspondences that it establishes and the theories that it utilizes.

When Foucault writes that 'the human body defines, by natural right, the space of origin and distribution of disease' (1988, 1; 1973, 3), he draws the imagery of geography through the atlas of anatomy, and he adds: 'the order of disease is simply a "carbon copy" of the world of life' (1988, 6; 1973, 7). For Foucault, 'the rationality of life is identical to the rationality of that which threatens it' (1988, 6; 1973, 7).

Furthermore, in citing Sydenham, who in 1736 defined what could be a historical and geographical consciousness of illness, Foucault retravelled the path of Hippocrates: 'The basis of this perception is not a specific type, but a nucleus of circumstances. The basis of an epidemic is not pestilence or catarrh: it is Marseille in 1721, or Bicêtre in 1780; it is Rouen in 1769, where "there occurred, during the summer, an epidemic among the children of the nature of bilious catarrhal fevers and bilious putrid fevers complicated by miliaria, and ardent bilious fevers during the autumn. This constitution degenerated into putrid biliousness towards the end of that season and during the winter of 1769 and 1770"' (1988, 22; 1973, 23). From this time on, the first task of the doctor is political and 'must begin with a war against bad government. Man will be totally and definitively cured only if he is first liberated' (1988, 33–4; 1973, 33). But here again the problem has already been posed by Hippocratus, who evokes the institutions and in particular the kings who, in Asia, would be responsible for the pusillanimity of the inhabitants.

Clinical thought, since then, is trained toward this form of correlation between the visible and the enunciable which leads to a doubly faithful description in relation to the object and in relation to language. It is therefore in the passage from *the totality of the visible to the group-structure of the enunciable* (1988, 114; 1973, 114) that the meaningful analysis of the perceived is accomplished. Here, Foucault draws out the enormous problem of the correlation between the gaze and language: 'The theoretical and practical problem confronting the clinicians was to know whether it would be possible to introduce into a spatially legible and conceptually coherent representation that element in the disease that belongs to a visible symptomatology and that which belongs to a verbal analysis' (1988, 113; 1973, 112).[2] The difficulty became apparent in the construction of the *table* that sought to correlate different variables through a system of abscissa and co-ordinates. In fact it is not a question of correlation but of the distribution of that which is given in a conceptual space defined beforehand. It is this description, 'or, rather, the implicit labour of language in description that authorizes the transformation of symptom into sign and the passage from patient to disease and from the individual to the conceptual. And it is there that is forged by the spontaneous virtues of description, the link between the random field of pathological events and the pedagogical domain in which they formulate the order of truth' (1988, 115; 1973, 114). In the end, all this comes down to the fact that to learn to see is to provide the key to a language that masters the visible: 'one now sees the visible only because one knows the language; things are offered to him who penetrated the closed world of words; and if those words communicate with things, it is because they obey a rule that is intrinsic to their grammar' (1988, 116; 1973, 115).

Above all this 'hovers the great myth of a pure Gaze that would be pure Language: a speaking eye' (1988, 115; 1973, 114). We find, here, the concept of Candillac, according to which the ideal of scientific knowledge is well-crafted language:

---

2 Raffestin misquotes Foucault here, reading 'visual analysis' instead of 'verbal analysis'.

'A hearing gaze and a speaking gaze: clinical experience represents a moment of balance between speech and spectacle' (1988, 116; 1973, 114).

In what way could Foucault, indirectly, have revolutionized geography in general and human geography in particular? In many ways, if geographers had dwelled on him as much as his thought merits. The work he did on the birth of the clinic could have been undertaken on the geographical gaze. No geographer, at least not to my knowledge, has preoccupied themself with the birth of this gaze.

It is, all the same, a fundamental question and one could legitimately express astonishment that it has not been approached, at some point, in a manner similar to that of Foucault. The astonishment is all the greater in that, since Alexander von Humboldt, geography has been entirely founded on what I would willingly call the totalitarianism of the eye. Despite that, the gaze has never been the object of analysis, for the good and simple reason that the object of geography has, for a long time and still now in some respects, never been researched in terms of relation, but in terms of external material objects, placed before an observer and lending itself to description as infinite as it is imprecise. The most obvious example, and moreover one pertaining to the same era as that chosen by Foucault for the birth of the clinic is supplied by landscape [*le paysage*],[3] which receives attention between the 17th and 18th centuries, doubtless because this notion sums up the new relation between man and nature and concretizes the immediate grasp of impressions that one gathers before the spectacle of the world, and which are as fleeting as the feelings that inspire and mobilize them.

Yet, whether one takes the landscapes described by writers such as Rousseau, Bernardin de Saint-Pierre and Chateaubriand, or naturalists such as Georg Forster or Humboldt, what matters is the table, the picture [*tableau*] that one can or does make of the external world, and not the relation, which is to say the practice and knowledge that one mobilizes for looking at and describing objects. There is, in sum, a sort of short-circuiting between the observer and the object, which leaves the practice of looking, of the gaze and of the language that grounds observation in the shadows and opacity of the unknown, the non-conceived and perhaps also the unwanted. Things pass as if the gaze goes without saying, as if it were not necessary that one stop before them. From the appearance of the notion of landscape, one places oneself before the object and one describes or one seeks to describe without truly knowing what happens upstream, in the relation that constructs the gaze that roots itself in a language with which one is moreover no longer preoccupied: the gaze and the language are there, at one's disposal, and are not questioned.

Emerging from painting, the notion of landscape played a nasty trick on geographers, who did not retain and hence could not transpose the lesson of the painter onto their own attempts at description. The work of the painter inscribes itself in a triangular process between his gaze on the object, his palette and his brush and the canvas. This dialogue between the gaze on one hand and the palette and the brush on the other does not occur in the approach of the geographer. While

---

3   This also means 'countryside'.

the painter makes explicit the practice and the knowledge that he has of material reality, in other words he grasps and projects what he sees inside him, the geographer is blinded by visible forms and only reproduces natural or human morphologies, without being entirely conscious of the practices and knowledges that condition his vision. Geographers have not understood that the landscape emerges as much from physics as from metaphysics, from the material as from the ideal; they have curiously sacrificed physics and the material, culminating in an analysis of the countryside that has led to the formation of a discipline more idiographic than nomothetical. Had they been more preoccupied by the dialogue between the gaze and language, they would have been more conscious of their intentionality and would probably have understood that the same practice could underlie diverse forms.

The failure of the landscape is illustrated by the German geographer, Passarge, who, in the aftermath of the First World War, published an enormous work on *Landschaftskunde,* in which he strives to catalogue all the elements suited for reviewing a description that seeks, without doubt, to be exhaustive, for what geography has for a long time sought to reproduce is nothing other than the image of an external world detached from the relations that made it possible. As it has often been described, the landscape is only a visual abstraction that from the outset evacuates the contributions of the other senses.

A visual end-in-itself, the geographical landscape is the product of a relation that, paradoxically, makes relation itself a dead-end, in order to concentrate almost exclusively on the result, which is to say the landscape-object over which it casts contemplative observation. Landscape geography, which sought to be an operation of unveiling, unveils in fact, only an image, a caricature of reality, whose development process remains to a very large degree hidden. The eye of the geographer is the only one not to be mobilized by an external intention, contrary to the eye of the engineer, to that of the architect, or even to that of the military man, though these are not the only examples. The geographical gaze is self-conditioned by a canonical tradition that finds legitimation in itself, without realizing that these practices are at fault. It is enough to say that, in these conditions, the geographical gaze would have necessitated an analysis, because, ultimately, in landscape, man is both subject and object, since he describes a proportion of reality in relation to an internal intention that is not clearly explained, and which moreover substitutes itself, at least in part, for the intentions from which the landscape itself results.

In his conclusion, Foucault touches on an essential point that will allow us to situate the position of the geography of the landscape: 'This medical experience is therefore akin even to a lyrical experience that has sought its language from Hölderlin to Rilke' (1988, 202; 1973, 198). The geographical gaze that is akin to a clinical gaze over the earth, a gaze that leans out over the earth, is in an intermediate situation, situated between medical experience and lyrical experience, at a sort of intersection devoid of intentionality. The back-and-forth that results from this leads to a purportedly scientific description, but one which, for multiple reasons, does not renounce lyricism, as various fragments from authors whom I cannot resist the necessary pleasure of citing can testify.

In his preface to *Cosmos*, Humboldt does not fail to note that 'the exact and precise description of phenomena is not absolutely irreconcilable with the animated and vibrant painting of the imposing scenes of creation' (1986, v). This is the explicit acknowledgement that he is playing across two registers that are in no way independent of one another in description, though their respective extents remain nonetheless unknown. Lyricism is perfectly identifiable a little further on: 'Another pleasure is that produced by the individual character of the landscape, the configuration of the world's surface in a determinate region. The impressions of this genre are livelier, better defined, more in conformity with certain situations of the soul. Sometimes it is the grandeur of the masses, the struggle of elements unleashed or the mournful barrenness of the steppes, as in the north of Asia, that excites our emotions; sometimes, under the inspiration of sweeter sentiments, it is the perspective on the fields brought out by bountiful harvests, the living of man on the edge of a torrent, the savage fecundity of soil vanquished by the plough' (Humboldt 1986, 5).

Rilke echoes Humboldt in one of his Valasian quatrains:

Landscape stopped halfway
between the earth and sky,
with voices of bronze and water,
ancient and new, tough and tender,

like an offering lifted
toward accepting hands:
lovely completed land,
warm, like bread!

Road that turns and plays
along the leaning vineyard
like a ribbon that we wind
around a summer hat.

Vineyard: hat on the head
that invents the wine
Wine: blazing comet
promised for next year. (Rilke 2002, 99, 111)

One thus discovers that the landscape of the writer composes a whole from elements mobilized by a gaze and a language that hark back to the situation of the painter, whereas the geographer only juxtaposes elements in the absence of a clearly defined gaze and language.

We find an analogous situation in Dardel: 'A love of the native soil or the pursuit of uprooting, a concrete relation forms a tie between man and the earth, a geographicity of man as a mode of his existence, of his destiny', and further on: 'It is to this primal amazement of man facing the world and to the initial intention of geographical reflection on this 'discovery' that we return here, by interrogating geography from a perspective that is proper to the geographer or more simply to the man interested in the surrounding world' (Dardel 1952). This quotation shows the extent to which the geographical gaze is a compromise between science and lyricism on one hand, and the extent to which it makes a dead-end of its own elucidation on the other.

The same goes for Vidal de la Blache when he begins to talk of the Vosges, whose forest seems to provide him with an identificatory key: 'All over, [the forest] haunts the imagination and sight. It is the natural clothing of the country. Under a heavy cloak, speckled by the light foliage of the beech trees, the undulations of the mountains are enveloped, as if dulled. The impression of height is subordinate to that of the forest' (Vidal de la Blache 1979, 189). Here too, the non-explication of the gaze speaks to the author as the forest haunts the imagination and sight, which is immediately banal in that everybody, and not just the geographer, can have this experience, but it is also extremely revelatory of a description that, in this case, is situated more on the side of lyricism than of science.

One of the rare 19th century authors who developed a geographical gaze was not a geographer. It was the little-known author George Perkins Marsh who sought to carry a sort of 'medical gaze' over the earth, after observing all the transformations provoked by anthropic action. This took the form of an ecological gaze *avant la lettre*, founded on the relation between men and spatial morphologies (1864). The intentionality of its gaze is truthfully explained, which signifies that the landscape is not an object of which one gives an image, but the outcome of a process – a set of practices – at the heart of which relation is present. Effectively, Marsh strove across history to rediscover the social practices responsible for the transformation of the landscape.

To try to understand the philosophical gaze – which can be defined as a gaze that apprehends the exteriority of the earth – requires the construction of a corpus composed of geogrammes elaborated in relation to a geostructure, or portion of reality. Whether or not articulated by a geographer, insofar as it is a process, the gaze does not grasp itself in the midst of its self-deployment, but after it has deployed itself and become concretized in discourses relative to a 'geographical reality'.

But perhaps it is convenient to return to the first sentence of Foucault's preface: 'This book is about space, about language, and about death; it is about the act of seeing, the gaze' (1988, v; 1973, ix). As always in his work, one must beware of words. The space is not that to which one habitually refers, language is taken in a very specific sense and death is that enunciated by autopsy. The whole comes together in the clinical gaze. He who would wish to write on the geographical gaze could begin in the same way as Foucault, and yet it would not consist in the same thing. It would nonetheless be inscribed in a relational approach that would accept as much generalization as singularization in the art of observing, which would be: 'a logic for these meanings which, more particularly, teach their operations and usages. In a word, it is the act of being in relation to relevant circumstances, of receiving impressions from objects as they are offered to us, and of deriving inductions from them that are their correct consequences' (Senebier, cited in Foucault 1988, 109; 1973, 108–9). If Senebier alluded to language, it would be, like Foucault, in the process that the latter called the syntax of language, spoken by things in-themselves in an originary silence.

What would our practices be, if the eye played no part in them? There would be no gaze, but the practices of hearing, of the sense of smell, of touch and taste – all those senses to which medicine makes recourse but to which the geographer has not paid much attention – doubtless because they concern circumstances whose fleetingness prevents them from being made the object of science, though they can in fact be irreplaceable, albeit neglected, resources.

Foucault will, ultimately, revolutionize geography, if we other geographers know how to engage ourselves in his philosophy of the relation, which we maladroitly brandish through the old relation of man and his surroundings, which all too often economizes the practices that we do not truthfully know how to analyze through the acts that collapse the practices and knowledges of things. Would this not be because the method of Foucault is as blinding as it is badly perceived? Or would it not be because geographers, privileging spatial forms independently of the meanings that actions have given to them over the course of history, are so strangely fascinated by a permanence that is denied them by the plasticity of practices that they have had real trouble hearing Foucault?

References

Dardel, Éric (1952) *L'homme et la Terre: Nature de la réalité géographique*. Paris: PUF.

Foucault, Michel (1973) *The Birth of the Clinic: An Archaeology of Medical Perception*. Translated by Alan Sheridan. London: Routledge.

Foucault, Michel (1976) *Histoire de la sexualité. 1. La volonté de savoir*. Paris: Gallimard.

Foucault, Michel (1978) *The History of Sexuality Volume I: The Will to Knowledge*. Translated by Robert Hurley. Harmondsworth: Penguin.

Foucault, Michel (1988 [1963]) *Naissance de la clinique*. Quadrige/PUF.

Hippocrate (1994) *De l'art medical, des airs, dea eaux et des lieux*. Paris.

Humboldt, Alexandre de (1986) *Cosmos, essai d'une description physique du monde*. Gide et Cie: libraries-éditeurs.

Perkins Marsh, George (1864) *Man and Nature or Physical Geography as Modified by Human Action*. New York.

Raffestin, Claude (1980) *Pour une géographie du pouvoir*. Paris: Litec.

Rilke, Rainer Maria (2002) *The Complete French Poems*. Translated by A. St. Paul: Poulin Graywolf Press.

Veyne, Paul (1978) *Comment on écrit l'histoire*, suivi de *Foucault révolutionne l'histoire*. Seuil.

Veyne, Paul (1997) 'Foucault Revolutionizes History.' Translated by Catherine Porter. In Arnold I. Davidson (Ed.) *Foucault and his Interlocutors*. Chicago: University of Chicago Press.

Vidal de la Blache, Paul (1979) *Tableau de la geographie de la France*. Tallandier.

PART 5
Texts

Chapter 15

The Incorporation of the Hospital
into Modern Technology

Michel Foucault, Translated by Edgar Knowlton Jr.,
William J. King, and Stuart Elden

Revista centro-american de Ciencias de la Salud, No. 10, mai-août 1978, 93–104. A
French translation appears in *Dits et écrits*, Vol. III, 508–21. This was the third lecture
given in the course of medicine which took place in October 1974 at The Institute of
Social Medicine, Rio de Janeiro, Brazil. The text of the first two lectures appeared in the
Revista centro-american de Ciencias de la Salud, Nos 3 & 6. They appear in English as
'Crisis of Medicine or Anti-Medicine?' Translated by Edgar C. Knowlton Jr., William
J. King, and Clare O'Farrell. *Foucault Studies*, No. 1, 5–19; and 'The Birth of Social
Medicine.' Translated by Robert Hurley. In *Power: The Essential Works of Michel
Foucault 1954–1984 Volume Two*. Edited by James Faubion and others. London: Allen
Lane, 2000, 134–56.

When did the hospital come to be considered as a therapeutic instrument, as an
instrument of intervention of illness for the patient, an instrument capable either
along or through its effects, of curing?

The hospital as a therapeutic instrument is a relatively modern concept, dating
from the end of the eighteenth century. Around 1760 the idea that the hospital can
and ought to be an instrument destined to cure the patient appears and is reflected
in a new practice: the investigation and systematic and comparative observation of
hospitals.

In Europe a series of investigations begin. Among these were the trips of the
Englishman [John] Howard, who went to hospitals, prisons, and poor houses of the
continent in the period of 1775–1780,[1] and that of the Frenchman [Jacques] Tenon,
at the request of the Academy of Sciences when the problem of the reconstruction of
the 'Hotel Dieu' of Paris was being posed.[2]

Those investigations had several characteristics:

---

1   John Howard (1726–1790), author of *State of the Prisons in England and Wales*.
Warrington: 1778, and *An Account of the Principal Lazarettos in Europe*. Warrington: William
Eyres, 1789. The Howard League for Penal Reform is named after him.

2   Jacques Tenon (1724–1816) author of *Mémoires sur les hôpitaux de Paris* (1788),
facsimile edition. Paris: AP-HP/Doin, 1998. A major Parisian hospital is named after him.

1. Their purpose consisted in defining, based on the inquiry, a program of reform or reconstruction of hospitals. When in France the Academy of Sciences decided to send Tenon to different countries in Europe to do research about the situation of hospitals, he expressed an important statement: 'It is the currently existing hospitals that enable the judging of the merits and defects of the new hospitals.'

No medical theory is sufficient by itself to define a hospital program. Moreover no abstract architectural plan can offer a formula for a good hospital. One is dealing with a complex problem of which the effect and consequences are not well known, which acts on illnesses and is capable of aggravating them, multiplying them, or by contrast attenuating them. Only an empirical investigation of that new object, the hospital, integrated and isolated in a similarly new manner, will be capable of offering a new program of construction of hospitals. The hospital then is no longer a simple architectural figure and comes to form part of a medical-hospital complex that must be studied the same way one studies climate, illness, etc.

2. These fact-finding missions afforded few details on the external aspect of the hospital and the general structure of the building. No longer were they descriptions of monuments, like those which were made by the classical travellers of the seventeenth and eighteenth centuries but functional descriptions. Howard and Tenon gave an account of the number of the patients per hospital, the number of beds, the useful space of the institution, the length and height of the rooms, the cubic units of air which each patient used, and the rate of mortality and cure.

They also tried to determine the relations that might exist between pathological phenomena and the state of cleanliness of each establishment. For example, Tenon investigated under what special conditions those hospitalized because of wounds were better cured and what were the most dangerous circumstances. Thus, he established a correlation between the growing rate of mortality among the wounded and the proximity to the patients with a malign fever, as it was called at that time. He also explained that the rate of mortality of those that were giving birth increased if they were located in a room situated above that of the wounded. As a consequence the wounded should not be placed below the rooms where those in labour were.

Tenon likewise studied journeys, dislocations and movements within the hospital, particularly in the room that the clean linen, sheets, dirty linen, rags utilized to treat the wounded, etc., were located. He tried to determine who transported that material and where it was taken, washed and distributed. According to him that route would explain several pathological facts interior to hospitals.

He analyzed why trephination, one of the operations practiced most frequently at this time,[3] had more satisfactory results in the English hospital of Bethlehem [Bedlam] than in the Hotel Dieu of Paris. Might there be internal factors of the hospital structure and distribution of patients to explain that circumstance? The problem is posed as a function of the interrelation of the location of the room, its ventilation and the transfer of dirty linen.

---

3 A surgical opening of the skull.

3. The authors of these functional descriptions of the medico-spatial organizations of the hospital were, however, not architects. Tenon was a doctor, and it was as such that the Academy of Sciences instructed him to visit hospitals; Howard was not a doctor, but rather a precursor of philanthropists and possessed an almost socio-medical competency.

There thus arises a new way of viewing the hospital, considered as a mechanism to cure, and of which the pathological affects it causes must be corrected. One might suggest that this is not new, since hospitals dedicated to curing patients have existed for millennia; that the only thing which perhaps may be affirmed is that in the seventeenth century it was discovered that hospitals do not cure as much as they ought; and that it is merely a question of refining the classically formulated requirements of the hospital as instrument.

I should like to express a series of objections to that hypothesis. The hospital which functioned in Europe from the Middle Ages on was not by any means a means of cure nor had it been conceived as such.

In the history of the care of the patient in the West, there were two distinct categories which did not overlap, which were sometimes paired but differed fundamentally: medicine and the hospital.

The hospital, as an important and even essential institution for urban life in the West from the Middle Ages on, is not a medical institution. At this time medicine is not a hospital profession. It is necessary to keep this situation in mind to understand the innovation that the introduction of hospital-medicine, or the medical-therapeutic hospital, represents in the eighteenth century. I shall try to show the divergences of those two categories in order to situate this innovation.

Before the eighteenth century the hospital was essentially the institution of assistance of the poor. It was at the same time an institution of separation and exclusion. The poor, as such, required assistance and as a patient, he was the carrier of disease and risked spreading them. In sum, he was dangerous. Hence the necessity of the existence of the hospital, as much to keep him apart as to protect others from the dangers he represented. Until the eighteenth century the ideal person of the hospital was not the patient, there to be cured, but the poor person on the point of death. It is a question of a person who needs help, material and spiritual, who has to receive final care and the last rites. This was the essential function of the hospital.

One used to say in those times – and with reason – that the hospital was the place where one went to die. The hospital personnel were not attempting the cure of the sick, but rather of attaining their salvation. It was the charitable personnel (comprised of religious or lay people) who were to perform works or mercy which would guarantee that person eternal salvation. As a consequence the institution served to save the soul of the poor in the moment of death and also save the soul of the staff members taking caring of him. He exercised a function in the transition of life to death, in the spiritual salvation more than the material one, all within the function of separating out the dangerous individual for the general health of the population.

For the study of the general significance of a hospital in the Middle Ages and Renaissance one must consider the text entitled *The Book of Active Life of the Hotel*

Dieu written by a parliamentarian who was an administrator of the Hotel Dieu in a language full or metaphors – a type of *Roman de la Rose* of hospitalization – which reflects clearly the mixture of functions of assistance and spiritual transformation which were incumbent upon the hospital.[4]

These were the characteristics of the hospital until the beginning of the eighteenth century. The General Hospital, a place of internment where the sick, the mad, prostitutes, etc., are jumbled and mixed up is still a place of the seventeenth century, a type of diverse instrument of exclusion, assistance, and spiritual transformation from which the medical function is absent.

As far as medical practice is concerned, none of the elements that it integrated and served as its scientific justification predestined it to be a hospital medicine. Medieval medicine and that of the seventeenth and eighteenth centuries were profoundly individualistic. Individualist on the part of the doctor who recognized this condition after an initiation guaranteed by the medical corporation itself, which comprised knowledge of texts and the more or less secret transmission of remedies. The hospital experience was not included in the ritual training of the doctor at that time. What authorized him was the transmission of remedies rather the experiences he would have assimilated and integrated.

The intervention of the doctor in the disease turned around the concept of 'crisis'. The doctor was to observe the patient and the disease from the appearance of the first symptoms to determine the moment at which the crisis was to occur. The crisis represented the moment in which the patient and disease confronted each other; the doctor was to observe the signs, to predict the evolution and to support, as far as possible, the triumph of health and nature over the disease. In the cure, nature, the disease and the doctor came into play. In this struggle, the doctor fulfilled a function of prediction, arbitrator and ally of nature against the disease. The type of battle whose cure took this form could only proceed through an individual relation between the doctor and the patient. The idea of a vast series of observations, collected within a hospital, which would have made it possible to raise the general characteristics of a disease and its particular elements, etc., did not form part of the medical practice.

Thus there was nothing in the medical practice of this period that permitted the organization of hospital knowledge, nor did the organization of the hospital permit the intervention of medicine. In consequence, up until the middle of the eighteenth century the hospital and medicine continued being two separated domains. How did the transformation occur, that is, how did the hospital become medicalized and how was hospital medicine achieved?

The principal factor in the transformation was not the search for a positive action of the hospital on the patient or the illness but simply the annulment of the negative effects of the hospital. It was not first a question of medicalizing the hospital but purifying it of its harmful effects, of the disorder that it created. And in this case one understands by disorder the illnesses which that institution might create in the

---

4   Jehan Henri (1480) *Le livre de vie active des religieuses de l'Hôtel-Dieu de Paris*. Paris.

interned people and propagate in the city in which it was located. It was thus that the hospital was a perpetual focal point of the economic and social disorder.

This hypothesis of the medicalization of the hospital through the elimination of disorder it produced is confirmed by the fact that the first great hospital organization of Europe is found in the seventeenth century, essentially in maritime and military hospitals. The point of departure of the hospital reform was not the civil hospital but the maritime one, which was a place of economic disorder. Through it one trafficked merchandise, precious objects, rare materials, spices, etc., proceeding from the colonies. The trafficker feigned illness and when he disembarked they would take him to the hospital. There he could distribute these goods avoiding the economic control of customs. The great hospitals of London, Marseilles and La Rochelle thus became places of an enormous traffic, against which the fiscal authorities protested.

Thus then the first regulation of the hospital that appears in the seventeenth century refers to the inspection of the coffers which the sailors, doctors and apothecaries retained in the hospital. From that moment on one could inspect the coffers and record their contents; if they found merchandise destined to be contraband their owners would be punished. Thus in this regulation appears an initial economic inquiry.

Moreover, another problem appears in these maritime and military hospitals: that of quarantine, that is to say the epidemic illnesses that can be carried by people disembarking ships. The lazarettos established in Marseilles and La Rochelle constitute a kind of perfect hospital. But it is essentially a type of hospitalization which does not conceive of the hospital as an instrument of cure, but rather as a means of preventing its constituting a focus of economic and medical disorder.

If military and maritime hospitals became a model as a point of departure for hospital reorganization, it is because with mercantilism economic regulations became stricter. But it is also because the value of a man increased more and more. It was in effect precisely in that period that the training of the individual, his capability and his aptitude began to have a value for society.

Let us examine the example of the army. Until the second half of the seventeenth century there was no difficulty in recruiting soldiers; it was sufficient to have financial means. Throughout the whole of Europe there were unemployed people, vagabonds, wretches ready to enter the army of any power, nationality or religion. At the end of the seventeenth century with the introduction of the rifle the army becomes more technical, subtle, and costly. To learn to wield a rifle exercise, manoeuvres, and training are required. This is how the price of a soldier exceeded that of a simple labourer and the cost of an army is converted into a budget entry for every country. Once trained, a soldier could not be permitted to die. If he dies, it has to be in a battle, as a soldier, not because of an illness. One must not forget that in the seventeenth century the index of mortality of a soldier was very high. For example, an Austrian army that left Vienna for Italy lost five sixths of the men before arriving at the field of combat. The losses because of illnesses, epidemic or desertion constituted relatively common phenomena.

From this technical transformation of the army on, the military hospital became an important technical and military matter. (1) It was necessary to oversee [*surveiller*] men in the military hospital so they did not desert because they had been trained at a considerable cost. (2) It was necessary to cure them so they did not die from illness. (3) It was necessary to ensure that having recovered they did not still pretend to be ill and remain in bed.

In consequence, an administrative and political reorganization, a new control of authority in the environs of the military hospital. And the same thing occurs in the maritime hospital, from the moment when the maritime technique become more complex and where similarly the person trained at a considerable cost also may not be lost.

How did this reorganization come to be carried out? The reorganization of the maritime and military hospitals did not stem from a medical technique but essentially from a technology which might be called political, namely discipline.

Discipline is a technique of exercising power, which was not so much invented but rather elaborated in its fundamental principles during the seventeenth century. It had existed throughout history, for example in the Middle Ages, and even in antiquity. For example, the monasteries constitute an example of a place of power of which a disciplinary system was at the heart. Slavery and the great slave companies existing in the Spanish, English, French, and Dutch colonies were also models of disciplinary mechanisms. We can go back to the Roman legion and in it we would similarly find an example of discipline.

Thus disciplinary mechanisms date from ancient times but in an isolated, fragmented manner, until the seventeenth and eighteenth centuries, when disciplinary power is perfected in a new technique with the management of men. We frequently speak of the technical inventions of the seventeenth century – chemical, metallurgical technology – yet we do not mention the technical invention of this new form of governing man, controlling his multiplicity, utilizing him to the maximum, and improving the useful products of his labour, of his activities thanks to a system of power which permits controlling them. In the great workshops which begin to appear, in the army, in schools, when we see throughout Europe great progress in literacy there also appear these new techniques of power which constitute the great inventions of the seventeenth century.

On the basis of the example of the army and school, what is it that arises in this period?

An art of spatial distribution of individuals. In the army of the seventeenth century individuals were herded together forming a conglomeration, with the stronger and most capable at the front. And those who did not know how to fight, the more cowardly or those who desired to flee, were at the flanks and at the middle. The power of a military body was rooted in the effect of the density of this human mass.

In the eighteenth century, on the contrary, beginning at the moment when a soldier receives a rifle, it is necessary to study the distribution of individuals and place them as they ought to be so their efficacy might reach the maximum. Military discipline begins at the moment when one teaches the soldier to locate himself and be at the place that is required.

In the same way, in the schools of the seventeenth century the students were grouped together. The teacher used to call one of them and for a few minutes gave him some instruction and then sent him back to his seat continuing the same operation with another, and so on in succession. Collective teaching works with all students and simultaneously demands a spatial distribution of the class.

Discipline is, above all, analysis of space; it is individualization through space, the placing of bodies in an individualized space that permits classification and combinations.

Discipline does not exercise its control on the results of an action but on its development. In the workshops of the corporate type of the seventeenth century what was required of the worker or master was the fabrication of a product of a determined quality. The mode of fabrication depended upon what was transmitted from one generation to another. The control did not affect the mode of production. In the same way one taught the soldier how to fight, to be stronger than the adversary in the individual fight or on the battlefield.

Beginning in the eighteenth century an art of the human body developed. Movements that are made begin to be observed, in order to determine which are the most efficacious, rapid and best adjusted. Thus the famous and sinister character of the supervisor or foreman appears in workshops, charged not with observing if the work was being done but how it would be done more quickly and with better-adapted movements. In the army appears the non-commissioned officer and with him the army exercises, manoeuvres and the breaking down of movements in time. The famous regulation of infantry that assured the victories of Frederick of Prussia comprises a series of mechanisms of the direction of the movement of the body.

Discipline is a technique of power, which contains a constant and perpetual surveillance of individuals. It is not sufficient to observe them occasionally or see if they work to the rules. It is necessary to keep them under surveillance to ensure activity takes place all the time and submit them to a perpetual pyramid of surveillance. There thus emerge a series of ranks in the army that go, without interruption, from the commander-in-chief to the simple soldier, as well as systems of inspection, reviews, parades, marches, etc., which permit each individual to be observed in a permanent manner.

Discipline supposes a continuous registration: annotations of the individual, relation of events, disciplinary elements, and communication of the information to the higher ranks, so that no detail escapes the top of the hierarchy.

In the classical system the exercise of power was confused and global and discontinuous. It was a question of the power of the sovereign over groups, integrated by families, cities, and parishes, that is by global units, not by the power which acted continuously on the individual.

Discipline is the collection of techniques by virtue of which systems of power have as their objective and result the singularization of individuals. It is the power of individualization whose basic instrument rests in the examination. The examination is permanent, classificatory surveillance, which permits the distribution of individuals, judging them, measuring or evaluating them and placing them so they can be utilized to the maximum. Through the examination, the individual is converted into an element for the exercise of power.

The introduction of the disciplinary mechanisms into the disorganized space of the hospital allowed its medicalization. Everything which has been set out, explains why the hospital is disciplined. Economic reasons, the value attributed to the individual, the desire to avoid the propagation of epidemics explains the disciplinary control to which the hospitals are subjected. But if this discipline acquires a medical character, if this disciplinary power is entrusted to the doctor, it is due to a transformation of medical knowledge. The formation of a hospital medicine has to be attributed, on one hand, to the introduction of discipline into hospital space, and on the other hand, to the transformation that the practice of medicine in that period was undergoing.

In the epistemological systems of eighteenth century, the great marvel of the intelligibility of illnesses is botany, the classification of [Carl von] Linné.[5] This means the necessity of understanding illnesses as a natural phenomenon. As in plants, in diseases there will be different species, observable characteristics, and courses of evolution. Disease is nature, but a nature due to a particular action of the environment on the individual. The healthy person, when he has submitted to certain actions of the environment, serves as a support to the disease, a phenomenon limited by nature. Water, air, food, and the general regimen constitute the bases on which the different types of diseases are developed in individuals.

In this perspective the cure is directed by a medical intervention which is no longer directed toward the disease itself, as in the medicine of crises, but precisely to the intersection of the disease and the organism, as it is in the surrounding environment: air, water, temperature, the regimen, food, etc. It is a medicine of the environment, which is being constituted, to the extent to which the disease can be conceived as a natural phenomenon that obeys natural laws.

In consequence it is in the articulation of those two processes – the displacing of medical intervention and the application of discipline to the space of the hospital – that one finds the origin of the medical hospital. Those two phenomena, of different origin, were going to be adjusted to the hospital discipline whose function would consist in guaranteeing the inquiry, surveillance, and application of disciplines into the disorganized world of the patients and of illness and in transforming the conditions of the environment which surrounds the patients. Likewise patients would be individualized and distributed in a space where one could oversee them and record the events that took place; one could also modify the air they breathed, the temperature of the environment, the water to drink, the regimen, so that the hospital panorama imposed by the introduction of discipline had a therapeutic function.

---

5   Carl von Linné or Carl Linneaus (1707–1778), founder of modern taxonomy, especially in biology, is discussed by Foucault in *The Order of Things*.

If one accepts the hypothesis that the hospital is born from techniques of disciplinary power and from the medicine of interventions on the environment, we can understand several characteristics possessed by that institution.

1. The localization of the hospital and the internal distribution of space. The question of the hospital at the end of the eighteenth century was fundamentally a question of space. In the first place it is a matter of knowing where to situate a hospital so that it does not continue to be a dark, obscure and confused place in the heart of the city where a person would arrive at the hour of death and spread dangerous miasma, contaminated air, dirty water, etc. It was necessary that the place in which the hospital was located conformed to the sanitary control of the city. The location of the hospital had to be determined within the overall medicine of urban space.

In the second place, one also had to calculate the internal distribution of the space of the hospital as a function of certain criteria: if it was certain that an action practiced in the environment would cure diseases, it would be necessary to create about each patient a small individualized space environment, specific to them and modifiable according to the patient, the disease, and its evolution. It is necessary to obtain a functional and medical autonomy of the space for survival of the patient. In this way the principle that beds should not be occupied by more than one patient is established, and thus ends the bed dormitory that at times would be filled by up to six people.

It would also be necessary to create around the patient a manageable environment, to allow the temperature to be increased, to cool the air, and to direct it toward a single patient. Because of this studies on the individualization of living space and the respiration of the patients would be undertaken, including in the collective wards. Thus for example, there was a project of isolating the bed of each patient employing screens at the sides and on the top that would permit the circulation of air but would block the propagation of miasmas.

All of this shows how, in a particular structure, the hospital constitutes a means of intervention on the patient. The architecture of the hospital must be the agent and instrument of cure. The hospital where patients were sent to die must cease to exist. Hospital architecture becomes an instrument of cure in the same category as a dietary regime, bleeding or other medical actions. The space of the hospital is medicalized in its purpose and its effects. This is the first characteristic of the transformation of the hospital at the end of the eighteenth century.

2. Transformation of a system of power in the heart of the hospital. Up to the middle of the seventeenth century religious personnel exercised power and rarely lay people. They were in charge of the daily life of the hospital, the salvation, and the feeding of interned persons. One called the doctor to attend to the most seriously ill, and rather than real action it was a question of a guarantee, a justification. The medical visit was a very irregular ritual, in principle it was performed once a day and for hundreds of patients. In addition, the doctor depended administratively on the religious personnel, who could even dismiss the doctor.

From the moment when the hospital was conceived as an instrument of cure and the distribution of space becomes a therapeutic means, the doctor assumes the main responsibility for the hospital organization. He is consulted as to how the hospital should be constructed and organized; for this reason Tenon realized the previously mentioned mission. Laws prohibited the cloister form of a religious community which had been employed to organize the hospital up to this point. Moreover, if the food regime, the ventilation, the frequency of beverages, were to be instruments of cure, the doctor, upon controlling the regime of the patient, takes charge to a certain point of the economic functioning the hospital, which up to then had been a privilege of the religious order.

At the same time, the presence of the doctor in the hospital is reaffirmed and intensified. The visits increase in an ever more accelerated rhythm during the eighteenth century. In 1680 at the Hotel Dieu of Paris the doctor would visit once a day; on the other hand in the eighteenth century several rules were established, which specify successively that there must be another visit at night for the more serious patients; that each visit should last two hours; and finally in about 1770, that a doctor must reside in the hospital to whom one could go at any hour of the day or night if necessary.

Thus appears the character of the doctor that did not exist before. Until the seventeenth century the great doctors did not appear in the hospital, there were doctors for private consultation that had acquired prestige thanks to a number of spectacular cures. The doctors to whom the religious community resorted for visits to the hospital were generally the worst ones in the profession. The great hospital doctor, the most competent with the greatest experience in those institutions is an invention of the end of the eighteenth century. Tenon, for example, was a hospital doctor, and the work achieved by [Philippe] Pinel at Bicêtre was possible thanks to his practice in the hospital.[6]

This inversion of the hierarchical order of the hospital with the exercise of power by the doctor is reflected in the ritual of the visit: the almost religious procession headed by the doctor, of the whole hierarchy of the hospital: assistants, students, nurses, etc., at the foot of the bed of each patient. This codified ritual of the visit, which signals the place of medical power, is found in the regulations of hospitals in the eighteenth century. It indicates the location of each person, and that the presence of the doctor must be announced by a bell, that the nurse must be at the door with a notebook in hand and accompany the doctor when he enters the room, etc.

3. The organization of a permanent and as far as possible complete records system, which registers whatever occurs. In the first place we must refer to the methods of identification of the patient. A small label will be tied to the wrist of each patient that will allow them to be distinguished if they live, but also if they die. In the upper part of the bed one will place an index card with the name of the patient and what they suffer from. Likewise one begins to utilize a series of records

---

6   Philippe Pinel's (1745–1826) work at Bicêtre is a key element in the founding of modern psychiatry. Foucault discusses his work extensively in *The History of Madness*.

which gather together and transmit information: the general records of admissions and discharges in which the name of the patient is written, the diagnosis of the doctor who admitted them, the ward in which they are located, and if they died or were given a discharge; the registry of each room prepared by the head nurse; the registry of the pharmacy in which are stated the prescriptions and for what patients they were issued; the records of what the doctor ordered during the visit, the prescriptions and the treatment prescribed, the diagnosis, etc.

Finally, it implanted the obligation of the doctor to confront their experiments and their records – at least once a month, in accord with the regulation of the Hotel Dieu in 1785 – to determine the different treatments administered, those that have turned out most satisfactory, the doctors that have the most success, or if epidemic illnesses are passing from one room to another, etc. Thus a collection of documents is formed in the heart of the hospital, and thus is constituted not only a place of cure but also a place of record and the acquisition of knowledge. Medical knowledge, which up until the eighteenth century was located in books, a type of medical jurisprudence concentrated in the great classical treatises of medicine, therefore begins to occupy a place which is not a text, but a hospital. It is no longer what was written and printed, but what every day was recorded in living, active and current actions which the hospital represents.

It is for these reasons that it can be asserted that the normative formation of the doctor in the hospital occurs in the period of 1780–1790. This institution, besides being a place of cure, is a place of medical training. The clinic appears as an essential dimension of the hospital. I understand by 'clinic' [*la clinique*] the organization of the hospital as a place of formation and transmission of knowledge [*savoir*]. But it happens also that, with an introduction of the discipline of the hospital space, it permits curing as well as the recording, capacitating and accumulating of knowledge [*connaissance*]. Medicine offers an immense field as an object of observation, limited on one side by the individual themselves and on the other by the population as a whole.

With the application of the discipline of medical space, and by the fact that it is possible to isolate each individual, install him in a bed, prescribe for him a regimen, etc., one is led toward an individualizing medicine. In effect it is the individual who will be observed, surveyed, known and cured. The individual thus appears as an object of medical knowledge and practice.

At the same time, through the same system of disciplined hospital space, one can observe a great number of individuals. The records obtained daily, when compared among hospitals and in diverse regions, permit the study of pathological phenomena common to the whole population.

Thanks to hospital technology, the individual and the population present themselves at the same time as objects of knowledge and medical intervention. The redistribution of those two medicines will be a phenomenon of the nineteenth century. The medicine that is formed in the course of the eighteenth century is simultaneously a medicine of the individual and the population.

Chapter 16

The Meshes of Power

Michel Foucault, Translated by Gerald Moore

This lecture was given at the University of Bahia, Brazil, on 1st November 1976. It was first published in two parts in *Barbárie*, No. 4, été 1981, 23–7 and No. 5, été 1982, 34–42. It first appeared in French in *Dits et écrits*, Vol. IV, 182–94, followed by a discussion to 201. The discussion is not translated here.

We are going to try to conduct an analysis of the notion of power. I am not the first, far from it, to try to bypass the Freudian schema that opposes instinct to repression, instinct and culture. A whole school of psychoanalysts tried, a few decades ago, to modify, to elaborate on this Freudian schema of instinct *versus* culture and of instinct *versus* repression – I refer to psychoanalysts in the English language as well as the French language, like Melanie Klein, [Donald] Winnicott and Lacan, who tried to show that repression, far from being a secondary, ulterior, delayed mechanism that would try to control any given game of instinct, is by nature part of a mechanism of instinct or, at least, of a process through which sexual instinct develops, unfurls and constitutes itself as drive [*pulsion*].

The Freudian notion of *Trieb* should not be interpreted as a simple natural given, a natural biological mechanism on which repression would come to lay its law of prohibition, but, according to psychoanalysts, as something that is already deeply penetrated by repression. Need, castration, lack, prohibition, the law are already elements through which desire constitutes itself as sexual desire, which therefore implies a transformation of the primitive notion of sexual instinct, such as Freud had conceived it at the end of the 19th century. We must therefore think instinct not as a natural given, but already as a whole development, a wholly complex game between the body and the law, between the body and the cultural mechanisms that ensure the control of the people.

I believe thus that psychoanalysts have considerably displaced the problem by bringing a new notion of instinct to the fore, or rather a new conception of instinct, of the drive, of desire. Nonetheless, what disturbs me, or at least what seems to me insufficient, is that, in this elaboration proposed by psychoanalysts, they perhaps change the conception of desire, but they nonetheless absolutely do not change the conception of power.

In these circles, they still continue to consider that the signified of power, the central point, that in which power consists, is still prohibition, the law, the fact of saying no, once again the form, the formula 'you must not'. Power is essentially what says 'you must not'. It seems to me that this is – and I will speak more of it presently – a totally insufficient conception of power, a juridical conception, a formal conception of power and that it is necessary to elaborate another conception of power that would allow us without doubt better to understand the relations that have established themselves between power and sexuality in Western societies.

I am going to try to develop, or better, to show in which direction one could better develop an analysis of power that would not simply be a negative, juridical conception of power, but a conception of a technology of power.

We frequently find amongst psychoanalysts, psychologists and sociologists this conception according to which power is essentially rule, the law, prohibition, that which marks the limit between what is permitted and what is forbidden. I believe that this conception of power was incisively formulated and broadly developed by ethnology at the end of the 19th century. Ethnology has always tried to detect systems of power, in societies different from our own, as systems of rules. And we, when we try to reflect on our society, on the way in which power exercises itself there, we do so essentially from a juridical conception: where power is, who holds power, what the rules are that govern power, what the system of laws is that power establishes over the social body.

We are thus always doing a juridical sociology of power for our society and, when we study societies different from our own, we do an ethnology that is essentially an ethnology of rules, an ethnology of prohibition. See, for example, in ethnological studies from Durkheim to Lévi-Strauss, what was the problem that would always reappear, perpetually re-worked: a problem of prohibition, essentially the prohibition of incest. And, from this matrix, from this core that is the prohibition of incest, we have tried to understand the general functioning of the system. And it was necessary to wait until more recent years to see new points of view on power appear, be they strictly Marxist or a point of view more distanced from classical Marxism. Anyway, from there we see appear, with the work of Clastres,[1] for example, a whole new conception of power as technology, which tries to break free from the primitive, from this privileging of rules and prohibition that had basically reigned over ethnology from Durkheim to Lévi-Strauss.

In any case, the question that I would like to pose is as follows: how is it that our society, Western society in general, has conceived power in such a restricted, such a poor and such a negative way? Why do we always conceive power as law and as prohibition, why this privileging? We can obviously say that it is due to the influence of Kant, to the idea according to which, in the last instance, the moral law, the 'you must not', the opposition 'you must'/'you must not' is at bottom the matrix of all regulation of human conduct. But, to speak truthfully, this explanation through the influence of Kant is obviously totally insufficient. The problem is of

---

1 Reference to the works of Pierre Clastres collected in *La société contre l'État: Recherches d'anthropologie politique*, Paris: Éditions de Minuit, coll. 'Critique', 1974. [*Society against the state*. Translated by Robert Hurley and Abe Stein. Oxford: Blackwell, 1977.]

knowing whether Kant had such an influence and why it was so strong. Why was Durkheim, a philosopher of vague socialist leanings at the beginning of the French Third Republic, able to rely in this way on Kant when it came to doing an analysis of the mechanism of power in a society?

I believe that we can roughly analyze the reason for this in the following terms: basically, in the West, the great systems established since the Middle Ages developed through the intermediary of the growth of monarchic power at the expense of feudal power, or better, feudal powers. Now, in this struggle between feudal powers and monarchic power, law had always been the instrument of monarchic power against institutions, mores, regulations, the forms of bondage and belonging characteristic of feudal society. I will give you two straightforward examples of this. On one hand monarchic power developed in the West by relying on judicial institutions and by developing these institutions; through civil war, it came to replace the old solution of private litigations by a system of tribunals, with laws, which in fact gave monarchic power the possibility of resolving disputes between individuals itself. In the same way, Roman law, which reappeared in the West in the 13th and 14th centuries, was a formidable instrument in the hands of the monarchy for coming to define the forms and mechanisms of its own power, at the expense of feudal powers. In other words, the growth of the State in Europe has been partly assured by, or in any case, utilized as an instrument, the development of juridical thought. Monarchic power, the power of the State, is essentially represented in law.

Yet it was found that the bourgeoisie, at the same time as broadly profiting from the development of royal power and the weakening, the regression of feudal systems, had every interest in developing this system of law that had allowed it, on the other hand, to shape the economic exchanges that assured its own social development. In such a way that the vocabulary, the form of law has been the system of representation of power common to the bourgeoisie and the monarchy. The bourgeoisie and the monarchy succeeded little by little in establishing, from the end of the Middle Ages up until the 18th century, a form of power that represented itself, that gave itself as a discourse, as a language, the vocabulary of law. And, when the bourgeoisie finally rid itself of monarchic power, it did so precisely by using this juridical discourse – which was nonetheless that of the monarchy – which it turned against the monarchy itself.

To give just one example: when Rousseau came up with his theory of the State, he tried to show how a sovereign, moreover a collective sovereign, a sovereign as social body or, better, a social body as sovereign, is born of the ceding of individual rights, their alienation and the formulation of laws of prohibition that each individual is obliged to recognize because it is he who has imposed the law on himself, to the extent that he is a member of the sovereign, to the extent that he is himself the sovereign. Consequently, this theoretical mechanism, through which the institution of the monarchy has been criticized, has been the instrument of law, which had been established by the monarchy itself. In other words, the West never had a system for the representation, the formulation and the analysis of power other than law and the system of law. And I believe that this is the reason for which, when it comes down to it, we have not had, until recently, other possibilities of analyzing power besides utilizing these elementary, fundamental, etc., notions that are those of law, of rules, of the sovereign, of the delegation of power, etc. I believe that it is this juridical conception of power, this conception of power derived from law and the sovereign, from rule and prohibition, of which we must now rid ourselves if we want to proceed to an analysis not just of the representation of power, but of the real functioning of power.

How could we try to analyze power in its positive mechanisms? It seems to me that we can find, in a certain number of texts, the fundamental elements for an analysis of this type. We can maybe find them in Bentham, an English philosopher from the end of the 18th and the beginning of the 19th century, who was ultimately the great theoretician of bourgeois power, and we can obviously also find them in Marx, essentially in Volume II of *Capital*. It is there, I think, that we can find several elements on which I can draw for the analysis of power in its positive mechanisms.

In sum, what we can find in Volume II of *Capital* is, in the first place, that there exists no *single* power, but several powers.[2] Powers, which means to say forms of domination, forms of subjection, which function locally, for example in the workshop, in the army, in slave-ownership or in a property where there are servile relations. All these are local, regional forms of power, which have their own way of functioning, their own procedure and technique. All these forms of power are heterogeneous. We cannot therefore speak of power, if we want to do an analysis of power, but we must speak of powers and try to localize them in their historical and geographical specificity.

A society is not a unitary body in which one power and one power only exercises itself, but in reality it is a juxtaposition, a liaising, a coordination, a hierarchy, too, of different powers which nonetheless retain their specificity. Marx continually insists, for example, on the simultaneously specific and relatively autonomous, in some way impermeable, character of the *de facto* power that the employer exerts in a workshop, in relation to the juridical type of power that exists in the rest of society. Thus the existence of regions of power. Society is an archipelago of different powers.

Secondly, it seems that these powers cannot and must not be understood simply as the derivation, the consequence of what would be a primordial, central type of power. The schema of jurists, be it that of Grotius, of Pufendorf or of Rousseau, consists in saying: 'In the beginning, there was no society, and then society appeared from the moment that there appeared a central point of sovereignty that organized the social body, and which then enabled a whole series of local and regional powers'; Marx, implicitly, does not recognize this schema. He shows on the contrary how, from the initial and primitive existence of these small regions of power – such as property, slavery, the workshop and also the army – great State apparatuses could

2　Karl Marx, *Das Kapital: Kritik der politischen Ökonomie*, Buch II: 'Der Zirkulationsprozess des Kapitals.' Hamburg: O. Meissner, 1867. [*Capital: A Kritique of Political Economy*, Volume Two. Translated by David Fernbach. Harmondsworth: Penguin, 1978.]

form, bit by bit. The unity of the State is essentially secondary in relation to these specific and regional powers, which come in the first place.

Thirdly, these specific, regional powers absolutely do not function primordially to prohibit, to prevent, to say 'you must not'. The primitive, essential and permanent function of these local and regional powers is, in reality, to be producers of an efficiency, an aptitude, producers of a product. Marx gave, for example, superb analyses of the problem of discipline in the army and in the workshops. The analysis that I will make of discipline in the army is not to be found in Marx, but no matter. What happened in the army, from the end of the 16th and the beginning of the 17th century until practically the end of the 18th century? A whole enormous transformation meant that, in the army, which up to this time had essentially been made up of relatively interchangeable small units of individuals, organized around a leader, these units were replaced by a great pyramidal unity, with a whole range of intermediary leaders, sub-officers, technicians too, essentially because of a technical discovery: the relatively quick-fire and aimable rifle.

From this moment on, the army could no longer be treated – it was dangerous to make it function in this way – in the form of small isolated units, composed of interchangeable elements. For the army to be efficient, it was necessary that these rifles be employed in the best possible way, that each individual be trained to occupy a determinate position in an extended front, to place himself simultaneously in harmony with a line that must not be broken, etc. A whole problem of discipline implied a new technique of power with sub-officers, subordinate and superior officers. And it is thus that the army could be treated as a very complex hierarchical unity, by ensuring its maximal performance through the unity of the whole in accordance with the specificity of the position and role of each individual.

Military performance was highly superior on account of a new procedure of power, whose function was absolutely not that of prohibiting anything. Of course, this led to prohibiting one thing or another, the goal was nonetheless absolutely not to say 'you must not', but essentially to obtain a better performance, a better production, a better productivity from the army. The army as the production of deaths – this is what has been perfected or, better, what has been ensured by this new technique of power. This was absolutely not prohibition. We can say the same thing about discipline in the workshops, which began to establish itself around the 17th and 18th centuries, when the replacement of small corporative-style workshops by great workshops with a whole series of workers – hundreds of workers – made it necessary simultaneously to oversee and coordinate their movements with one another through the division of labour. The division of labour was, at the same time, the reason for which this new workshop discipline had to be invented; but inversely we can say that this workshop discipline was the condition for the division of labour being able to take hold. Without this workshop discipline, which is to say without the hierarchy, without the overseeing, without the supervisors, without the chronometric control of movements, it would not have been possible to obtain the division of labour.

Finally, a fourth important idea: these mechanisms of power, these procedures of power, must be considered as techniques, which is to say procedures that have been invented, perfected and which are endlessly developed. There exists a veritable technology of power or, better, powers, which have their own history. Here, once again, one can easily find between the lines of Volume II of *Capital* an analysis, or at least the sketch of an analysis, which would be the history of the technology of power as it has been exercised in the workshops and in the factories. I will therefore follow these essential indications and I will try, where sexuality is concerned, not to envisage power from a juridical point of view, but from a technological one.

It seems to me, in fact, that if we analyze power by privileging the State apparatus, if we analyze power by considering it as a mechanism of conservation, if we consider power as a juridical superstructure, we basically do no more than return to the classical theme of bourgeois thought, when it essentially envisaged power as a juridical fact. To privilege the State apparatus, the function of conservation, the juridical superstructure, is to 'Rousseau-ize' Marx. It is to reinscribe it in the bourgeois and juridical theory of power. It is not surprising that this supposedly Marxist conception of power as State apparatus, as agent of conservation, as juridical superstructure, finds itself in the European social democracy of the end of the 19th century, when the problem was precisely that of knowing how to make Marx function on the inside of the juridical system of the bourgeoisie. So, what I would like to do in revisiting that which is found in Volume II of *Capital*, and in distancing from it everything that has subsequently been added to it and rewritten on the privileges of State apparatus, the function of the reproduction of power, the character of the juridical superstructure, would be to try to see how it is possible to do a history of powers in the West, and essentially of the powers that have been invested in sexuality.

From this methodological principle, how then would we be able to do a history of the mechanisms of power in relation to sexuality? I believe that, in a very schematic way, we would be able to say the following: the system of power that the monarchy had succeeded in organizing since the end of the Middle Ages presented two major disadvantages for the development of capitalism. Firstly, political power, such as it was exercised in the social body, was a very discontinuous power. The mesh of the net was too large, an almost infinite number of things, elements, conducts and processes escaped the control of power. If we take for example a precise point: in the importance of contraband across Europe until the end of the 18th century, we note a very important economic flow, a flow almost as important as any other, a flow that entirely escaped power. It was, moreover, one of the conditions of the existence of people; if there had been no maritime piracy, commerce would not have been able to function, and people would not have been able to live. In other words, illegality was one of the conditions of life, but at the same time it signified that there were certain things that escaped power, and over which power had no control. Consequently, economic processes that after a fashion remained out of control required the establishment of a continuous power, to be precise, of a certain atomistic manner; to pass from lacunary, global power to a continuous, atomistic and individualizing power: that each one, every individual himself, in his body, in his movements, could be controlled, in the place of global and mass controls.

The second great disadvantage of the mechanisms of power as they functioned in the monarchy, is that they were excessively onerous. And they were onerous precisely because the function of power – that in which power consisted – was essentially the power of taking away, of having the right and the force to perceive something – a tax, or a tithe, when it came to the clergy – in what had been harvested: the obligatory perception of such and such a percentage for the master, for royal power, for the clergy. Power was thus essentially perceiver and predator. To this extent, it always operated an economic subtraction and, as a consequence, far from favouring and stimulating economic flows, it was perpetually an obstacle, a break on them. Whence this second preoccupation, this second necessity: to find a mechanism of power that, at the same time as controlling things and people up to the finest detail, is neither onerous nor essentially predatory on society, that exercises itself in the very sense of the economic process.

With these two objectives, I believe that we can roughly understand the great mutation of technological power in the West. We have the habit – once again conforming to a more or less primary spirit of Marxism – of saying that the great invention, as everybody knows, was the steam engine, or some other invention of this type. It is true that this was very important, but there was a whole other series of technological inventions equally important as this and which, in the last instance, were the condition of the functioning of others. This was the case with political technology; there was a glut of invention at the level of forms of power right across the 17th and 18th centuries. As a consequence, we must undertake not only a history of industrial techniques, but also of political techniques, and I believe that we can group the inventions of political technology into two large chapters, for which we must credit the 17th and 18th centuries foremost. I would group them into two chapters because it seems to me that they developed in two different directions. On one hand, there is this technology that I would call 'discipline'. Discipline is basically the mechanism of power through which we come to control the social body in its finest elements, through which we arrive at the very atoms of society, which is to say individuals. Techniques of the individualization of power. How to oversee someone, how to control their conduct, their behaviour, their aptitudes, how to intensify their performance, multiply their capacities, how to put them in the place where they will be most useful: this is what discipline is, in my sense.

I just cited you an example of discipline in the army. It is an important example, because it has truly been the point at which the great discovery of discipline was made and developed almost in the first place. One moreover linked to that other invention of technico-industrial order, namely the relatively quick-fire rifle. From this moment on we can basically say the following: that the soldier ceased to be interchangeable, ceased to be purely and simply flesh with a gun and a simple individual capable of hitting. To be a good soldier, it was necessary to know how to shoot, it was therefore necessary to have passed through a process of apprenticeship. It was necessary that the soldier knew equally how to move, how to coordinate his movements with those of the other soldiers, in sum: the soldier becomes something of skill, and therefore valuable. And the more valuable he was, the more it was necessary to preserve him; the more it was necessary to preserve him, the more it became necessary to teach him the techniques capable of saving his life in battle; and the more he was taught techniques, the longer his apprenticeship, the more valuable he was. And suddenly you have a sort of take-off of the military techniques of dressage that culminated in the famous Prussian army of Frederick II, which spent the most part of its time doing drills. The Prussian army, the model of Prussian discipline, is precisely the perfection, the maximal intensity of this corporal discipline of the soldier, which was, up to a point, the model for other disciplines.

Another point through which we see this new disciplinary technology appear is education. It is initially in the schools and then in the primary schools that we see appear these new disciplinary methods through which the multiplicity of individuals are individualized. The school brings together tens, hundreds and sometimes thousands of schoolchildren, students and it is as such a question of exercising over them a power that is precisely much less onerous than the power of the private tutor, one which could only exist between the pupil and the master. There we have a master for dozens of disciples; it is therefore necessary, despite this multiplicity of pupils, that there is an individualization of power, a permanent control, an overseeing of every moment. Whence the appearance of this person known to all those who have studied in school, namely the invigilator, who, in the pyramid, corresponds to an army sub-officer; equally the appearance of quantitative marking, the appearance of exams, the appearance of competitions, the possibility, consequently, of classing individuals in such a way that each one is exactly in their place, under the eyes of the master, or even in the qualification and in the judgment that we hold over them.

See for example how you sit in a row before me. It is a position that perhaps appears natural to you, but it is worth recalling however that it is relatively recent in the history of civilization, and that it is still possible at the beginning of the 19th century to find schools where the pupils present themselves standing upright in a group, around a teacher who gives them a lesson. And that implies, obviously, that the teacher cannot really and individually oversee them: there is a group of pupils and then the teacher. Nowadays, you are placed like this in a row, the gaze of the professor can individualize each one, can call them to know that they are present, what they do, if they dream, if they yawn... There are trivialities there, nonetheless very important futilities, because in the end, at the level of a whole series of exercises of power, it is these little techniques that these new mechanisms of power could invest in and were able to make work. What happened in the army and in the schools could equally be seen in the workshops throughout the 19th century. What I will call the individualizing technology of power, a power that basically targets individuals right up to their bodies, in their behaviour; it is *grosso modo* a type of political anatomy, an anatomy that targets individuals to the point of anatomizing them.

This is one family of technologies of power that appeared in the 17th and 18th centuries; we have another family of technologies of power that appeared a bit later, in the second half of the 18th century, and which was developed (it must be said that the first, to the shame of France, was developed primarily in France and in Germany) primarily in England: technologies that did not target individuals as individuals, but which on the contrary targeted the population. In other words, the 18th century discovered this principal thing: that power is not simply exercised over subjects; this was the fundamental thesis of the monarchy, according to which there is the sovereign and then subjects. We discover that that on which power is exercised is the population. And what does population mean? It does not simply mean to say a numerous group of humans, but living beings, traversed, commanded, ruled by processes and biological laws. A population has a birth rate, a rate of mortality, a population has an age curve, a generation pyramid, a life-expectancy, a state of health, a population can perish or, on the contrary, grow.

Now all this began to be discovered in the 18th century. We see, consequently, that the relation of power to the subject or, better, to the individual must not simply be this form of subjection that permits power to take from the subject goods, riches and eventually its body and blood, but that power must be exercised on individuals insofar as they constitute a species of biological entity that must be taken into consideration, if we want precisely to utilize this population as a machine for producing, producing riches, goods, producing other individuals. The discovery of population is, alongside the discovery of the individual and the body amenable to dressage, the other great technological core around which the political procedures of the West transformed themselves. At this moment, what I will call 'bio-politics', in opposition to the anatomo-politics I mentioned a moment ago, was invented. It is at this moment that we see appear problems like those of housing, of the conditions of life in the city, of public hygiene, of the modification of the relation between birth and mortality. It is at this moment that there appeared the problem of knowing how we can bring people to have more children, or at any rate how we can regulate population flux, how we can equally regulate migrations and the growth rate of a population. And, from this, a whole series of techniques of observation, including statistics, obviously, but also all the great administrative, economic and political organisms, are charged with this regulation of the population. There were two great revolutions in the technology of power: the discovery of discipline and the discovery of the regulation and perfection of an anatomo-politics and the perfection of a bio-politics.

Life has now become, from the 18th century onwards, an object of power. Life and the body. Once, there were only subjects, juridical subjects from whom one could take goods, life too, moreover. Now, there are bodies and populations. Power has become materialist. It ceases to be essentially juridical. It must deal with these real things that are bodies and life. Life enters into the domain of power: a crucial mutation, without doubt one of the most important in the history of human societies; and it is evident that one can see how sex was able to become from this moment, which is to say beginning precisely in the 18th century, an absolutely crucial theatre; for, basically, sex is very exactly placed at the point of articulation between individual disciplines of the body and regulations of the population. Sex is that point from which the overseeing of individuals can be ensured, and we understand how, in the 18th century, and precisely in schools, the sexuality of adolescents became a medical problem, a moral problem, almost a political problem of the highest order, because through – and under the pretext of – this control of sexuality, schoolgoers and adolescents, could be overseen throughout their lives, at every instant, even during sleep. Sex thus goes on to become an instrument of 'disciplinarization', it comes to be one of the essential elements of this anatomo-politics of which I have spoken; but, on the other hand, it is sex that ensures the reproduction of populations, it is with sex, with the politics of sex that we can change the relation between birth and death; in any case, the politics of sex comes to be integrated into the interior of this whole politics of life, which will become so important in the 19th century. Sex is the hinge between anatomo-politics and bio-politics, it is at the intersection of disciplines and regulations, and it is in this function that it has become, at the end of the 19th century, a political drama of first importance for making society a machine of production.

Chapter 17

The Language of Space

Michel Foucault, Translated by Gerald Moore

'Le langage de l'espace.' *Dits et écrits*. Vol I, 407–12. This piece originally appeared in the journal *Critique* in 1964.

Writing, over the centuries, has been coordinated with time. Narrative, be it fictional or real, was not the only form of this belonging [to time], nor the one that is most essential to it; it is even probable that it has concealed the depths and law of writing in the movement that seemed best to exhibit them. At the point of liberating writing from narrative, from its linear order, from the great syntactical game of the concordance of times, it was believed that the act of writing was relieved of its old obedience to time. In fact the rigour of time did not exercise itself over writing through the leanings of what it wrote, but in its dense layering, in that which constituted its singular being – incorporeal. Whether or not addressing itself to the past, submitting to the order of chronologies, or applying itself to unravelling them, writing was caught in the fundamental curve of the Homeric return; but also that of the accomplishment of Jewish Prophecies. Alexandria, which is our birthplace, had prescribed this circle to all Western language: to write was to make return, it was to return to the origin, to re-capture oneself in the primal moment; it was to be new every morning. From this the mythical function, up until the present, of literature; from this the relation of literature to the ancient; from this the privilege that literature accorded to analogy, to the same, to all the marvels of identity. From this, above all, a structure of repetition that designates its being.

The 20th century is perhaps the era when such kinships were undone. The Nietzschean return closed once and for all the curve of Platonic memory, and Joyce closed that of the Homeric narrative. This does not condemn us to space as the only other possibility, for too long neglected, but reveals that language is (or, perhaps, became) a thing of space. That it might describe or pass through space is no longer what is essential here. And if space is, in today's language, the most obsessive of metaphors, it is not that it henceforth offers the only recourse; but it is in space that, from the outset, language unfurls, slips on itself, determines its choices, draws its figures and translations. It is in space that it transports itself, that its very being 'metaphorizes' itself.

The gap, distance, the intermediary, dispersion, fracture and difference are not the themes of literature today; but in which language is now given and comes to us: what makes it speak. Language has not, like the verbal model, removed these dimensions from things in order to reinstate something analogous. These dimensions are common to things and to language itself: the blind spot where things and words come to us in the moment where they go toward their meeting point. This paradoxical 'curve', so different from the Homeric return or from the fulfilment of the Promise, is without doubt for the moment the unthinkable of Literature. Which is to say that which makes it possible in the texts where we can read it today.

***

La Veille, by Roger Laporte,[1] *clings tightly* to this simultaneously pallid and awesome 'region'. It is designated here as an ordeal: a danger and probation, an opening that instantiates but remains gaping, an approach and a distancing. What imposes its imminence in this way, but also immediately turns away, is not language. But a neutral subject, a faceless '*it*' through which all language is possible. Writing is given only if *it* does not withdraw in the absolute of distance; but writing becomes impossible when it threatens with the full weight of its extreme proximity. In this gap that is full of perils, there can be neither Midst, nor Law, nor Measure (no more so than in Hölderlin's *Empedokles*[2]). For nothing is given but distance and the night watch [*la veille*] of the lookout who opens his eyes on the day that is not yet there. In an enlightened but absolutely reserved way, this *it* states the excessive, unmeasured measure of the distance that keeps vigil, where language speaks. The experience recounted by Laporte as the past of an ordeal is exactly where the language that recounts it is given; it is the fold where language redoubles the empty distance from where it comes to us and separates itself from itself in the approach of this distance over which it is proper to language, and to language alone, to keep watch.

In this sense, the work of Laporte, in proximity to Blanchot, thinks the unthought of Literature and approaches its being through the transparency of a language that seeks not so much to join with it as to receive and host it.

***

An Adamite novel, *Le Procès-Verbal*,[3] is also a watch [*une veille*], but in the full light of day. Stretched out across 'the diagonal of the sky', Adam Pollo is at the point where the aspects of time fold in on one another. At the beginning of the novel, he is perhaps an escapee of the prison in which he will be enclosed at its end; perhaps he comes from the hospital whose black paint, metal and mother-of-pearl shell he finds in the final pages. And the breathless old woman who climbs toward him with the whole world as a halo around her head is undoubtedly, in the discourse of madness,

---

1   Roger Laporte (1963) *La Veille*. Paris: Gallimard, collection 'Le Chemin'. The title means 'eve', 'sleeplessness', 'keeping wake' or 'night watch'.

2   Friedrich Hölderlin (1798) *Der Tod des Empedokles*. The German text and English translation can be found in *Hölderlin: Selected Verse*. Edited and translated by Michael Hamburger. Harmondsworth: Penguin, 1961, 40–65.

3   J.-M.G. Le Clézio (1963) *Le Procès-Verbal*. Paris: Gallimard, collection 'Le Chemin'. *The Statement*, literally 'the trial of the Word', it can also mean a transcript of a discussion.

the young girl who, at the beginning of the text, climbs up to his abandoned house. And in this refolding of time an empty space is born, an as yet unnamed distance where language precipitates. This distance is steepness itself and at its summit Adam Pollo is like Zarathustra: he descends toward the world, the sea, the town. And when he climbs back up to his den, it is not the solar circle, the inseparable enemies of the eagle and the serpent, who await him; but only the dirty white rat that he tears apart with a knife, and which he sends to rot on a sun of thorns. Adam Pollo is a prophet in a singular sense; he does not announce Time; he speaks of this distance that separates him from the world (from the world that 'came to him from his head by dint of being watched'), and, by the tide of his discourse contradicts; when the world flows back to him, like a big fish swimming against the current he will swallow it and hold it closed for an indefinite and immobile time in the quartered bedroom of an asylum. Closed in on itself, time now redistributes itself on this chessboard of bars and the sun. A grid that is perhaps the puzzle of language.

***

The entire work of Claude Ollier is an investigation of the space that is common to language and things; in appearance, an exercise for adjusting long and patient sentences, undone, resumed and fastened in the movements of a simple gaze or a stroll, to the complex spaces of towns and countrysides. To speak truthfully, the first novel of Ollier, *La Mise en scène*,[4] already revealed a deeper relation between language and space than that of a description or a sublimation: in the blank circle of an unmapped region, narrative had given birth to a precise space peopled and furrowed by events in which he who described them (in giving birth to them) found himself immersed, as if lost; because the narrator had a 'double' who, himself inexistent in this same inexistent place, had been killed by a sequence of factual events identical to those that wove around the narrator: so much so that this hitherto undescribed space had been named, recounted and measured up only at the cost of a murderous redoubling; space acceded to language by a 'stuttering' that abolished time. Space and language were born together, in *Le Maintien de l'ordre*,[5] of an oscillation between a gaze that saw itself being overseen and an obstinate and mute double gaze that oversaw it and surprised the overseer with a game of constant retrospection.

Été indien[6] obeys an octagonal structure. The axis of abscissa is the car that, from the tip of its hood, cuts the expanse of the landscape in two; it is the stroll on foot or by car through the city; it is tramways and trains. On the vertical axis of coordinates, there is the climb up the side of the pyramid, the elevator in the sky-scraper, the panoramic view that hangs over the city. And in the space opened up by these perpendiculars, every composite movement unfurls: the gaze that turns, the one that plunges over the expanse of the city as if studying a plan; the curve of the air train

---

4   Claude Ollier (1958) *La Mise en scène*. Paris: Éditions de Minuit. *The Staging*, or *Direction*, literally 'the putting in place'.

5   Claude Ollier (1961) *Le Maintien de l'ordre*. Paris: Éditions de Minuit. *The Maintenance of Order*.

6   Claude Ollier(1963) *Été indien*. Paris: Éditions de Minuit. *Indian Summer*.

that propels itself beyond the bay and then descends again toward the suburbs. What is more, some of these movements are prolonged, reverberated, sent back or forth or fixed by photos, fixed points of view and fragments of film. But all are redoubled by the eye that follows them, relates them or completes these movements itself. For the gaze is never neutral; it gives the impression of leaving things there where they are; in fact, it 'removes' them, virtually detaching them from their depths and layers, in order to enter them into the composition of a film that is yet to exist and whose screenplay has not been determined. These are the 'views' that are not decided upon, but 'under option', and which, between the things that are no longer and the film that is yet to be, form with language the weaving plot of the book.

In this new place, that which is perceived abandons its consistency, detaches itself from itself, floats in a space and in accordance with improbable combinations, acquires the gaze that detaches them and knots them, so much that it enters inside them, creeps into this strange impalpable distance that separates and unites their place of birth and their screen grand finale. Entering the aircraft that leads them back toward the reality of film (producers and authors) as if he had entered into this slender space, the narrator disappears with it – with the fragile distance established by his gaze: the plane falls into a tide that closes in on all the things seen in this 'removed' space, leaving only the red flowers 'under no gaze' beyond the now-calm perfect surface, and this text that we read – the floating language of a space that has devoured itself along with its creator, but which still and forever remains present in all these words that no longer have a voice to pronounce them.

***

Such is the power of language: that which is woven of space elicits space, gives itself space through an originary opening and removes space to take it back into language. But again it is devoted to space: where else could it float and posit itself, if not on this place that is the page, with its lines and its surfaces, if not in this *volume* that is the book? Michel Butor has, on several occasions, formulated the laws and paradoxes of this space so visible that language ordinarily encompasses it without protest. The *Description de San Marco*[7] does not seek to restore in language the architectural model of that which the gaze can traverse. But it systematically, and of its own accord, makes use of all the spaces of language that are subsidiary to the edifice of stones: anterior spaces that language recovers (the sacred texts illustrated by the frescos), spaces immediately and materially superimposed on painted surfaces (inscriptions and legends), ulterior spaces that analyze and describe elements of the Church (commentaries in books and guides), neighbouring and correlative spaces that grasp at us somewhat accidentally, caught up in words (the reflections of watching tourists), nearby spaces whose gazes are turned elsewhere (fragments of dialogues). These spaces have their own proper place of inscription: rolls of manuscripts, the surfaces of walls, books, magnetic tapes that one cuts with

---

7   Michel Butor (1963) *Description de San Marco*. Paris: Éditions de Minuit, 'Collection blanche'.

scissors. And this three-fold game (the basilica, verbal spaces, and the place of their writing) distributes its elements in accordance with a double system: the usual route (which is itself the entangled outcome of the space of the basilica, the strolling of the walker and the movement of his gaze), and that which is prescribed by the great white pages on which Butor had his text printed, where strips of words are cut up by no more than the law of margins, some laid out in verse and others in columns. And this organization perhaps brings us back to yet another space, which is that of photography… An immense architecture along the lines of the basilica, but differing absolutely from its space of stones and paintings – directed toward it, clinging to it, traversing its walls, opening the trove of words buried inside it, bringing back to it the whole murmur of that which escapes it or turns away from it, making the games of verbal space, in its grappling with things, surge up with methodological rigour.

'Description' here is not a reproduction, but more a deciphering: the meticulous undertaking for untangling this mess of the diverse languages that are things, in order to restore each to its natural place and make the book a white place where everything, after de-scription, can find a universal place of inscription. And this, without doubt, is the being of the book, the object and place of literature.

Chapter 18

The Force of Flight

Michel Foucault, Translated by Gerald Moore

'La force de fuir.' *Dits et écrits*. Vol. II, 401–5. This was originally published in March 1973 in *Derrière le miroir*, No. 202, 1–8 to accompany a series of paintings by Paul Rebeyrolle (1926–2005) known as *Dogs*.

You have entered. Here you are surrounded by ten paintings, which run the length of a room whose every window has been carefully closed.

Are you not, in turn, in prison, like the dogs you see priming themselves and pushing up against the bars?

Unlike the *Birds* from the Cuban sky, the *Dogs* belong to neither a determinate time nor a specific place.[1] It is not about the prisons of Spain, of Greece, of the USSR, of Brazil or Saigon; it is about *prison*. But prison – Jackson has given testimony to this[2] – is today a political place, which is to say a place where forces are born or become manifest, a place where history is formed and where time surges up.

The *Dogs* are not as such a variation on a form, on colours, a movement like the *Frogs* were. They form an irreversible series, an interruption that cannot be mastered. Do not say: a history appears thanks to the juxtaposition of canvases; but rather: the movement that initially trembles, then breaks free from the canvas, really passes beyond its limits to inscribe itself, to continue itself in the following canvas and to make all the canvases shudder with a great movement that ends up escaping them and leaving them there in front of you. The series of paintings, instead of recounting what has happened, gives rise to a force whose history can be recounted as the ripple of its flight and its freedom. Painting has at least this much in common with discourse: when it gives rise to a force that creates history, it is political.

Observe: the windows are white, so much so that enclosure reigns. Neither sky nor light: nothing of the interior can be glimpsed; nothing risks penetrating it anymore. Rather than an exterior, there is pure outside, neutral, inaccessible and without form. These white squares do not indicate a sky and an earth that one could see from afar,

---

1   *Birds*, *Frogs* and *Guerrillas* are previous series of Rebeyrolle's paintings. The other titles are those of individual paintings in the *Dogs* series.

2   Foucault is referring to Bruce Jackson, and his book *In the Life: Versions of the Criminal Experience*. New York: Holt, Rinehart & Winston, 1972. Foucault wrote a preface for the French edition, *Leurs prisons: Autobiographie de prisonniers et d'ex-détenus américains*. Paris: Plon 1975, reprinted in *Dits et écrits*, Vol. II, 687–91.

they denote that one is here and nowhere else. In classical painting, windows allow an interior to be re-placed in an external world; these unseeing eyes fix, nail and anchor shadows to walls that would otherwise know only night. Emblems of stark impotence.

Power, obstinate and immobile power, rigid power: such are the woods in the paintings of Rebeyrolle. Woods superimposed on canvas, glued to it by the strongest glue that one could find ('one cannot uproot them without uprooting the canvas'), they are simultaneously in the painting and outside its surface. In the middle of these hourless nights, in this darkness without direction, fragments of truncheons are like clockhands, but which mark height and depth: a timepiece of verticality. When the dogs are at rest, the batons hang straight; they are the immobile guards of *The Jail*, the single watchman of the sleeping *Condemned*, the pikes of *Torture*; but when the dog is primed, the wood lengthens and becomes a bar; it is the formidable lockdown of the *The Cooler*; against whose door *The Enraged* presses up; against the window of the *Prisoners*, still and always the horizontal baton of power.

In the world of prisons, as in the world of dogs ('lying down' and 'upright'), the vertical is not one of the dimensions of space, it is the dimension of power.

It dominates, rises up, threatens and flattens; an enormous pyramid of buildings, above and below; orders barked out from up high and down low; you are forbidden to sleep by day, to be up at night, stood up straight in front of the guards, to attention in front of the governor; crumpled by blows in the dungeon, or strapped to the restraining bed for having not wanted to go to sleep in front of the warders; and, finally, hanging oneself with a clear conscience, the only means of escaping the full length of one's enclosure, the only way of dying upright.

The window and the baton oppose one another yet form a couple, as power and impotence. The baton, which is external to the painting, which, with its miserable straightness has come to be stuck to the painting, breaks the darkness and the body until bloody. The represented window, by contrast, with the limited means of painting, is incapable of opening onto any space. The straightness of the one bears on and underlines the powerlessness of the other: they intertwine in the bars. And, by these three elements (bars-window-baton), the splendour of this painting is wilfully pulled back from the aesthetics and forces of enchantment and onto politics – the struggle of forces and power.

When the white surface of the window shines out against an immense blue, it is the decisive moment. The canvas whereupon this mutation takes effect has the title *Inside*: the division takes place and the inside begins to open itself, despite itself, to the birth of a space. The wall splits from top to bottom: one would say divided by a great blue sword. The vertical, which, in the foreground of the baton, once marked power, now digs for freedom. The vertical batons that hold up the grill do not prevent the wall beside them from cracking. A muzzle and paws throw themselves into opening it with an intense joy, an frisson of electricity. In the world of men, nothing big has ever happened through these windows, but everything, always, through the triumphant bringing down of these walls.

The futile window has anyhow disappeared in the subsequent canvas (*The Enclosure*): pulled up against the ledge of a wall, the dog, erect but already somewhat drawn into himself, pulled back to pounce, looks out on the blue and infinite surface in front of him, from which only two driven in stakes and a half-battered grill separate him.

One leap and the surface swings around. Inside outside. From an inside that had no exterior to an outside that leaves no interior standing. Field and reverse angle [*champ et contrechamp*]. The white window is obscured and the blue that lay before us becomes a white wall that one leaves behind. This leap, this irruption of force (which is not represented on the canvas, but which produces itself unspeakably *between* two canvases, in the lightening-burst of their proximity) was enough for all signs and values to be inverted.

The abolition of verticals: henceforth everything takes flight in accordance with rapid horizontals. In *The Beautiful* (the most 'abstract' of the series: because it is pure force, the night rising up from the night and carving itself out as a vibrant form in the light of day), the impotent baton this time designates a forced portico. Surging forth from the obscure, which still seems to impregnate it and form a body with it, a beast takes flight, feet first, penis raised.

And the grand finale, the great last canvas unfurls and spreads out a new space, hitherto absent from the whole series. It is the charting of transversality; it is divided by halves between the black fortress of the past and the clouds of future colour. But, across its whole length, the traces of a gallop – 'the sign of an escapee'. It seems that the truth comes softly, in the steps of a dove. Force, too, leaves on the earth the claw-marks, the signature of its flight.

There were, in Rebeyrolle, three grand series of animals: first the trout and the frogs; then the birds; and here the dogs. Each one corresponds not only to a distinct technique, but to a different act of painting. The frogs and the trout weave in and out of weeds, pebbles and swirling streams. Movement is achieved through reciprocal displacements: the colours slip over their original forms and constitute, beside them, a bit further on, floating and liberated flecks; the forms displace one another under colours and cause the line of a nervous twitch or behaviour to rise up between two immobile surfaces. In such a way that it produces a leaping in some green, a darting amid the transparent, a furtive burst of speed through blue reflections. Animals from down below, animals of the water, of earth, the humid earth, formed within water and earth and broken down in them (a bit like Aristotle's rats), frogs and trout can only be painted as linked to and dispersed through them. They carry with them the world that eludes them. The painter apprehends them where they hide themselves only in order to liberate them and make them disappear in the movement that traces them.

The bird, like power, comes from on high. It beats down against the force that also comes from on high, and which it wants to *master*. But, in the moment it approaches this terrestrial force, yet livelier and more burning than the sun, it breaks down and falls, dislocated. In the series of *Guerillas*, birds-helicopters-parachutists swing toward the sun, head first, already struck by death, which they sew around them in a final somersault. In Bruegel, a miniscule Icarus falls, struck by the sun: this happens amidst the indifference of a working and everyday countryside. The bird in the green beret, in Rebeyrolle, falls in an enormous clatter from which beaks, claws, blood and feathers fly out. It is tangled up in the soldier into whom it clatters, but who kills it; red fists and arms thrown out. The contours from which the frogs and the trout furtively free themselves are rediscovered here, but in fragments, and on the periphery of a struggle where the violence of colour crushes forms. The act of painting is beaten back onto the canvas where it will thrash out for a long time yet.

The dogs, like the frogs, are animals from down below. But animals of force that rage. Form, here, is entirely reconstructed; despite the gloomy colours and overtone, the silhouettes carve themselves out with precision. However, the contour is not obtained by a line that runs neatly the length of the body; but by thousands of perpendicular strokes, blades of straw that form a general bristling, a gloomy electric presence in the night. It is less a question of form than of energy; less of a presence than an intensity, less of a movement and a behaviour than an agitation, a trembling contained only with difficulty. Mistrustful of language, Spinoza feared that in the word 'dog' one might confuse 'barking animal' and 'celestial constellation'.[3] The dog of Rebeyrolle is resolutely both barking animal and terrestrial constellation.

Here, the painting of form and the unleashing of force come together. Rebeyrolle has found, in a single movement, the means of bringing out the force of painting in the vibrancy of the painting. Form is no longer charged with representing force in its distortions; the latter no longer has to jostle with form to realize itself. The same force passes directly from painter to canvas, and from one canvas to the next; from trembling dejection and supported grief to the glimmering of hope, to the leap, to the endless flight of this dog, who, turning right around you, has left you alone in the prison where you find yourself now enclosed, high on the passing of this force which is now already far from you and whose traces you no longer see before you – the traces of one who 'saves oneself'.

---

3   Le Grand/Petit Chien or The Great/Little Dog are star constellations, also known as *Canis Major* and *Canis Minor*.

Chapter 19

Questions on Geography

Michel Foucault, Translated by Colin Gordon

Reprinted from Colin Gordon, Ed. (1980) *Power/Knowledge: Selected Interviews and Other Writings, 1972–1977*. New York: Pantheon, 63–77.

Interviewers: the editors of the journal *Hérodote*.

> Your work to a large extent intersects with, and provides material for, our reflections about geography and more generally about ideologies and strategies of space. Our questioning of geography brought us into contact with a certain number of concepts you have used – knowledge *(savoir)*, power, science, discursive formation, gaze, *episteme* – and your archaeology has helped give a direction to our reflection. For instance the hypothesis you put forward in *The Archaeology of Knowledge* – that a discursive formation is defined neither in terms of a particular object, nor a style, nor a play of permanent concepts, nor by the persistence of a thematic, but must be grasped in the form of a system of regular dispersion of statements – enabled us to form a clearer outline of geographical discourse. Consequently we were surprised by your silence about geography. (If we are not mistaken, you mention its existence only once in a paper about Cuvier, and then only to number it among the natural sciences.) Yet, paradoxically, we would have been astounded if you had taken account of geography since, despite the example of Kant and Hegel, philosophers know nothing about geography. Should we blame for this the geographers who, ever since Vidal de la Blache, have been careful to shut themselves off under the cover of the human sciences from any contact with Marxism, epistemology or the history of the sciences? Or should we blame the philosophers, put off by a discipline which is unclassifiable, 'displaced', straddling the gulf between the natural and the social sciences? Is there a 'place' for geography in your archaeology of knowledge? Doesn't archaeology here reproduce the division between the sciences of nature (the inquiry and the table) and the human sciences (examination, discipline), and thereby dissolve the site where geography could be located?

First let me give a flatly empirical answer; then we can try and see if beyond that there is more that can be said. If I made a list of all the sciences, knowledges and domains which I should mention and don't, which I border on in one way or another, the list would be practically endless. I don't discuss biochemistry, or archaeology. I haven't even attempted an archaeology of history. To me it doesn't seem a good method to take a particular science to work on just because it's interesting or important or because its history might appear to have some exemplary value. If one wanted to do a correct, clean, conceptually aseptic kind of history, then that would be a good method. But if one is interested in doing historical work that has political meaning, utility and effectiveness, then this is possible only if one has some kind of involvement with the struggles taking place in the area in question. I tried first to do a genealogy of psychiatry because I had had a certain amount of practical experience in psychiatric hospitals and was aware of the combats, the lines of force, tensions and points of collision which existed there. My historical work was undertaken only as a function of those conflicts. The problem and the stake there was the possibility of a discourse which would be both true and strategically effective, the possibility of a historical truth which could have a political effect.

> That point connects up with a hypothesis I would put to you: if there are such points of collision, tensions and lines of force in geography, these remain on a subterranean level because of the very absence of polemic in geography. Whereas what attracts the interest of a philosopher, an epistemologist, an archaeologist is the possibility of either arbitrating or deriving profit from an existing polemic.

It's true that the importance of a polemic can be a factor of attraction. But I am not at all the sort of philosopher who conducts or wants to conduct a discourse of truth on some science or other. Wanting to lay down the law for each and every science is the project of positivism. I'm not sure that one doesn't find a similar temptation at work in certain kinds of 'renovated' Marxism, one which consists in saying, 'Marxism, as the science of sciences, can provide the theory of science and draw the boundary between science and ideology'. Now this role of referee, judge and universal witness is one which I absolutely refuse to adopt, because it seems to me to be tied up with philosophy as a university institution. If I do the analyses I do, it's not because of some polemic I want to arbitrate but because I have been involved in certain conflicts regarding medicine, psychiatry and the penal system. I have never had the intention of doing a general history of the human sciences or a critique of the possibility of the sciences in general. The subtitle to *The Order of Things* is not *'the* archaeology', but *'an* archaeology of the human sciences'.

It's up to you, who are directly involved with what goes on in geography, faced with all the conflicts of power which traverse it, to confront them and construct the instruments which will enable you to fight on that terrain. And what you should basically be saying to me is, 'You haven't occupied yourself with this matter which isn't particularly your affair anyway and which you don't know much about'. And I would say in reply, 'If one or two of these "gadgets" of approach or method that I've tried to employ with psychiatry, the penal system or natural history can be of service to you, then I shall be delighted. If you find the need to transform my tools or use others then show me what they are, because it may be of benefit to me'.

> You often cite historians like Lucien Febvre, Braudel and Le Roy Ladurie, and pay homage to them in various places. As it happens these are historians who have tried to open up a dialogue with geography, in order to found either a geo-history or an anthropogeography. There might have been occasion for you to make contact with geography through these historians. Again in your studies of political economy and natural history you were

verging on the domain of geography. Your work seems to have been constantly bordering on geography without ever taking it explicitly into account. This isn't a demand for some possible archaeology of geography, nor even really an expression of disappointment, just a certain surprise.

I hesitate to reply only by means of factual arguments, but I think that here again there is a will to essentiality which one should mistrust, which consists in saying, 'If you don't talk about something it must be because you are impeded by some major obstacle which we shall proceed to uncover'. One can perfectly well not talk about something because one doesn't know about it, not because one has a knowledge which is unconscious and therefore inaccessible. You asked if geography has a place in the archaeology of knowledge. The answer is yes, provided one changes the formulation. Finding a place for geography would imply that the archaeology of knowledge embraces a project of global, exhaustive coverage of all domains of knowledge. This is not at all what I had in mind. Archaeology of knowledge only ever means a certain mode of approach.

It is true that Western philosophy, since Descartes at least, has always been involved with the problem of knowledge. This is not something one can escape. If someone wanted to be a philosopher but didn't ask himself the question, 'What is knowledge?', or, 'What is truth?', in what sense could one say he was a philosopher? And for all that I may like to say I'm not a philosopher, nonetheless if my concern is with truth then I am still a philosopher. Since Nietzsche this question of truth has been transformed. It is no longer, 'What is the surest path to Truth?', but, 'What is the hazardous career that Truth has followed?' That was Nietzsche's question, Husserl's as well, in *The Crisis of the European Sciences*. Science, the constraint to truth, the obligation of truth and ritualized procedures for its production have traversed absolutely the whole of Western society for millennia and are now so universalized as to become the general law for all civilizations. What is the history of this 'will to truth'? What are its effects? How is all this interwoven with relations of power? If one takes this line of enquiry then such a method can be applied to geography; indeed, it should be, but just as one could equally do the same with pharmacology, microbiology, demography and who knows what else. Properly speaking there is no 'place' in archaeology for geography, but it should be possible to conduct *an* archaeology of geographical knowledge.

If geography is invisible or ungrasped in the area of your explorations and excavations, this may be due to the deliberately historical or archaeological approach which privileges the factor of time. Thus, one finds in your work a rigorous concern with periodization that contrasts with the vagueness and relative indeterminacy of your spatial demarcations. Your domains of reference are alternately Christendom, the Western world, Northern Europe and France, without these spaces of reference ever really being justified or even precisely specified. As you write, 'Each periodization is the demarcation in history of a certain level of events, and conversely each level of events demands its own specific periodization, because according to the choice of level different periodizations have to be marked out and, depending on the periodization one adopts, different levels of events become accessible. This brings us to the complex methodology of discontinuity'. It is possible, essential even, to conceive such a methodology of discontinuity for space and the scales of spatial magnitude. You accord a *de facto* privilege to the factor of time, at the cost of nebulous or nomadic spatial demarcations whose uncertainty is in contrast with your care in marking off sections of time, periods and ages.

We are touching here on a problem of method, but also on a question of material constraint, namely the possibility available to any one individual covering the whole of this spatio-temporal field. After all, with *Discipline and Punish* I could perfectly well call my subject the history of penal policy in France – alone. That after all is essentially what I did, apart from a certain number of excursions, references and examples taken from elsewhere. If I don't spell that out, but allow the frontier to wander about, sometimes over the whole of the West, that's because the documentation I was using extends in part outside France, and also because in order to grasp a specifically French phenomenon I was often obliged to look at something that happened elsewhere in a more explicit form that antedated or served as a model for what took place in France. This enabled me – allowing for local and regional variations – to situate these French phenomena in the context of Anglo-Saxon, Spanish, Italian and other societies. I don't specify the space of reference more narrowly than that since it would be as warranted to say that I was speaking of France alone as to say I was talking about the whole of Europe. There is indeed a task to be done of making the space in question precise, saying where a certain process stops, what are the limits beyond which something different happens – though this would have to be a collective undertaking.

This uncertainty about spatialization contrasts with your profuse use of spatial metaphors – position, displacement, site, field; sometimes geographical metaphors even – territory, domain, soil, horizon, archipelago, geopolitics, region, landscape.

Well, let's take a look at these geographical metaphors. *Territory* is no doubt a geographical notion, but it's first of all a juridico-political one: the area controlled by a certain kind of power. *Field* is an economico-juridical notion. *Displacement:* what displaces itself is an army, a squadron, a population. *Domain* is a juridico-political notion. *Soil* is a historico-geological notion. *Region* is a fiscal, administrative, military notion. *Horizon* is a pictorial, but also a strategic notion.

There is only one notion here that is truly geographical, that of an *archipelago*. I used it only once, and that was to designate, via the title of Solzhenitsyn's work, the carceral archipelago: the way in which a form of punitive system is physically dispersed yet at the same time covers the entirety of a society.

Certainly these notions are not geographical in a narrow sense. Nonetheless, they are the notions which are basic to every geographical proposition. This pinpoints the fact that geographical discourse produces few concepts of its own, instead picking up notions from here, there and everywhere. Thus landscape is a pictorial notion, but also an essential object for traditional geography.

But can you be sure that I am borrowing these terms from geography rather than from exactly where geography itself found them?

> The point that needs to be emphasized here is that certain spatial metaphors are equally geographical and strategic, which is only natural since geography grew up in the shadow of the military. A circulation of notions can be observed between geographical and strategic discourses. The *region* of the geographers is the military region (from *regere*, to command), a *province* is a conquered territory (from *vincere*). *Field* evokes the battlefield...

People have often reproached me for these spatial obsessions, which have indeed been obsessions for me. But I think through them I did come to what I had basically been looking for: the relations that are possible between power and knowledge. Once knowledge can be analyzed in terms of region, domain, implantation, displacement, transposition, one is able to capture the process by which knowledge functions as a form of power and disseminates the effects of power. There is an administration of knowledge, a politics of knowledge, relations of power which pass via knowledge and which, if one tries to transcribe them, lead one to consider forms of domination designated by such notions as field, region and territory. And the politico-strategic term is an indication of how the military and the administration actually come to inscribe themselves both on a material soil and within forms of discourse. Anyone envisaging the analysis of discourses solely in terms of temporal continuity would inevitably be led to approach and analyze it like the internal transformation of an individual consciousness. Which would lead to his erecting a great collective consciousness as the scene of events.

Metaphorizing the transformations of discourse in a vocabulary of time necessarily leads to the utilization of the model of individual consciousness with its intrinsic temporality. Endeavouring on the other hand to decipher discourse through the use of spatial, strategic metaphors enables one to grasp precisely the points at which discourses are transformed in, through and on the basis of relations of power.

> In *Reading Capital*, Althusser poses an analogous question: 'The recourse made in this text to spatial metaphors (field, terrain, space, site, situation, position, etc.) poses a theoretical problem: the problem of the validity of its *claim* to existence in a discourse with scientific pretensions. The problem may be formulated as follows: *why* does a certain form of scientific discourse necessarily need the use of metaphors borrowed from scientific disciplines?' Althusser thus presents recourse to spatial metaphors as necessary, but at the same time as regressive, non-rigorous. Everything tends on the contrary to suggest that spatial metaphors, far from being reactionary, technocratic, unwarranted or illegitimate, are rather symptoms of a 'strategic', 'combative' thought, one which poses the space of discourse as a terrain and an issue of political practices.

It is indeed, war, administration, the implantation or management of some form of power which are in question in such expressions. A critique could be carried out of this devaluation of space that has prevailed for generations. Did it start with Bergson, or before? Space was treated as the dead, the fixed, the undialectical, the immobile. Time, on the contrary, was richness, fecundity, life, dialectic.

For all those who confuse history with the old schemas of evolution, living continuity, organic development, the progress of consciousness or the project of existence, the use of spatial terms seems to have the air of an anti-history. If one started to talk in terms of space that meant one was hostile to time. It meant, as the fools say, that one 'denied history', that one was a 'technocrat'. They didn't understand that to trace the forms of implantation, delimitation and demarcation of objects, the modes of tabulation, the organization of domains meant the throwing into relief of processes – historical ones, needless to say – of power. The spatializing decription [sic] of discursive realities gives on to the analysis of related effects of power.

> In *Discipline and Punish,* this strategizing method of thought advances a further stage. With the Panoptic system we are no longer dealing with a mere metaphor. What is at issue here is the description of institutions in terms of architecture, of spatial configurations. In the conclusion you even refer to the 'imaginary geopolitics' of the carceral city. Does this figure of the Panopticon offer the basis for a description of the State apparatus in its entirety? In this latest book an implicit model of power emerges: the dissemination of micro-powers, a dispersed network of apparatuses without a single organizing system, centre or focus, a transverse coordination of disparate institutions and technologies. At the same time, however, you note the installation of State control over schools, hospitals, establishments of correction and education previously in the hands of religious bodies or charitable associations. And parallel with this is the creation of a centralized police, exercising a permanent, exhaustive surveillance which makes all things visible by becoming itself invisible. 'In the eighteenth century the organization of police ratifies the generalization of disciplines and attains the dimensions of the State.'

By the term 'Panoptism', I have in mind an ensemble of mechanisms brought into play in all the clusters of procedures used by power. Panoptism was a technological invention in the order of power, comparable with the steam engine in the order of production. This invention had the peculiarity of being utilized first of all on a local level, in schools, barracks and hospitals. This was where the experiment of integral surveillance was carried out. People learned how to establish dossiers, systems of marking and classifying, the integrated accountancy of individual records. Certain of the procedures had of course already been utilized in the economy and taxation. But the permanent surveillance of a group of pupils or patients was a different matter. And, at a certain moment in time, these methods began to become generalized. The police apparatus served as one of the principal vectors of this process of extension, but so too did the Napoleonic administration. I think in the book I quoted a beautiful description of the role of the Attorneys-General under the Empire as the eyes of the Emperor; from the First Attorney-General in Paris to the least Assistant Public Prosecutor in the provinces, one and the same gaze watches for disorder, anticipates the danger of crime, penalizing every deviation. And should any part of this universal gaze chance to slacken, the collapse of the State itself would be imminent. The Panoptic system was not so much confiscated by the State apparatuses, rather it was these apparatuses which rested on the basis of small-scale, regional, dispersed Panoptisms. In consequence one cannot confine oneself to analyzing the State apparatus alone if one wants to grasp the mechanisms of power in their detail and complexity. There is a sort of schematism that needs to be avoided here – and which incidentally is not to be found in Marx – that consists of locating power in the State apparatus, making this into the major, privileged, capital and almost unique instrument of the power of one class over another. In reality, power in its exercise goes much further, passes through much finer channels, and is much more ambiguous, since each individual has at his disposal a certain power, and for that very reason can also act as the vehicle for transmitting a wider power. The reproduction of the relations of production is not the only function served by power. The systems of domination and the circuits of exploitation certainly interact, intersect and support each other, but they do not coincide.

Even if the State apparatus isn't the only vector of power, it's still true, especially in France with its Panoptico-prefectoral system, that the State spans the essential sector of disciplinary practices.

The administrative monarchy of Louis XIV and Louis XV, intensely centralized as it was, certainly acted as an initial disciplinary model. As you know, the police was [sic] invented in Louis XV's France. I do not mean in any way to minimize the importance and effectiveness of State power. I simply feel that excessive insistence on its playing an exclusive role leads to the risk of overlooking all the mechanisms and effects of power which don't pass directly via the State apparatus, yet often sustain the State more effectively than its own institutions, enlarging and maximizing its effectiveness. In Soviet society one has the example of a State apparatus which has changed hands, yet leaves social hierarchies, family life, sexuality and the body more or less as they were in capitalist society. Do you imagine the mechanisms of power that operate between technicians, foremen and workers are that much different here and in the Soviet Union?

You have shown how psychiatric knowledge presupposed and carried within itself the demand for the closed space of the asylum, how disciplinary knowledge contained within itself the model of the prison, Bichat's clinical medicine the enclave of the hospital and political economy the form of the factory. One might wonder, as a conceit or a hypothesis, whether geographical knowledge doesn't carry within itself the circle of the frontier, whether this be a national, departmental or cantonal frontier; and hence, whether one shouldn't add to the figures of internment you have indicated – that of the madman, the criminal, the patient, the proletarian – the national internment of the citizen-soldier. Wouldn't we have here a space of confinement which is both infinitely vaster and less hermetic?

That's a very appealing notion. And the inmate, in your view, would be national man? Because the geographical discourse which justifies frontiers is that of nationalism?

Geography being together with history constitutive of this national discourse: this is clearly shown with the establishment of Jules Ferry's universal primary schools which entrust history-geography with the task of implanting and inculcating the civic and patriotic spirit.

Which has as its effect the constitution of a personal identity, because it's my hypothesis that the individual is not a pre-given entity which is seized on by the exercise of power. The individual, with his identity and characteristics, is the product of a relation of power exercised over bodies, multiplicities, movements, desires, forces. There is much that could be said as well on the problems of regional identity and its conflicts with national identity.

> The map as instrument of power/knowledge spans the three successive chronological thresholds you have described: that of measure with the Greeks, that of the inquiry during the Middle Ages, that of the examination in the eighteenth century. The map is linked to each of these forms, being transformed from an instrument of measurement to an instrument of inquiry, becoming finally today an instrument of examination (electoral maps, taxation maps, etc.). All the same the history (and archaeology) of the map doesn't correspond to 'your' chronology.

A map giving numbers of votes cast or choices of parties: this is certainly an instrument of examination. I think there is this historical succession of the three models, but obviously these three techniques didn't remain isolated from each other. Each one directly contaminates the others. The inquiry used the technique of measure, and the examination made use of inquiry. Then examination reacted back on the first two models, and this brings us back to an aspect of your first question: doesn't the distinction between examination and inquiry reproduce the distinction between social science and science of nature? What in fact I would like to see is how inquiry as a model, a fiscal, administrative, political schema, came to serve as a matrix for the great surveys which are made at the end of the eighteenth century where people travel the world gathering information. They don't collect their data raw: literally, they inquire, in terms of schemas which are more or less clear or conscious for them. And I believe the sciences of nature did indeed install themselves within this general form of the inquiry; just as the sciences of man were born at the moment when the procedures of surveillance and record-taking of individuals were established. Although that was only a starting-point. And because of the effects of intersection that were immediately produced, the forms of inquiry and examination interacted, and as a consequence the sciences of nature and man also overlapped in terms of their concepts, methods and results. I think one could find in geography a good example of a discipline which systematically uses measure, inquiry and examination.

> There is a further omnipresent figure in geographical discourse: that of the inventory or catalogue. And this kind of inventory precisely combines the triple register of inquiry, measure and examination. The geographer – and this is perhaps his essential, strategic function – collects information in an inventory which in its raw state does not have much interest and is not in fact usable except by power. What power needs is not science but a mass of information which its strategic position can enable it to exploit.

This gives us a better understanding both of the epistemological weakness of geographical studies, and at the same time of their profitability (past more than present) for apparatuses of power. Those seventeenth-century travellers and nineteenth-century geographers were actually intelligence-gatherers, collecting and mapping information which was directly exploitable by colonial powers, strategists, traders and industrialists.

I can cite an anecdote here, for what it's worth. A specialist in documents of the reign of Louis XIV discovered while looking at seventeenth-century diplomatic correspondence that many narratives that were subsequently repeated as travellers' tales of all sorts of marvels, incredible plants and monstrous animals, were actually coded reports. They were precise accounts of the military state of the countries traversed, their economic resources, markets, wealth and possible diplomatic relations. So that what many people ascribe to the persistent naïveté of certain eighteenth-century naturalists and geographers were in reality extraordinarily precise reports whose key has apparently now been deciphered.

Wondering why there have never been polemics within geography, we immediately thought of the weak influence Marx has had on geographers. There has never been a Marxist geography nor even a Marxist current in geography. Those geographers who invoke Marxism tend in fact to go off into economics or sociology, giving privileged attention to the planetary or the medium scale. Marxism and geography are hard to articulate with one another. Perhaps Marxism, or at any rate *Capital* and the economic texts in general, does not lend itself very readily to a spatializing approach because of the privilege it gives to the factor of time. Is that what is at issue in this remark of yours in an interview: 'Whatever the importance of their modification of Ricardo's analyses, I don't believe Marx's economic analyses escape from the epistemological space established by Ricardo'?

As far as I'm concerned, Marx doesn't exist. I mean, the sort of entity constructed around a proper name, signifying at once a certain individual, the totality of his writings, and an immense historical process deriving from him. I believe Marx's historical analysis, the way he analyzes the formation of capital, is for a large part governed by the concepts he derives from the framework of Ricardian economics. I take no credit for that remark, Marx says it himself. However, if you take his analysis of the Paris Commune or *The Eighteenth Brumaire of Louis Bonaparte,* there you have a type of historical analysis which manifestly doesn't rely on any eighteenth-century model.

It's always possible to make Marx into an author, localizable in terms of a unique discursive physiognomy, subject to analysis in terms of originality or internal coherence. After all, people are perfectly entitled to 'academize' Marx. But that means misconceiving the kind of break he effected.

If one re-reads Marx in terms of the treatment of the spatial his work appears heterogenous. There are whole passages which reveal an astonishing spatial sensibility.

There are some very remarkable ones. Everything he wrote on the army and its role in the development of political power, for instance. There is some very important material there that has been left practically fallow for the sake of endless commentaries on surplus value.

I have enjoyed this discussion with you because I've changed my mind since we started. I must admit I thought you were demanding a place for geography like those teachers who protest when an education reform is proposed, because the number of hours of natural sciences or music is being cut. So I thought, 'It's nice of them to ask me to do their archaeology, but after all, why can't they do it themselves?' I didn't see the point of your objection. Now I can see that the problems you put to me about geography are crucial ones for me. Geography acted as the support, the condition of possibility for the passage between a series of factors I tried to relate. Where geography itself was concerned, I either left the question hanging or established a series of arbitrary connections.

The longer I continue, the more it seems to me that the formation of discourses and the genealogy of knowledge need to be analyzed, not in terms of types of consciousness, modes of perception and forms of ideology, but in terms of tactics and strategies of power. Tactics and strategies deployed through implantations, distributions, demarcations, control of territories and organizations of domains which could well make up a sort of geopolitics where my preoccupations would link up with your methods. One theme I would like to study in the next few years is that of the army as a matrix of organization and knowledge; one would need to study the history of the fortress, the 'campaign', the 'movement', the colony, the territory. Geography must indeed necessarily lie at the heart of my concerns.

PART 6
Development

Chapter 20

Geographies of Governmentality

Margo Huxley

Introduction

The first English version of Foucault's (1979a) lecture 'On governmentality' appeared in the journal *Ideology and Consciousness* in 1979. Over the course of the next decade, the influence of this essay became visible in sociological studies of disciplines in the human sciences and their connections to practices of subject formation (Dean 1999, 1). Papers and books by Miller (1987), Miller and Rose (1986; 1990) and Rose (1985; 1988; 1990), for example, brought Foucauldian perspectives to bear on 'psy' sciences, accountancy, auditing and the government of the economy. These studies explored how diverse and disparate projects sought to govern the conduct of others and to foster individual practices of self-government.

By 1991, when Foucault's essay was reprinted in the collection *The Foucault Effect* (Burchell et al. 1991), there was a burgeoning field of English language studies of governmentality that provided fertile ground into which this collection could be planted, and from which governmentality studies proliferated into increasing numbers of areas. By the end of the decade, very few academic disciplines had not employed analytic lenses that owed at least something to the idea of governmentality (Dean 1999, 1–4). Geographic studies also increasingly drew on this framework, developing earlier work in historical geography in examining the role of space in disciplining, fostering, managing and monitoring the conducts of individuals and the qualities of populations.

Nevertheless, while there have been many responses to the substantive historical spatial and geographic analyses to be found in Foucault's work (Philo 1992), there are still further possibilities for developing the insights offered by analyses of mentalities of rule. Among these is the further exploration of the rationalities that underpin programmes and practices of government, focusing on logics that attribute causal effects to space and environment and that seek to manipulate these towards governmental ends.

The aim of this chapter is, therefore, to explore some of the ways space and environment can be seen as rationalities of government. It begins with developments of the notion of governmentality, particularly in relation to analyses focusing on the governmental production of subjects, and how mentalities and rationalities of government subtend such attempts. In sociological and political studies of liberal governmentality, space and environment are suggestive, but under-developed, presences.

The discussion then moves to a consideration of the ways governmentality has been taken up in geographic literature – whether implied in examinations of the spaces of discipline and control or in explicit elaborations of spatial logics of mapping and enumeration. The geographic emphasis on spatial aspects of the conduct of conduct is marked (in reciprocal relation to sociological approaches) by relative under-development of genealogies of liberal government and rationalities of rule in the name of freedom.

Finally, I argue that governmental reason involves chains of *causal* spatial and environmental logic that include, but are distinct from, logics of spatial arrangements and visibilities of measurement, calculation or discipline. Governmental thought that postulates spatial and environmental causes shaping comportments and moral states of individuals, and influencing the bio-social conditions of populations, underpins diverse programmes and practices of the conduct of conducts. To illustrate this, I provide brief examples of some spatial and environmental rationalities – 'diagrams' of government from a spatial point of view.

Governmentality

Although the explicit use of the term 'governmentality' is traceable to the 1979 and 1991 English publications of Foucault's essay, the ways in which studies of governmentality have used and developed this perspective are derived from sources scattered throughout Foucault's work: from general discussions of power (Foucault 1982) to the micro-politics of disciplinary institutions (Foucault 1979b); from delineations of bio-power and the management of populations (Foucault 1981; 2003) to explorations of technologies of the self (Foucault 1988b,c; 1990); and from outlines of different forms of governmental thought (disciplinary, pastoral, liberal: e.g. Foucault 1988a) to indications of neo-liberal forms of government by the state (Lemke 2001).[1]

Inevitably and productively, there are differences of exegesis and interpretation (see for instance, the differences between Dean 1999; and Rose 1999). But in general, 'government' is understood as 'le conduire des conduits' – the 'conduct of

---

1   Key lecture series in which Foucault elaborated the notion of 'governmentality' – *Securite, territoire et population* 1978, and *La naissance de la biopolitique*' 1979 – are not yet available in English. The way the concept has been taken up by English language researchers has inevitably been influenced in part by their ability to access the French publications. For English translations of Foucault's summaries of these lectures, see Rabinow 2000 and Faubion 2000; for an exegeses of some of the series, see e.g. Elden 2002 (race and the constitution of the state); and 2007 (security, territory, population); Lemke 2001 (neo-liberal governmentality); Rabinow 1982 (town planning, regulation and discipline). The treatment of discipline, bio-politics and race in the 1975–1976 lectures, translated as *Society Must be Defended* (2003) foreshadow some of the themes taken up in Foucault's later discussions of governmentality.

conducts' (Gordon 1991, 2; Foucault 1982, 220–1) – a generalized power that seeks to fashion and guide the bodily comportments and inward states of others and of the self; a form of action on the actions and capacities of the self and of others. In this sense, government is the form of power 'by which, in our culture, human beings are made subjects' (Foucault 1982, 221), and thus 'entails any attempt to shape with some degree of deliberation aspects of our behaviour ... for a variety of ends' (Dean 1999, 10).

Governmentality includes the exercise of discipline over bodies (anatomo-politics, Foucault 1981, 139); 'police' supervision of the inhabitants of the sovereign's territory (1979b, 213–17; 1986a, 241–2; 1991b); the bio-political regulation of the 'species life' of a population (Foucault 1981, 139–45); and self-formation through ethical care of the self (Foucault 1990; 1997).

Studying 'governmentality', however, involves not only examination of practices and programmes aiming to shape, guide and govern the behaviour of others and the self, or the calculations, measurements and technologies involved in knowing and directing the qualities of a population; but also pays attention to the aims and aspirations, the mentalities and rationalities intertwined in attempts to steer forms of conduct. These mentalities or rationalities of government are framed within 'regimes of truth' that inform the 'thought' secreted in projects of rule.

The problem is thus, 'to see how men govern (themselves and others) by the production of truth' (Foucault 1991a, 79). Or as Rose (1999, 19) puts it, analyses of governmentality are studies of:

> ... the emergence of particular 'regimes of truth' concerning the conduct of conduct, ways of speaking the truth, persons authorized to speak truths, ways of enacting truths and the costs of so doing. Of the invention and assemblage of particular apparatuses and devices for exercising power and intervening upon certain problems.

This interest in the 'truths' expressed in the aims and aspirations of government links governmentality to Foucault's analyses of discursive formations of the human sciences to be found in his earlier writings (e.g. 1974a, b), to power and struggles over discourses of truth (1984; 2003, 167–87), and to discussions of the Enlightenment (Foucault 1986b, 32–50), and of Nietzsche and genealogy (Foucault 1986c, 76–100).

Thus, in tracing the emergence of mentalities of government and liberal rationalities of rule that presuppose the freedom of the subject, Foucault (e.g. 1981; 1991b) is not outlining a simple historical progression from despotism to liberal democracy. Rather, he shows how possibilities for, and exercises of, government emerge at particular points of confluence between thought and practices, and how these ways of framing the problem of government, including government by the state, persist, become transformed and converge or conflict with other rationalities and practices of subjectification.

In this way, rule conceived as the power of the sovereign over territory, does not disappear with the concerns with the art of government of 'men and things'; and equally, disciplinary practices focused on the body, the minutiae of the knowledge derived from police power, and attention to the management of populations (bio-power) give rise to, and persist in liberal formulations of government (Dean 1999; 2002; Fontana and Bertani 2003). That is, there is no necessary opposition between discipline and liberal government: indeed, the study of disciplinary practices, institutions, spaces in *Discipline and Punish* (Foucault 1979b), or practices and logics of nineteenth century concerns with the control and direction of sexuality (Foucault 1981) are also studies of 'how the subjection of the body forms the subject' (Elden 2001, 104) – a subject which is produced, not repressed, and is governed and governs itself as having capacities to act (Patton 1998).

It is these suggestions about the subjectification of subjects in tandem with the increasing 'governmentalization of the state' – the contingent history of the state as a locus in which various forms of government are taken up and reconfigured – that have been developed in sociological and political studies of governmentality.

Dean (1994) usefully distinguishes three ways in which governmentality can be understood: as 'political subjectification'; as 'governmental self-formation'; and as 'ethical self-formation' (Dean 1994, 154–8). 'Political subjectification' involves practices and discourses that see individuals as political subjects of various kinds, but particularly under regimes of liberal rule, as 'sovereign subjects or citizens within a self-governing political community under the conditions of liberal democracy' (155; Hindess 1993; 1997).

'Governmental self-formation' relates to the ways in which assorted agencies, authorities, organizations and groups seek to shape and incite the self-formation of the comportments, habits, capacities and desires of particular categories of individuals towards particular ends. And 'ethical self-formation' concerns the government of the self by the self 'by means of which individuals seek to know, decipher, and act on themselves' (Dean 1994, 156; Foucault 1990).

The focus of the present discussion is not primarily concerned with problems of ethical self-formation, although they are important elements in Foucault's understanding of the political (Foucault 1988a, b; Simons 1995). Rather, the focus is on the development of notions of rationalities of governmental self-formation and political subjectification.

Governmental self-formation is fostered through a myriad of practices and disciplines that encourage individuals to govern themselves in accordance with various regulatory, moral or spiritual or pastoral norms (Foucault 1988a, 57–85). As examples, Rose's (1985; 1990) studies of psychology and psychiatry show how these disciplines are implicated in projects of governing individuals through inciting them to govern themselves. He extends these analyses to the ways in which government by the state converges with self-help and managerial discourses attempting to bring about subjects exhibiting self-reinforcing entrepreneurial social and economic ('advanced liberal') behaviours; and shows how government by the state is backed up by measures for controlling or segregating those failing to attain sufficient degrees of acceptable self-direction (Miller and Rose 1990; Rose 1999; see also Cruikshank 1999; Dean 1999; 2002; Hindess 2001).

Hence, governmental aspirations for the self-formation of subjects linked to the state have convergences with aims for the political subjectification of liberal subjects, for the creation of appropriately autonomous subjects (Burchell 1996; Dean 1999; 2002; Dean and Hindess 1998; Barry et al. 1996; Hindess 1993; 1997). Liberal political reason, in particular, expects of:

> ... the governed that they freely conduct themselves in a certain rational way ... that requires the proper use of liberty. Individual freedom, in appropriate forms, is here a technical condition of rational government rather than the organizing value of a Utopian dream. (Burchell 1996, 24)

This formulation of mentalities of government involves examination of rational and technical precepts embedded in practices of rule; the mobilization of truths about what and who should be governed, how and by whom (Dean 1998; Rose 1999), and what it means to see government as 'technical', that is, as attempting to render aspects of conduct continuous, calculable, measurable and comparable (Foucault 1979b, 26, 224–8; Dean 1996, 63). In liberal government, these rationalities and technical practices operate with the aspiration of producing conditions in which (regulated) freedoms can be exercised, at least by some of the population some of the time (see Dean 2002; Hindess 2001).

The emphasis here is on the rationalities and technologies traversing diverse, dispersed, reinforcing and contradictory projects and programmes for the conduct of conducts. 'In seeking to reconstruct the logic underlying particular programmes of government, and the specific articulation of means, ends and objects' (Bennett 2004, 11; Dean 2002, 121), studies of mentalities of government provide important insights into the truths that animate governmental aspirations.

However, focus on rationalities can, at times, obscure struggles over discourses of truths and the messy, contingent and haphazard fashion in which localized practices of regulation get hooked up with, modified by, and in turn modify, rationales for projects of government. Accounts of the 'programmers' view' (Dean 2002, 121; Bennett 2004, 11) can appear to suggest that governmental aims and rationales are capable of automatically producing the reality for which they hope.[2] But for Foucault, the study of rationalities is inextricably entangled with governmental practices, not only because this entanglement renders outcomes and effects problematic, but also because practices and truths are mutually constitutive:

---

2 Both Dean, e.g. *The Constitution of Poverty* (1991) and Rose, e.g. *The Psychological Complex* (1985) have written detailed historical accounts of particular, contingent and contested instances of the installation of projects government. But their influential overviews of governmentality (Dean 1999; Rose 1999), focusing as they do on the productive logics of government, have tended to eclipse the precarious uncertainties surrounding the achievement of governmental aims. See Elden 2001, 133–50, for an analogous critique of readings of *Discipline and Punish* 'through Bentham's eyes' as the untrammelled achievement of discipline and repression.

If I have studied 'practices' … it was in order to study [the] interplay between a 'code' which rules ways of doing things … and a production of true discourses which serve to found, justify and provide reasons and principles for these ways of doing things. (Foucault 1991a, 79)

Nevertheless, the study of governmental rationalities in themselves is a constructive focus as an analytical device that singles out only one aspect of complex interplays of thought and practices. An analytics of government that renders up presupposed 'postulates of thought' for re-examination (Foucault 1996, 423–4) can indeed contribute to bringing into question the power congealed in taken-for-granted ways of doing things.

The purpose of this overview of some of the (non-geographical) developments of perspectives on governmentality has been to suggest that, despite some dangers of generalization and loss of the detailed specificity that marks Foucault's historical studies, these approaches highlight important aspects of the rationalities underpinning aspirations of government to produce various comportments, behaviours and qualities in bodies and populations, individuals and political subjects. Sociological and political studies of governmentality call attention to different modes of subjectification (political subjectification, governmental self-formation, ethical practices of the self); they point to the continuities between discipline and police power and the making up of individuals who conduct themselves as if they were free; and they direct attention to the 'thought' that resides in government – the 'operative rationales that animate aspirations aimed at shaping the conduct of others' (Osborne and Rose 1999, 738).

But, in ways not unconnected with the emphasis on a generalized 'programmers' view', the specificities of spatial aspects of governmentality tend to be relatively under-developed in studies of political and sociological rationalities and technologies of government. And so, in the next section I examine some of the ways governmentality has been developed in the geographic literature, in order to argue for the productive possibilities of an analytics of spatial rationalities of government.

Space, Geography, Government

As Foucault (1986a, 252), famously said: 'Space is fundamental in any form of communal life; space is fundamental in any exercise of power.' Yet, as noted above, much of the development of governmentality in sociological and political frames barely touches on the question of space, possibly because of these disciplines' long-standing ambivalence about the place of space in social and political relations.

Nevertheless, there are hints at the role of space in governmental projects. For example, Rose (1996, 143–4; see also 1999, 31–40) suggests that:

> … we need to render being intelligible in terms of the localization of routines, habits and techniques within specific domains of action and value: libraries and studies; bedrooms and bathhouses; courtrooms and school rooms; consulting rooms and museum galleries; markets and department stores … To the apparent linearity, unidirectionality

and irreversibility of time, we can counterpose the multiplicity of places, planes and practices.

However, these hints are seldom followed through in any detail (except perhaps, in examinations of the government of cities – see below): and in general, space seems to be conceived as a series of surfaces and containers upon which governmental aims can be projected and within which certain practices can be enacted. In contrast, the spatial aspects of practices of the conduct of conducts have been more thoroughly explored in Geography.

In English language geography, there was little explicit use of governmentality frameworks until the mid-1990s, when governmentality perspectives began to inform numerous policy critiques and assessments of the reconfigurations of the Welfare State (especially in Britain and Australia) that included spatial reconfigurations of cities and regions – for instance, urban regeneration projects, public-private development partnerships and attempts to foster community (e.g. Murdoch 2000; Raco 2003; Raco and Imrie 2000).[3]

Whilst governmentality provides policy studies with crucial insights into the aims of subjectification contained in disparate and seemingly unrelated projects, ungrounded and a-historical policy analyses tend to dilute the critical force of genealogical approaches which might otherwise trace unsettling 'histories of the present'. Instead, policy analyses run the risk of becoming ideology critiques, aimed at unmasking the true and repressive intentions of the state and, at times, uncritically celebrating 'resistance' (see Keith 1997 for a discussion of the romanticization of resistance).

Policy critiques such as these parallel views that position Foucault as the prophet of repression and surveillance. There has been a prevalent Lefebvrian attitude in Geography that sees Foucault as an analyst of domination, offering no escape from ubiquitous control, failing to engage with the multiplicities of lived experience, and dealing only in spatial metaphors (Lefebvre 1991, 4; Smith and Katz 1993; Thrift 2000).

This last accusation of metaphorical ungroundedness (Smith and Katz 1993) may also have contributed to a neglect of Foucault by those geographers searching for an explicit Theory of Space (with capitals). But as Philo (1992) points out, 'geographies' are present in the 'spaces of dispersion' in Foucault's (1974b) analysis of discourse – planes 'across which all of the events and phenomena relevant to a substantive study are dispersed' – so that:

> ... a researcher would envisage such things as asylums, upland environments, dirty towns, ardent reformers, the 1807 Select Committee, John Conolly, the *Asylum Journal*, a parliamentary debate, country walks, and Bentham's Panopticon all being scattered over the space available. (Philo 1992, 148)

---

3 Perspectives on governmentality now influence almost every area of geographic study. In calling attention to 'spatial rationalities', in this chapter I am only able to trace a narrow path through these expanding domains.

At the same time, Philo (1992) argues, 'substantive geographies' can be discovered in the arrangements, visibilities and the particularities of location and place described in detail in Foucault's histories of the asylum, the prison, the spaces of medical practice in hospitals and in the city (Philo 1992, 155–8).

So, what we might call 'substantive geographies of government' are found in the historical geographies that develop and extend Foucauldian approaches to the prison and the workhouse system (Driver 1985; 1993), the asylum (Philo 1987; 1989) and the regulation of urban behaviours (Ogborn 1992; 1998). Despite the absence of 'governmentality' as an explicit framework, these studies give a sense of the rationalities, the practices and the spaces involved in attempts to produce or control certain kinds of subjectivities. In particular, Driver's (1993) examination of local and national negotiations around, and oppositions to, the installation of the workhouse system from 1834 to 1884, goes beyond a simplistic history of the installation of a disciplinary complex, and argues that 'the 1834 [Poor Law] reform set the stage for a dramatic shift in policy; but it could not dictate the script of subsequent practice' (Driver 1993, 57).

Historical studies (in Geography, History and Historical Sociology) show how spaces of the institution, the city and the territory are implicated in projects of government. They examine experimental and unsecured assemblages of disparate elements – sanitary technologies, philanthropic practices, architecture, government regulations, aesthetic and moral theory, methods of measurement and calculation (such as the census), mapping, medical knowledge, transport and communications – constituting discursive and material arrangements aimed at the making up and government of particular kinds of individuals, populations, locations and territories.[4] These *dispositifs* (ensembles, apparatus, constructs, grids of intelligibility) (Foucault 1980a, 194–5, 158–9; Elden 2001, 110–11) express configurations of relations that are connected to the 'thought' of government in solutions to problems of rule.

Paul Rabinow's (1989) French Modern is an exemplary working through of these conceptions. Knowledges – statistics, medicine, biology, architecture and building, regional geography; the 'technicians of general ideas' or 'specific intellectuals' – whose writings addressing problems of government are located somewhere between high culture and everyday life (9); and practices of architecture, colonial rule, working class reform, sanitary infrastructure, are intertwined in French colonial experiments with 'urbanism', attempting to create *milieux* and built 'forms' in order to induce, express and maintain specific and differentiated 'norms'. And although Rabinow does not use the terms, his study of the interconnections and

---

4   Examples of such studies (not all geographic) include: Dean 1991 on the 'constitution of poverty'; Braun 2000 on the cartographic production of Canadian territory; Hannah 2000 on territory, masculinity, race and the US census in the nineteenth century; Hunt 1996 on liberal modes of governing the city; Joyce 2003 on the city as an expression of liberal rule; Matless 1998 on making English bodies in the landscape; Osborne 1996 on drains and liberalism; Otter 2002; 2004 on 'liberalism made durable' in the material technologies of the city; Rose 1994 on technologies of medical truth in the city.

interplays between thought, rationales, practices, regulation, environment, built form and populations perfectly illustrates the intricacies of 'governmentality' and the discursive materiality of '*dispositifs*'. These assemblages are not solely the result of individual inspiration, but are enmeshed in a complex matrix of rationalities and practices; nor are they equivalent to the reality they try to manipulate, nor to the effects they seek to produce.

For this reason, it is important to distinguish between rationalities of government in themselves, and their implication in assembling *dispositifs*. To the extent that arrangements of built form, urban spaces and drains and trains partially embody causal reasonings that posit chains of determinations between 'men and things', and these logics require critical examination of their implication in attempts to fabricate spaces and places as elements in governmental projects of subjectification.

So, for example, Hannah's (2000) examination of the US Census and its protagonist, Francis A. Walker, at the end of the nineteenth century, works within an explicit governmentality framework, focusing on the historical geography of the modern American nation-state and the Census as a technology in a cycle of social control, to demonstrate the inherent spatiality of government as the conduct of conducts and the making up of people. Governmentality has also figured as a framework for analyses of maps, cartographic visualizations and practices, and the disciplinary and regulatory impulses informing cartographic reason or rationality (Crampton 2003; Harley 2001; Pickles 2004).

However, the emphasis on rationalities of discipline and control, visualizations and attempted arrangements of space tends to privilege social control over the productive elements of attempts to 'make up' liberal, autonomous political, or self-forming social, subjects that are the focus of sociological and political studies of governmentality.

In summary, in geographic work on disciplinary and regulatory spaces, and on governmentality, there are fruitful historical studies of institutional spatial arrangements, built form, and more generally, urban and rural environments (joined by work in history and historical sociology), which explore material discursive struggles around regulatory regimes and spaces actually produced. And there are studies that examine forms of disciplinary and regulatory logics that inform technologies of mapping and geographic visualization. But many of these approaches rely on a view of an essentialized autonomous subject as the target of domination, and it is only recently that conceptions of the productive rationalities associated with the making of liberal freedom have been connected to the spatial concerns of geography. Elden (2001), in particular, makes the case for seeing Foucault's concerns with space as integral to the formation, rather than the suppression, of the 'modern soul'.

In order to broaden understandings of the way space figures in rationalities of government, we might begin to examine the *causal and productive* powers attributed to spaces and environments in aspirations to catalyze appropriate comportments and subjectivities. These rationalities can then be investigated: but not as false or mistaken ascriptions of spatial, physical or environmental determinism; nor as ontological questions about the nature of space; nor in order to formulate a general theory of how such 'irreal' spaces come to be fabricated (Rose 1999, 32–3). Rather, 'space' and 'environment' can be posed as 'analytical and political' questions (Rabinow 1982, 269) that need to be asked anew of each instance of governmental problematizations and attempts to assemble solutions.

And while such reasoning appears in different guises according to the ways problems of government are contingently and specifically formulated, it is nonetheless possible to identify certain recursive spatial rationalities that posit causal relations within an 'individual-population-environmental complex' (Burchell 1991, 142).

Space as a Rationality of Government

Spatial rationalities postulate causal qualities of 'spaces' and 'environments' as elements in the operative rationales of government, and these postulates can be examined as truths having histories. Thus, the writing of histories of 'spaces' and 'powers' (in the plural) is also the examination of the logics contained in 'strategies' and 'tactics' of power/government that seek to use space for particular ends (Foucault 1980b, 149).

Such an approach develops the notion of the 'diagram' or 'theorem' that Foucault uses to characterize the rationality of Bentham's plans for the Panopticon. The Panopticon is not a description of a real prison: for, as Foucault (1991a, 81) said:

> If I had wanted to describe 'real life' in the prisons, I wouldn't indeed have gone to Bentham. But the fact that this real life isn't the same thing as the theoreticians' schema doesn't entail that these schemas are therefore utopian, imaginary, etc. One could only think that if one had a very impoverished notion of the real ... It is absolutely correct that the actual functioning of the prisons ... was a witches' brew compared to the beautiful Benthamite machine.

Instead, the Panopticon is 'a diagram of a mechanism of power reduced to its ideal form ... it is in fact a figure of a political technology' (Foucault 1979b, 205). Schemes and 'diagrams' such as the Panopticon, serve as models, tests and ongoing aims against which programmes of government are evaluated and adjusted, with the continuous (but seldom attained) aspiration that reality can be made to conform to the truth of these schemes (Elden 2001, 145–50). They are, at the same time, distillations of underlying logics of multiple and dispersed practices for the conduct of conducts, and this is what gives such 'diagrams' their place in the thought of government.

Osborne and Rose (1999; 2004) extend this idea of the 'diagram' to the analysis of nineteenth and twentieth century aspirations for reforming and governing cities. They identify 'diagrams' and logics of order, of health, of happiness, of progress, as operative rationales in technologies for visualizing and governing urban spaces. As an analytical focus, the identification such diagrams or rationalities seeks to 'individuate the regularities that are giving form to the multitude of local, fluid, fleeting endeavours, stratagems, and tactics that characterize the forces seeking to govern this or that aspect of urban existence' (Osborne and Rose 1999, 758).

In their paper, Osborne and Rose (1999) trace how 'the city' is variously constituted as an object of government in the face of its persistent 'ungovernability' (758). While the causal logics mobilized in attempts to 'govern existence in urban form' (758, emphasis removed) are not unpacked in any detail, these ideas are provocations to further explorations of the spatial rationalities informing projects of government (see also Osborne and Rose 2004 on 'spatial phenomenotechnics').

Thus, spaces can be delineated for various purposes: to produce grids of classification, order and discipline; but equally to foster particular kinds of environmental qualities (cleanliness, beauty); or to concentrate or fragment the effects of broader social processes found to be present in particular localities (social progress/regress). And 'sometimes space has certain magical qualities in that there is a kind of interpenetration of space and that which inhabits it' (Osborne and Rose 2004, 213).

It is this 'magical interpenetration' of space and inhabitants that I argue is not 'sometimes' but centrally implicated in spatial rationalities of government. The causal qualities of space and environment act as technologies for 'clearings' in which to 'gather together' and release the potentials of the individuals and populations subjected to their influences (see Dean 1996, 59–63; Elden 2001, 84–92, discussing Heidegger on technology; and Osborne and Rose 2004).

That is, spaces and environments are not simply delineated or arranged for purposes of discipline or surveillance, visibility or management. In projects of political subjectification or governmental self-formation, appropriate bodily comportments and forms of subjectivity are to be fostered through the positive, catalytic qualities of spaces, places and environments. These productive spatial rationalities operate in different modes, making use of different combinations of, for instance, geometric, biological, medical, environmental or evolutionary causalities. In order to illustrate the different causal logics of spatial rationalities, I provide three brief examples of 'diagrams' of spatial thought: dispositional, generative and vitalist (Huxley 2006).[5]

Dispositional Spatial Rationality

Discourses and practices aiming to mobilize the efficacy of geometric ordering and spatial grids of classification emerge, as Foucault (1979b) has shown in *Discipline and Punish*, from conjunctures between Descartian philosophies of a universal order (*mathesis universalis*), practices of incarceration of the criminal and the insane, and the exercise of police power in knowing and managing the 'men and things' of a territory. Order and visibility operate as moral registers in calls to combat the chaos and evil of the city at the end of the eighteenth century (see Rabinow 1982).

---

5   In Huxley (2006) I explore three exemplary 'diagrams' of such spatial rationalities in more detail.

Here, the problem for government is the dangerous mob and insurrection that threaten to replace order with bedlam. The new urban masses, plunged in obscurity and debauchery, beyond the influence of moral example, must be brought into the light of reason through re-placement in a visible spatial order and social hierarchy. This spatial, social and moral order is exemplified – 'diagrammed' – in James Silk Buckingham's (1849) (unrealized) plan for the hierarchically ordered, quadrilateral model town of Victoria.

Victoria's layout reflects a social hierarchy, in which the labouring classes are housed around the perimeter, the supervisors and bourgeoisie occupy the intermediate quadrilaterals, and the governing elite are positioned at the centre. The relative proportions of inhabitants, residences, trades and occupations of each class are meticulously calculated. However, these groups are not entirely cut off from interaction with each other, but mingle in covered shopping arcades, libraries, galleries and open spaces. Thus, mutual visibility is ensured, both through the spatial layout and through social interaction, so that the lower orders will be exposed to the examples of the higher, while the upper classes can monitor and guide the lower.

Buckingham's Victoria provides a 'diagram' of police power and perfect visibility (similar to that of a military camp, Foucault 1986a, 255), but it also seeks a spatial catalyst for the *production* of social and moral order.

These productive logics of geometrical order and spatial clarity are available as technologies that can be deployed in liberal modes of governmentality, fostering the exercise of liberty and choice on the part of autonomous political subjects. Grids do not only control – they enable. For example, Paul Carter (1987, Chapter 3) describes the liberal possibilities of choice, direction and development contained in the grid plan for the new settlement of Melbourne; and Rose-Redwood (2003) analyzes the liberal aspirations for self-perpetuating order and rationality secreted in the 1811 plan for New York.

A second form of spatial rationality, in contrast, draws on medical and sanitary ideas of the generative qualities of environments and milieux in the production of bodily and moral health.

Generative Spatial Rationality

In projects of reform of the nineteenth century city, Hippocratic theories of diseases generated in swamps, fens and miasmas join with sanitarian projects to cleanse the city with running water and canalized sewerage systems, and with the identification of areas of the city as breeding grounds for sin and degradation. The free circulation of air and water, people and things, and the importance of sunlight and visibility are not only necessities for the medical health of bodies, but are fundamental requirements for moral and spiritual health (Driver 1988; Joyce 2003; Valverde 1991). Rationalities of spatial order and sanitary reform and regulation converge with philanthropic practices of 'training urban man' [sic] (Schoenwald 1973) in attempts to create conditions for the production and maintenance of appropriate comportments and behaviours.

The 'diagram' of the ideal city described in Benjamin Richardson's *Hygeia, a city of health* (1876/1985) crystallizes the problems that generative rationalities both identify and propose solutions to. The layout of Hygeia is on a grid pattern that promotes visibility and order, but more important are the sanitary technologies that cleanse the city streets, the houses and the bodies of the inhabitants. Richardson describes in minute detail the methods for provision of clean water and the carrying away of waste in arched tunnels beneath the houses and the streets which also carry water, gas, sewerage and rail transport. In Hygeia, abattoirs and factories are located outside the city perimeter and the sale of alcohol is prohibited.

The houses are designed and constructed for maximum efficiency in cleaning, with tiled exterior and interior walls, kitchens in the upper stories (so that heavy plates are carried down stairs and empty ones up); floors are polished wood to prevent the retention of dust; rubbish is disposed of through chutes connected to the underground tunnels; and the city and its residents are subject to regular inspection by the Health authorities. Hygeia is thus an environment that is positively generative of both healthy bodies and moral beings who will comport themselves in appropriate sanitary, civil and political ways.

Generative rationalities presume specific areas that are susceptible to reform and specific populations that are the subjects of government. Although health and morality will be produced for all, it is the unhealthy and unsanitary 'slums' or 'rookeries' that pose the danger of the spread of disease and degradation. If these diseased areas and their inhabitants can be cured and improved, the body of the city, the social body, and the proper relations between its parts and processes, will be restored to normal, healthy equilibrium (see Foucault 2003, 246).

But with the increasing 'governmentalization of the state' (Foucault 1991b, 103) and the emergence of 'government from a social point of view' (Rose 1999, Chapter 3) from the end of the nineteenth century, a form of spatial rationality can be discerned that aims not only to govern spatial aspects of the whole sphere of the social within a national territory, but posits a spiritual or vitalist dimension to the problems of social government and the incitement of political subjectivities and self-forming subjects.

Vitalist Spatial Rationality

In this last example of spatial rationality, problematizations of chaos, evil, disease, and immorality as inhering specific, disorganized spaces and being generated by particular 'abnormal' environments become translated into the governmentalized state's assumption of pastoral care for the whole population, constituted as the sphere of 'the social'.

Problematic spaces and environments still occupy the thought and programmes of government, but come to be relocated within a general 'individual-population-environment' relationship that includes bio-social evolutionary causalities. The spread of practices of social management and the growth of social science expertise in government contribute to the making up of 'the social' as an object of governmental problematization; but a mix of eugenics, neo-Lamarckian theories of acquired characteristics and ideas of possibilist geographies also play their parts in New Liberal, Fabian and US Progressive Era politics that converge in the vision of universal social progress (Campbell and Livingstone 1983; Dean 1999; Huxley 2006; Rose 1999).

In addition to concerns with the bio-politics of population, these aspirations to create conditions of health and happiness for all (a Benthamite 'eudaemonic diagram'; Osborne and Rose 1999) are also infused with spiritualist and vitalist elements, both as wide-spread practices and experiments with séances and mediums, and as philosophical explorations of vitalism and 'creative evolution', such as those of Henri Bergson (Burwick and Douglass 1992; Roe 1984). These philosophies and practices link social reform and good government to biological evolutionary progress and, further, to the intellectual and spiritual advance of the human race, by suggesting that humanity (or at least the most advanced members of it) can guide and direct the course of bio-social, intellectual-spiritual development.

The writings of Patrick Geddes are emblematic of vitalist spatial rationalities. For Geddes, the unique qualities of individual cities and regions, influenced by their physical and natural environments, are the locations and conditions of human evolution. The city-region – with its *genius loci*, its layers of experience and memory embodied in the natural and built environment – is the medium from which higher evolution emerges. Geddesian town planning aims to foster conscious engagement by individuals and populations in the positive guidance and direction of their own environments (see Mercer 1997; Osborne and Rose 2004), and thus contributes to their bio-social, civic-political and creative evolution.[6]

For Geddes and his colleague, Victor Branford (Branford and Geddes 1917, 143):

> A town becomes a true city in the measure that it develops new and higher powers to enrich and enhance the inner life of its citizens, to combine their diverse interests into an ethical polity, and to evoke those high gifts of personality which master circumstance, transcend tradition, and rise on the wings of the spirit into the realm of creative culture.

Others have noted spiritualist aspects of Geddes' thought (e.g. Matless 1992; 1998; Welter 2002; see also Osborne and Rose 2004), but these preoccupations are not peculiar only to him. 'The vitalist' and 'the non-material' are integral to truths of the effects of space and environment on individuals, populations and human evolution,

---

6   Osborne and Rose (2004) compare Charles Booth's mapping of social space by class with Geddes open-ended vitalist conceptions of the city. They draw on the concept of 'phenomenotechnics' and (suitably vitalist Deleuzean) ideas of 'striated' (Booth) and 'smooth' (Geddes) space, but this analysis seems to underplay the governmental elements of both Booth's and Geddes' aims to reform problematic aspects of the city and society.

and inflect the thought of government more generally. There is no necessary contradiction between these aims to foster the non-material aspects of creative evolution and the production and management of particular kinds of political subjectivities and self-forming subjects. As Joyce (2003, 172) points out, Geddes' participatory surveys are exercises in the production of, and depend upon, subjects of regulated governmental self-formation.

The modes of spatial rationality outlined here are only three possible ways in which the causal qualities of space and environment might be configured in the thought of government. There is further potential for an 'analytics of government' to explore the multiple, taken-for-granted spatial presuppositions that inform technologies and practices of regulation and subjectification.

Conclusion

Governmentality as the conduct of conducts and the aspiration to fabricate and guide appropriate comportments and behaviours, comprises diverse and disparate series and assemblages of practices, regulations, philosophies, texts, buildings and authorities. Governmentality is also indelibly spatial, both in terms of the spaces it seeks to create and in the causal logics that imbue such attempts with their rationality.

But the major developments of Foucauldian perspectives on the rationalities and technologies of government have taken place in disciplines that, while acknowledging the importance of space, generally have not explored the implications of space/power relations in great detail. Conversely, in Geography, there have been meticulous historical studies of the spaces of institutions and cities as sites of discipline and/or subjectification, but relatively less explicit concern with rationalities and technologies of government or with the spatial as an element of governmental thought.

Where space has been examined as a rationality of government, it is often taken as the technology of a disciplinary order of visuality, surveillance and control. However, if power and government are seen not only as forms of control, but also as *productive* of political subjectivities and self-forming subjects; and space is taken as integral to the exercise of power and the conduct of conducts; then spatial and environmental causality can be examined as central elements in the thought of government.

Three examples or 'diagrams' of spatial and environmental rationalities illustrate different governmental mobilizations of causal relations within an individual-population-environment nexus. The schemes for ideal cities or city-citizen relations within and through which appropriate comportments and subjectivities can be fostered are exemplars of dispositional/geometrical, generative/bio-medical and vitalist/evolutionary logics that characterize the taken-for-granted spatial presuppositions of disparate programmes of urban and social reform.

This chapter has made a case for space and environment as rationalities of government, to take seriously Foucault's (1986a, 254) statement that he was concerned with 'spatial techniques, not metaphors'. By investigating the ways in which spaces and environments are invested with causal powers in programmes, projects and plans for the government of individuals and populations, we can begin to trace how particular specifications of spaces, buildings, environments, suburbs, cities and regions enter into unstable, heterogeneous, assemblages of technologies of rule. And in so doing, we might also reflect on how, and with what effects, the truths produced by Geography are implicated in these rationalities of governmentality.

References

Barry, A., T. Osborne and N. Rose, Eds (1996) *Foucault and Political Reason: Liberalism, Neo-liberalism and Rationalities of Government*. London: UCL Press.

Bennett, T. (2004) *Pasts beyond Memory: Evolution, Museums, Colonialism*. London: Routledge.

Branford, V. and P. Geddes (1917) *The Coming Polity: A Study in Reconstruction*. London: Williams and Norgate.

Braun, B. (2000) 'Producing Vertical Territory: Geology and Governmentality In Later Victorian Canada.' *Ecumene*, 7(1), 7–46.

Buckingham, J. (1849) *National Evils and Practical Remedies, with the Plan of a Model Town*. London: Peter Jackson.

Burchell, G. (1991) 'Peculiar Interests: Civil Society and Governing "the System of Natural Liberty".' In G. Burchell et al. (Eds), 119–50.

Burchell, G. (1996) 'Liberal Government and Techniques of the Self.' In A. Barry et al. (Eds), 19–35.

Burchell, G., C. Gordon and P. Miller (1991) *The Foucault Effect: Studies in Governmentality*. London: Harvester/Wheatsheaf.

Burwick, F. and P. Douglass, Eds (1992) *The Crisis in Modernism: Bergson and the Vitalist Controversy*. Cambridge: Cambridge University Press.

Campbell, J. and D. Livingstone (1983) 'Neo-Lamarckism and the Development of Geography in the United States and Great Britain.' *Transactions of the Institute of British Geographers*, NS 8, 267–94.

Carter, P. (1987) *The Road to Botany Bay: An Essay in Spatial History*. London: Faber & Faber.

Crampton, J.W. (2003). 'Cartographic Rationality and the Politics of Geosurveillance and Security.' *Cartography and Geographic Information Systems*, 30(2), 131–44.

Cruikshank, B. (1999) *The Will to Empower: Democratic Citizens and Other Subjects*. Ithaca, NY: Cornell University Press.

Dean, M. (1991) *The Constitution of Poverty: Toward a Genealogy of Liberal Governance*. London: Routledge.

Dean, M. (1994) '"A Social Structure of Many Souls": Moral Regulation, Government, and Self-formation.' *Canadian Journal of Sociology*, 19(2), 145–68.

Dean, M. (1996) 'Putting the Technological into Government.' *History of the Human Sciences*, 9(3), 47–68.

Dean, M. (1998) 'Questions of Method.' In I. Velody and R. Williams (Eds) *The Politics of Constructionism*. London: Sage, 182–99.

Dean, M. (1999) *Governmentality: Power and Rule in Modern Society*. London: Sage.

Dean, M. (2002) 'Liberal Government and Authoritarianism.' *Economy and Society*, 31(1), 37–61.

Dean, M. and B. Hindess, Eds (1998) *Governing Australia: Studies in Contemporary Rationalities of Government*. Melbourne: Cambridge University Press.

Driver, F. (1985) 'Power, Space and the Body: A Critical Assessment of Foucault's *Discipline and Punish*.' *Environment and Planning D: Society and Space*, 3, 435–46.

Driver, F. (1988) 'Moral Geographies: Social Science and the Urban Environment in Mid-nineteenth Century England.' *Transactions of the Institute of British Geographers*, NS, 13, 275–87.

Driver, F. (1993) *Power and Pauperism: The Workhouse System, 1834–1884*. Cambridge: Cambridge University Press.

Elden, S. (2001) *Mapping the Present: Heidegger, Foucault and the Project of a Spatial History*. London: Continuum.

Elden S. (2002) 'The War of Races and the Constitution of the State: Foucault's 'Il faut defendre la societe' and the Politics of Calculation.' *boundary 2*, 29(1), 125–51.

Elden, S. (2007) 'Governmentality, Calculation, Territory.' *Environment and Planning D: Society and Space*, 25, 4.

Faubion, J., Ed. (2000) *Aesthetics, Methods and Epistemology. Essential Works of Foucault 1954–1984*. Volume 2. London: Penguin.

Fontana, A. and M. Bertani (2003) 'Situating the Lectures.' In M. Foucault, *Society Must be Defended*, 273–94.

Foucault, M. (1974a) *The Order of Things: An Archaeology of the Human Sciences*. London: Routledge.

Foucault, M. (1974b) *The Archaeology of Knowledge*. London: Tavistock.

Foucault, M. (1979a) 'On governmentality.' *Ideology and Consciousness*, 6, 5–21.

Foucault, M. (1979b) *Discipline and Punish; The Birth of the Prison*. New York: Vintage Books.

Foucault, M. (1980a) 'The Confessions of the Flesh.' In C. Gordon (Ed.), 194–228.

Foucault, M. (1980b) 'The Eye of Power.' In C. Gordon (Ed.), 146–65.

Foucault, M. (1981) *The History of Sexuality: Volume I, An Introduction*. London: Penguin.

Foucault, M. (1982) 'The Subject and Power, Afterword.' In H. Dreyfus and P. Rabinow, *Michel Foucault: Beyond Structuralism and Hermeneutics*. London: Harvester Wheatsheaf. 208–26.

Foucault, M. (1984) 'The Order of Discourse.' In M. Shapiro (Ed.) *Language and Politics*. Oxford: Basil Blackwell, 108–38.

Foucault, M. (1986a) 'Space, Knowledge and Power.' In P. Rabinow (Ed.), 239–56.

Foucault, M. (1986b) 'What is Enlightenment?' In P. Rabinow (Ed.), 32–50.

Foucault, M. (1986c) 'Nietzsche, Genealogy, History.' In P. Rabinow (Ed.), 76–100.

Foucault, M. (1988a) 'Politics and Reason.' In L. Kritzman (Ed.) *Michel Foucault: Politics, Philosophy, Culture*. London: Routledge. 57–85.

Foucault, M. (1988b) 'Technologies of the Self.' In L. Martin et al. (Eds), 16–49.
Foucault, M. (1988c) 'The Political Technologies of Individuals.' In L. Martin et al. (Eds), 145–62.
Foucault, M. (1990) *The Care of the Self: The History of Sexuality*. Volume 3. London: Penguin.
Foucault, M. (1991a) 'Questions of Method.' In G. Burchell et al. (Eds), 73–86.
Foucault, M. (1991b) 'Governmentality.' In G. Burchell et al. (Eds), 87–104.
Foucault, M. (1996) 'Clarifications on the Question of Power.' In S. Lotringer (Ed.) *Foucault Live: Michel Foucault, Collected Interviews 1961–1984*. New York: Semiotext(e), 255–63.
Foucault, M. (1997) 'The Ethics of Concern for the Self as a Practice of Freedom.' In P. Rabinow (Ed.) *Ethics: Essential Works of Foucault 1954–1984*. Harmondsworth: Penguin, 281–302.
Foucault, M. (2003) *Society Must be Defended: Lectures at the Collège de France 1975–1976*. London: Allen Lane/Penguin.
Gordon, C., Ed. (1980) *Michel Foucault: Power/Knowledge*. Brighton: The Harvester Press.
Gordon, C. (1991) 'Governmental Rationality: An Introduction.' In G. Burchell et al. (Eds), 1–52.
Hannah, M. (2000) *Governmentality and the Mastery of Territory in Nineteenth-century America*. London: Cambridge University Press.
Harley, J. (2001) *The New Nature of Maps: Essays in the History of Cartography*. Baltimore: The Johns Hopkins University Press.
Hindess, B. (1993) 'Liberalism, Socialism and Democracy: Variations on a Governmental Theme.' *Economy and Society*, 22(3), 300–13.
Hindess, B. (1997) 'Politics and Governmentality.' *Economy and Society*, 26(2), 257–72.
Hindess, B. (2001) 'The Liberal Government of Unfreedom.' *Alternatives*, 26, 93–111.
Hunt, A. (1996) 'Governing the City: Liberalism and Early Modern Modes of Governance.' In A. Barry et al. (Eds), 167–88.
Huxley, M. (2006) 'Spatial Rationalities: Order, Environment, Evolution and Government.' *Social and Cultural Geography*, 7(5), 771–87.
Joyce, P. (2003) *The Rule of Freedom: Liberalism and the Modern City*. London: Verso.
Keith, M. (1997) 'Conclusion: A Changing Space and a Time for Change.' In S. Pile and M. Keith (Eds) *Geographies of Resistance*. London: Routledge, 277–86.
Lefebvre, H. (1991) *The Production of Space*. Oxford: Blackwell.
Lemke, T. (2001) '"The Birth of Bio-politics": Michel Foucault's Lecture at the Collège de France on Neo-liberal Governmentality.' *Economy and Society*, 30(2), 190–207.
Martin, L., H. Gutman and H. Hutton, Eds (1988) *Technologies of the Self: A Seminar with Michel Foucault*. London: Tavistock Publications.

Matless, D. (1992) 'Regional Surveys and Local Knowledges: The Geographical Imagination in Britain, 1918–39.' *Transactions of the Institute of British Geographers*, NS 17, 464–80.

Matless, D. (1998) *Landscape and Englishness*. London: Reaktion Books.

Mercer, C. (1997) 'Geographies for the Present: Patrick Geddes, Urban Planning and the Human Sciences.' *Economy and Society*, 26(2), 211–32.

Miller, P. (1987) *Domination and Power*. London: Routledge.

Miller, P. and N. Rose, Eds (1986) *The Power of Psychiatry*. Cambridge: Polity Press.

Miller, P. and N. Rose (1990) 'Governing Economic Life.' *Economy and Society*, 19(1), 1–31.

Murdoch, J. (2000) 'Space Against Time: Competing Rationalities in Planning for Housing.' *Transactions of the Institute of British Geographers*, 25, 503–19.

Ogborn, M. (1992) 'Local Power and State Regulation in Nineteenth Century Britain.' *Transactions of the Institute of British Geographers*, NS 17, 215–26.

Ogborn, M. (1998) *Spaces of Modernity: London's Geographies 1680–1780*. New York: Guilford Press.

Osborne, T. (1996) 'Security and Vitality: Drains, Liberalism and Power in the Nineteenth Century.' In A. Barry et al. (Eds), 99–122.

Osborne, T. and N. Rose (1999) 'Governing Cities: Notes on the Spatialisation of Virtue.' *Environment and Planning D: Society and Space*, 17, 737–60.

Osborne, T. and N. Rose (2004) 'Spatial Phenomenotechnics: Making Space with Charles Booth and Patrick Geddes.' *Environment and Planning D: Society and Space*, 22, 209–28.

Otter, C. (2002) 'Making Liberalism Durable: Vision and Civility in the Late Victorian City.' *Social History*, 27(1), 1–15.

Otter, C. (2004) 'Cleansing and Clarifying: Technology and Perception in Nineteenth-century London.' *Journal of British Studies*, 43(1), 40–64.

Patton, P. (1998) 'Foucault's Subject of Power.' In J. Moss (Ed.) *The Later Foucault: Politics and Philosophy*. London: Sage. 64–77.

Philo, C. (1987) '"Fit Localities for an Asylum": The Historical Geography of the Nineteenth Century "Mad Business" in England as Viewed Through the Pages of the Asylum Journal.' *Journal of Historical Geography*, 13(4), 398–415.

Philo, C. (1989) '"Enough to Drive One Mad": The Organisation of Space in Nineteenth-century Lunatic Asylums.' In J. Wolch and M. Dear (Eds) *The Power of Geography: How Territory Shapes Social Life*. London: Unwin Hyman. 258–90.

Philo, C. (1992) 'Foucault's Geography.' *Environment and Planning D: Society and Space*, 10(2), 137–62.

Pickles, J. (2004) *A History of Spaces: Cartographic Reason, Mapping and the Geocoded World*. New York: Routledge.

Rabinow, P. (1982) 'Ordonnance, Discipline, Regulation: Some Reflections on Urbanism.' *Humanities in Society*, 5(3 & 4), 267–78.

Rabinow, P., Ed. (1986) *The Foucault Reader*. Harmondsworth: Penguin.

Rabinow, P. (1989) *French Modern: Norms and Forms of the Social Environment.* Chicago: The University of Chicago Press.

Rabinow, P., Ed. (2000) *Ethics, Subjectivity and Truth. Essential works of Foucault 1954–1984.* Volume 1. London: Penguin.

Raco, M. (2003) 'Governmentality, Subject Building and Discourses and Practices of Devolution.' *Transactions of the Institute of British Geographers,* 28, 75–95.

Raco. M. and R. Imrie (2000) 'Governmentality and Rights and Responsibilities in Urban Policy.' *Environment and Planning A,* 32(12), 2187–240.

Richardson, B. (1876) *Hygeia: a city of health.* London: Macmillan & Co. (Reproduced by Garland Publishing, New York, 1985.)

Roe, M. (1984) *Nine Australian Progressives: Vitalism and Bourgeois Social Thought 1890–1960.* St. Lucia: University of Queensland Press.

Rose, N. (1985) *The Psychological Complex: Psychology, Politics and Society in England, 1869–1939.* London: Routledge and Kegan Paul.

Rose, N. (1988) 'Calculable Minds and Manageable Individuals.' *History of the Human Sciences,* 1(2), 179–200.

Rose, N. (1990) *Governing the Soul: The Shaping of the Private Self.* London: Routledge.

Rose, N. (1994) 'Medicine, History and the Present.' In C. Jones and R. Porter (Eds) *Reassessing Foucault: Power, Medicine and the Body.* London: Routledge, 48–72.

Rose, N. (1996) 'Identity, Genealogy, History.' In S. Hall and P. du Gay (Eds) *Questions of Cultural Identity.* London: Sage, 128–50.

Rose, M. (1999) *Powers of Freedom: Reframing Political Thought.* Cambridge: Cambridge University Press.

Rose-Redwood, R. (2003) 'Disciplined Minds, Rational Souls: The Manhattan Grid and the Rationalization of the Landscape.' Unpublished paper to the Association of American Geographers Conference, New Orleans, April.

Rycroft, S. and D. Cosgrove (1995) 'Mapping the Modern Nation: Dudley Stamp and the Land Utilisation Survey.' *History Workshop Journal,* 40, 91–105.

Schoenwald, R. L. (1973) 'Training Urban Man.' In J. Dyos and M. Wolff (Eds) *The Victorian City: Images and Realities, Vol. 1.* London: Routledge, 669–92.

Sharp, J., P. Routledge, C. Philo and R. Paddison, Eds (2000) *Entanglements of Power: Geographies of Domination and Resistance.* London: Routledge.

Simons, J. (1995) *Foucault and the Political.* London: Routledge.

Smith, N. and C. Katz (1993) 'Grounding Metaphor: Towards a Spatialised Politics.' In M. Keith and S. Pile (Eds) *Place and the Politics of Identity.* London: Routledge, 67–83.

Thrift, N. (2000) 'Entanglements of Power: Shadows?' In J. Sharp et al. (Eds), 269–78.

Valverde, M. (1991) *The Age of Light, Soap, and Water: Moral Reform in English Canada, 1885–1925,* McCelland and Stewart, Toronto.

Welter, V. (2002) *Biopolis: Patrick Geddes and the City of Life.* Cambridge, MA: MIT Press.

Chapter 21

The History of Medical Geography after Foucault

Gerry Kearns

Medical geography is a small sub-discipline of academic geography. Its presence within histories of geography depends in part upon whether geography is considered as a discipline or whether there is a broader understanding of geographies as forms of practical and popular knowledge. If we take the second approach, medical geography would seem to be quite important in the development of geographical imaginations. There are connections of cause and responsibility between these popular and academic geographies. Quite often, popular knowledges get treated as precursors of academic disciplines yet the connections are reciprocal and continuing.

Medical topics appear in histories of geography under two main guises: as an occasion of environmentalism or as a field of spatial analysis (Mayer 1990). These two form the backbone of this chapter. I begin by indicating some of the ways these two aspects of medical geography have featured in histories of geography. Then, I consider these aspects in the light of three tactics from Foucault's historical writings (see Table 21.1). I take up the question of the relations between discourses and practices. I look at the processes whereby subjectivities are shaped. Finally, I look at the political issues raised in history writing, illustrating my account with reference to the 'imaginative geographies' (Said 1978, 55) of AIDS.

Medical Topics in Histories of Geography

Hartshorne's sectarian *The Nature of Geography* (1939) is largely a polemic against Geography considered as a form of environmentalism. Medical matters are absent both from his account of those who came to prepare the way for Humboldt and Ritter and of the contemporary heresies which geographers following the classical model should denounce as deviations. Glacken's broad church in *Traces on the Rhodian Shore* (1967) deals only with Hartshorne's Old Testament and stops before the nineteenth-century era of Hartshorne's Classical Geography.

By virtue of his treatment of Geography as a science of dynamic distributions, Sauer, Glacken's mentor, was one of Hartshorne's heretics. For Hartshorne, the true geographer did not dabble in change. Historians and geographers should be good neighbours, comfortable in their good fences. Glacken appeared to ignore these

Table 21.1   Some links between medical geography and the historical writings of Michel Foucault

Styles of Medical Geography	Tactics and themes in Foucault's historical writings		
	Discourses and Practices	Governmentality and Subjectivation	Critical and Effective Histories
Environmentalism	'Of Airs, Waters and Places'	Dietetics and 'Of Airs, Waters and Places' (Subjectivation)	Africanization of AIDS
Spatial Analysis	Exemplary institutional Archipelago	Ecological analysis (Governmentality)	Diffusion and Diffusionism in AIDS studies

debates and wrote a biography of a set of intellectual ideas about the environment; not properly geographical ideas at all according to Hartshorne. Up to the late eighteenth century, environmentalists, according to Glacken, took up three questions: is the earth a fit home for people; has it actively shaped cultures; and has it been irreparably damaged by people? The last of these was a central concern of Sauer's, particularly after the Second World War. His anxiety about the environmental impact of industrialism informed both his optimistic reading of the carrying capacity of pre-industrial (and pre-Conquest) Meso-American agriculture, and also the doomwatch tone of the conference he instigated on 'Man's Role in Changing the Face of the Earth' (Thomas 1956). Glacken writes the prehistory of his mentor's moral vision. Environmental medical themes receive careful attention. Disease bore upon both the question of the fitness of the earth for habitation and on how natural forces had shaped settlement and society. Hippocrates' works, such as 'Of Airs, Waters and Places', and their rediscovery by Bodin were treated carefully in Glacken's book. There is nothing on human modification of the environment as a precondition of population increase. Like Sauer, Glacken is sceptical about human mastery of nature.

Glacken's work has been criticized of late. His studies of environmental ideas fail to address their context. This means at least two things. At one point, before 1989 perhaps, it seemed as if 'We're all Marxists now'; no doubt on grounds which would have led Karl Marx himself to repeat his claim to Paul Lafargue: 'As for me, I am no Marxist' (Draper 1978, 5). Since the landmark collection, *Geography, Ideology and Social Concern* (Stoddart 1981), historians of geography have been invited to ask of any geographical idea: what was the social or economic demand for this view of the world? By and large, this has served to make geography a moment in the history of imperialism. For example, one of Glacken's former research assistants returned to the issue of environmentalism and presented it as capitalism's alibi for the dirty business of class exploitation (Peet 1985). The second use of context brings me to the central issue of this chapter for in some, equally superficial, senses we are all Foucauldians now insofar as we treat geography as a discourse which establishes its own truth conditions anew in each context: 'He possessed that rare capacity as a thinker to open us to an optic ... whose perspective now seems so familiar that it is difficult to see how we previously failed to bring it to bear' (Soper 1995, 21).

Livingstone's *The Geographical Tradition* (1992) is a somewhat unstable marriage between these two versions of context. As Latour (1993 [1991]) suggests of the approach of the Edinburgh School to the history of science, a conventionalist reading of science has been placed alongside a realist reading of material determination. Livingstone is concerned both with how environmentalist ideas work and with what needs they serve in each period. The geographical tradition turns into a relay race in which each generation finds its own environmental voice to address the current needs of the hegemonic social order: comes the time, comes the geographer. At least, that is, until the present, when the relay race breaks down with runners wandering off in postmodern indifference, each thinking they have the baton but uninterested in where to take it.

Livingstone takes up Glacken's reading of medical geography as a branch of environmentalism. He goes back to Hippocrates, back to Bodin and on to the nineteenth-century debates over the natural limits to white settlement in the tropics. The two senses of context are laid alongside each other. Bodin's writings are considered in the light of the Age of Discovery but also against attempts to read God through his creation, specifically to produce an astrological anthropology. Likewise, the nineteenth-century moral climatology is presented as an ethical reflection on racial responsibility and difference, although here the voice of Marx prevails over Foucault's since the instrumental value of these writings for commercial imperialism is repeatedly stressed. I do not want to adjudicate this ambivalence, although I find Latour's discussion of what Haraway has called hybrids a useful way to think about objects of knowledge which are yet embedded in practices and in networks of verification and replication.

If we turn to the second of my styles of medical geography, that is as spatial science, we find an even more fragmented conception of its history. In the first place, the spatial scientists were disinclined to take up Hartshorne's challenge and write a story of the emergence of geography which justified their approach. Bunge (1966 [1962]) suggested that the history of the discipline was more or less irrelevant to current practice because modern geography contained the accumulated wisdom of geographies past. Haggett (1965) was a little more guarded and proposed that a version of geography as locational analysis could call upon a long tradition of geometrical studies going back to the Greeks. Yet neither Haggett, nor any other 'scientific' geographer made any serious attempt to document or use that tradition. In the second place, when the quantitative geographers spoke of diffusion studies in the 1960s they almost always meant the diffusion of innovations and not of diseases. This reflected the strongly economic nature of the topics that models in human geography were intended to address. To some it extent, it was the attacks (Blaikie 1978) on the validity of innovation diffusion as a model for economic growth which shifted the attention of the modellers from an area (innovation adoption) where there seemed to be a need to complicate models with a whole set of inaccessible behavioural postulates to an area (the spreading of diseases) where there seemed to be no such need. Modelling now had a different set of antecedents, and modern works in quantitative geography now make reference to John Snow and the Broad-Street pump among other early studies in 'medical geography' (Cliff and Haggett 1988, ch.1; Gould 1985, ch.19). Although Haggett (1965) referred at first to these medical studies as the application of geographical methods outside geography, they are now admitted more readily to the canon. It might be possible to take up Haggett's invitation to consider a history of the geometric tradition in geography, and to document its connections with popular and scientific understandings of disease diffusion or other medical matters. In this way Philo (1995) has provided a contextual account of the so-called Jarvis Law describing the distance decay function covering the use of mental health institutions. However, I am not going to follow up that work here.

Instead, I want to return to Foucault and excavate some different medical geographies that we might historicize. Then, I shall turn to modern medical geography. There is a connection between the particular vision of medical geography current in histories of geography and some of the limitations of modern medical geography. Livingstone is always looking for versions of popular geographies around which the discipline can heroically, if embarrassingly, institutionalize by the late nineteenth-century. There is value, I think, in recovering a more dispersed medical geography by continuing with the study of the connections between popular and academic geographies beyond the period when university departments were created.

Discourse and Practice

In 1984, Foucault suggested that his work had covered three topics: 'I tried to locate three major types of problems: the problem of truth, the problem of power and the problem of individual conduct' (1988a, 243). If we take Foucault at his word, it is possible to locate some of his central obsessions in terms of the relations between truth, power and the individual (see Figure 21.1). It would perhaps be more consistent with Foucault's suspicions to speak of truth-effects, power-effects, individual-effects and social effects detectable in the practices they inflect and allow.

Works such as *The Birth of the Clinic* (1973 [1963]), *Madness and Civilisation* (1965 [1961]) and *Discipline and Punish* (1977 [1975]) concern the relations between power and truth. In *Madness and Civilisation*, Foucault, as is well known, connects discourses about madness to the disciplines practised upon the mad. In the name of having the true understanding of madness, people locked up the mad, first in all-purpose *hôpitaux généraux* and later in specialist asylums. Madness was rendered as a fit subject for incarceration by being seen as homologous with other forms of refusing the universal imperative to labour. The result of madness was an unfitness to labour and thus incarceration was appropriate. Through incarceration a new subjectivity could be formed in which, by physical and moral means, regularity and a moderation of habits could be made second nature. Individuals, now in control

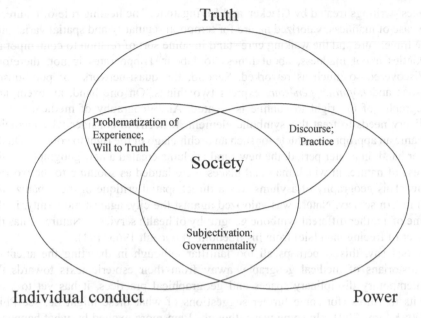

Figure 21.1 Some themes in the work of Michel Foucault

of themselves, could be trusted to take up again their obligation to earn their living by the sweat of their brow. Foucault insisted upon the interrelations between the discourses of madness and the practices of institutions such as the asylum. The conditions of existence of the propositions of madness discourses include both adjacent discourses (on poverty, on the body) and specific institutions (the asylum, the church, the sovereign). Some things can only be thought because other things can be done, while some things can only be done because they can be thought that way. Lest this seem overly deterministic, Foucault also insisted, more fully in interviews and lectures than in the text of *Madness and Civilisation*, that there are counter-discourses and counter-conducts (Bouchard 1977). Yet these contests are not even. The sovereign appropriation of Truth is an attempt to censor counter-discourses and to block counter-conducts.

In some ways this is the Foucault who is closest to the Edinburgh School and to David Livingstone. This is the Foucault who proliferates distinctions specifying the mechanics of the formation of discourses (1991a). Those practical and discursive conditions of existence of practices and discourses that are sedimented in earlier times can be isolated and linked in temporal genealogies. Likewise, discursive systems can be made comprehensible through mapping their interrelations as an episteme of promiscuously bifurcating divergences and dependencies. The archive of any earlier episteme can be approached as an archaeology.

Madness and Civilisation opens up a rich field for histories of medical geography. In terms of my earlier distinction between environmentalism and spatial science, let me mention but two. First, it gives a particular inflection to the 'Of Airs, Waters and

Places' writings treated by Glacken and Livingstone. The healing role of nature, in the case of madness, valorized nature for temporal regularity and spatial variability. The travel cure and the walking cure stand in some sort of relation to contemporary anxieties about madness, about fitness to labour. Hippocrates is not, therefore, rediscovered so much as reworked. Second, the quasi-networks of poorhouses, prisons and *hôpitaux généraux* express two things. On one hand, an exemplary geography of the right to confine is created. Any geography of medical service delivery needs to treat this symbolic element. It needs to consider why medicine became an appropriate vehicle for such an archipelago of the right to confine. On the other hand, in a later period, the new rural asylums created a new geography of the caress of nature, in which marginal places were lauded as soothing to the worried brow. This geography of asylums was a direct spatial critique of the urbanization of modern society. Nature was valorized against the city, against modernity, in the name of a rather different symbolic geography of health services: 'Nature ... has the power of freeing man [sic] from his freedom' (Foucault 1965, 194).

As I say, this is perhaps all too familiar although in directing the attention of historians of medical geography away from their esoteric texts towards the contemporary disciplinary spaces and geographical practices, it has yet to yield all its treasures (for some further suggestions of what these might be, see Philo 2000; Kearns 2001). In some ways, though, I am more excited by what happened to Foucault's work when he triangulated his mistrust of power/knowledge systems with a new concern for individual conduct.

Governmentality and Subjectivation

Governmentality was treated by Foucault (2004a; 2004b) in two of his annual series of lectures (1978 and 1979) to the Collège de France. In his published works, notably the second and third volumes of the *History of Sexuality* (1985, 1986; published in France in 1984), he turned instead to subjectivation. The two are related but not identical. Governmentality produces more than subjects. Subjectivation involves more than just the actions of the state.

Modern states rely upon power rather than mere violence. In some respects they require the consent of the ruled. The practices of government presuppose people with choices, people as decision-making subjects. The modern state, suggested Foucault (1991c; see also Pasquino 1991; Burchell 1991; Procacci 1991; Donzelot 1991; Castel 1991), was essentially created in the seventeenth and eighteenth centuries when the population became the focus and instrument of government. Formerly, the acquisition of estates paying taxes had been the goal of a successful monarch. Now, population became the focus of government, replacing territory as the thing that defined the well-being of the state. Intellectually, population was increasingly seen as having its own laws of motion, detectable in swings relating to national prosperity. The good of the state was seen as best served by a numerous and contented population. Foucault argued that this period saw the emergence of a form of governmentality,

Discourses	Problematize one's relations with	Subjectivation: Practices of the Self
Dietetics, Health	One's own body	Aphrodisis/Ontology: Identify the part of the self which is subject to moral conduct
Economics, Household Management	Women, Spouse	Chresis/Deontology: Identify the ways one must relate to the rules of good conduct
Erotics	Men, Boys	Enkrateia/Ascetic: Elaborate exercises which make one ethically fit, and develop self control
Philosophy	Truth, Wisdom	Soprhosyne/Teleology: Recognise the broader sense of moderation to which cultivation of the self must tend

Figure 21.2   **The dimensions of the subjectivation of adult male citizens in fourth-century (BCE) Greek society**

of a governmental rationality based on police theory. This combined the religious, pastoral concern for individuals with the civic, abstract management of the many. The police state individualizes, normalizes and totalizes.

The individualizing dimension of the policing state centres on good conduct and the problems of the moral order of the city. When Foucault tried to specify what this involved he began to describe the forms of self-examination that we learn and to show how these relate to notions of individual good conduct in social life. He had already described how madness served to reinforce an injunction to labour, but in his later works he was increasingly concerned to explore how our obsessive regulation of sexuality might teach and reinforce forms of good conduct that make us governable subjects. In *The Use of Pleasure* (1985), Foucault took fourth-century Greek society as an example of one form of this self-regulation of private conduct in the public good, suggesting in addition that there were significant respects in which early Christian writers drew upon these ideas. Foucault identified four dimensions of the self along which Greek male citizens were encouraged to examine their conduct, particularly with regard to sexuality (see Figure 21.2). Greek male citizens were asked to examine their relations with their own body, to problematize their use of it. This discourse of care for one's body was called dietetics. In part, it depended upon recognizing the part of one's self that should be subject to moral conduct.

One was expected to be other-regarding in various situations among which Foucault examined those areas covering the relations between the male citizen and his spouse and between the male citizen and his boy lovers. The first of these referred in particular to various forms of courtship and was known as erotics. In each case, the forms of self-reflection took on a particular character relating, respectively, to ethics and ascetics. Integrating and overseeing all these areas of self-examination was the question of the Greek male citizen's relations with truth and the problematization of this set of relations constituted philosophy. Cultivating this last dimension of wisdom involved submitting to a particular moral regimen. Moderation and self-control allowed the cultivation of pleasure and the pursuit of that moral perfection which was most consistent with fulfilling the obligations laid upon one as a citizen.

These inter-related concerns with governmentality and subjectivation invite extension and elaboration. In the course of this, new histories of medical geography might be broached. Clearly Foucault's account of dietetics provides a further way of exploring the 'Of Airs, Waters and Places' discourses in terms of how climate affected the body and how the prudent management of one's exposure to different sorts of weathers and places might be part of a regimen for good living. The anxieties and promptings of these discourses suggest a range of things that might be at stake in what Livingstone terms moral climatology. Stoler's *Race and the Education of Desire* (1995) develops some of these.

Stoler argues that European bourgeois identity was framed by the experience of being colonizers and that it was this colonizing European who defined domestic nervousness around cleanliness and health. These agitating worries were learned as part of an attempt to define civilized conduct in a place where race trumped class, and where the colonial encounter proliferated a whole set of subject positions which seemed to be not quite European, not quite native. A regimen of domestic cleanliness set against a background of environmentally-derived moral danger was charged with holding the European line in the face of what Spivak would term hybridity. McClintock shows in *Imperial Leather* (1995) that health and cleanliness became central concerns of bourgeois subjectivation. Indeed, the race experience of the colonies increasingly served as a grid for articulating other dimensions of difference in nineteenth-century Britain. Soap became a fetish. Strict spatial distinctions created pools of purity in dangerous seas. Here we have a medical geography that brought the empire back home and used health and cleanliness as a way of racializing the domestic sense of bourgeois superiority. Moral climatology was more than just part of the practice of colonialism; it was also part of the practices of the self.

These practices of the self were elaborated not only with regard to a temporal sensibility, as Foucault notes (the play of the seasons in dietetics, of the fleeting moment in erotics), but also as a spatial sensibility around environment, and across public and private spaces. This spatial sensibility carried its own dangers relating to the elements of the environment, to shame in public space, and to scarcity in the domestic household. The moral topography of disease and dirt articulated these concerns in the form of a medical geography for the British imperial bourgeoisie and its associates.

Governmentality also disposes a new set of fields for histories of medical geography. The concentration on population gave rise, argued Foucault, to a state technology he termed biopower (Legg 2005). To a large extent this involved the management by the state of the ecology of populations located in various spaces; in particular in either urban or rural spaces. The aggregate description of this managed population in census and vital registers was both an instrument of policy as well as an objectification of the state's new field of competence.

In his book on the decline of British fertility in the late nineteenth century, *Fertility, Class and Gender in Britain, 1860–1940*, Szreter (1996) shows just how tenaciously a certain strand of liberalism held onto the ecological method, with its urban-rural divide, as an instrument of knowledge which allowed certain policies and not others. Thus, when a class analysis threatened to legitimize eugenics policies, the liberals at the General Register Office fought tooth and nail to retain control of the census. In this way they hoped to preserve an ecological treatment of the British social structure that could be married to their geographical framing of the collection of vital statistics. More importantly, perhaps, this geographical framework was ill-suited to articulating anything other than an ecological account of how populations should be managed. Medical geography as spatial science was a very distinct dimension of governmentality.

Critical and Effective Histories

I want to conclude this discussion of Foucault's tactics by turning to the politics of history. Kritzman finds in Foucault's work 'an unquestionable suspicion toward any order through which knowledge is transformed into power and vice versa' (1988b: xvii). This seems right, as does Dean's (1994) account of the relations between critical and effective histories in Foucault. By showing the odd nature of earlier links between knowledge and power, Foucault could denaturalize the present, undermine its self-evidence (1991b, 76; Bouchard 1977), create space for thinking how it might be different. Thus a 1984 interview found him proposing that:

> The work of the intellectual is not to shape other's political will; it is, through the analyses that he carries out in his own field, to question over and over again what is postulated as self-evident, to disturb people's mental habits, the way they do and think things, to dissipate what is familiar and accepted, to reexamine rules and institutions and on the basis of this reproblematization (in which he carries out his specific task as an intellectual) to participate in the formation of a political will (in which he has his role as a citizen to play). (1988b, 265)

Foucault's tactics in moving from discourses to practices and from institutions to subjects have a political edge to them that he explored in several lectures and in many interviews.

In some ways, simply dispersing medical geography away from the relay race of esoteric texts problematizes our relation to its genealogy and alerts us to the possibility that it might indeed exist as a discourse of power. The ways historical work problematizes the present, its fatal self-evidence, is ultimately an imaginative and political question. It is not dictated by or guaranteed by historical scholarship. With that important caveat, I turn to modern medical geography and to AIDS (surveyed in Kearns 1996).

What are the connections between modern medical geography and the old environmentalist discourses of 'Of Airs, Waters and Places' or the population discourses of biopower? In some ways they remain close. The environmental focus is still important and valuable work is still written under the sign of the ecology of disease. Secondly, the practices of biopower are still present in two ways. On one hand, we have pure epidemiological models retooled as a form of spatial science. On the other, we have the managerial perspective of service delivery with its account of hierarchical regionalization, health care systems in space, and so on. This caricature makes medical geography sound rather old-fashioned in the light of current trends in modern geography. This is unfair, and I shall return to this point in my conclusion, but it is certainly striking how far medical geography retained an environmental focus long after determinism coughed its last in human geography more generally. It is also striking how far medical geography has retained its status as a purely spatial science long after the behaviourist critique had undermined the neoclassical foundations of geographical models more generally. All this is to say that medical geography remains concerned with some of the persistent questions about health, society, space and place that animate popular as well as academic discourse.

This is where the matter of AIDS becomes relevant and this is where a work such as Gould's (1993) *The Slow Plague*, written, so he tell us, pro bono publico, requires more extended commentary than I can give here. Let me take the three dimensions of medical geography to which I drew attention in the previous paragraph and see how they correlate with the dispersed medical geographies of the popular imaginary when it comes to AIDS. Take environmentalism. In films such as *Outbreak* (Petersen 1995) or at the beginning of works such as Shilts' (1987) *And the Band Played On*, in television documentaries which treat African AIDS as an expression of the same identity of nature and culture which brought famine in the Sahel, the early deaths displayed in charity commercials, inter-tribal violence, and so on; in all these places dark Mother Africa sweats disease, tragedy and threat (Packard and Epstein 1991; Watney 1989). This is a threat that is racialized in ways recalling directly its colonial referents. It is an imaginary environmentalism equating Black people with African nature, and African nature with evil, in a discourse of origins that is actually a discourse of blame. In the case of Haiti, in the case of the inner-cities of the United States, this is the race discourse of AIDS as WOGS (the wrath of god syndrome). Geography can counter this by producing non-environmentalist discourses of poverty in Africa, by questioning the significance accorded the holy grail of origins.

The objective space of epidemiological models runs many of the same dangers. Instead of the conflation of nature and culture we get their almost complete discounting. Diseases are autonomous spatial processes and may be figured as diffusions. As Brown (1995) argues powerfully, this amounts to a decontextualization of the spread of the virus erasing the sufferings and achievements of the specifically gay communities which first confronted HIV and AIDS. The diffusion model is a dangerously seductive and powerful figuring of disease. Disease comes from somewhere else; it is external to our society at least. Disease moves across space and down urban hierarchies, it is contagious. The population ecology is understood as risk groups rather than risk behaviours (Oppenheimer 1992). Yet AIDS is largely a function of internal arrangements, what Blaikie et al. (1994) term the social distribution of vulnerability (see also Barnett and Blaikie 1992). It is not contagious. It is not spread by casual contact. It poses no threat to everyday sociability. Finally, a non-stigmatizing emphasis on education for behavioural change may be the only effective preventive measure available for some time. The lazy, dangerous geographies of the public imagination need challenging, not encouraging; pro bono publico.

Lastly, the technocratic focus on health service delivery emphasizes intelligent direction from the centre and implementation in the periphery; a conceptualization that Blaut criticizes as 'diffusionism' (1977; 1987). Against this, I would urge us to acknowledge the striking effectivity of place not just the disciplinary regulation of space. Let me give two examples. Shilts' *And the Band Played On* documents such indifference on the part of the government of the United States towards what they saw as a gay plague between 1981 and 1987 that Kramer (1989) is certainly justified in concluding that the conservative Right, many of whom purport not to believe in natural selection, saw gay people as expendable. It was a gay man, Michael Callen, who preached the gospel of safe sex at a time when the Jeremiah's of the 'Just Say No' campaign were urging an insulting and deadly double standard: thou shalt have no sex other than marital heterosexual sex. It was a gay voluntary organization in New York, Gay Men's Health Crisis, which developed the techniques of buddying and of domestic support for the sick. It was the gay men in San Francisco who humanized Ward 5 (Wolf 1991) of the San Francisco General Hospital, turning it into a beacon of hospice care which has eased suffering in a myriad of other Western cities. It was from the counter-discourses and counter-conducts articulated around the solidarities of place and community (Geltmaker 1992), that effective and dignified health care sprang. Kayal (1993) describes this as an ethic of voluntary communalism and while I would look more at the political structures than render communalism more or less effective from place to place, I think Kayal is right. The technocratic discourses of centralized health care management can be murderous. The lesson will only be learned if we find ways to talk about and valorize these resistances.

Finally, let me turn briefly to the question of the passivity of the Third World. Within the same technocratic discourse of diffusionism, the West is presented as the source of all wisdom and Third World countries are berated for their fatalism in not recognizing the urgency of the Western agenda. Again, there are counter discourses and counter practices to recover (Kearns 2006). There have been, as Farmer (1992) shows in *AIDS and Accusation*, Haitian scientists resisting the racialized construction by scientists in the United States of Haitians as a risk group for AIDS sui generis. The Haitian scientists conducted their own research. They re-questioned HIV-positive

Haitians and found, contrary to the answers given to epidemiologists by illegal aliens in the United States, that many had indeed engaged in relevant (illegal) risk practices. They showed that HIV cases clustered in a part of the Haitian capital that served as a red-light district for American tourists. They questioned the lazy acceptance of speculative reports about blood rituals in voodoo and about green monkey business in Haitian brothels. Haitians in the United States demonstrated against and occupied the headquarters of the Food and Drugs Administration, urging that the embargo on people of Haitian origin giving blood be rescinded. The government of Haiti lobbied the government of the United States on the same issue. This is enlightened David confronting bigoted Goliath.

Much the same is true of certain Asian and African countries. Rejecting the expense and false security of testing, they have tried to develop appropriate technologies of prevention including comics, theatre and advertising (Reid, 1995). On occasion, these have shown a frankness that should shame the West with its own rather coy advertising campaigns. Yet when it comes to thinking about poor countries, the Western geographical imagination is willing to entertain almost any bizarreness given only that the picture presented is bad, fatalistic and bestial. Geographers should not collude with this, they should not, as did Gould (1993), offer inappropriately sweeping generalizations about, say, sexual mores sustained by little more than anecdote. In short, we can only counter Eurocentrism by looking for the agency of others, not by imagining it away (see Coronil 1996; Dussel 2000).

Conclusion

To some extent, of course, I am pushing against an open door in arguing that the works of Foucault be taken seriously by those writing historical studies of the subject matter and approaches of medical geography. Ogborn (1993a, 1993b) has taken up Foucault's work on the law, Driver that on poverty (1993) and Philo (1992; 2004) that on madness. These works are particularly strong on the links between geographical discourses and geographical practices in these areas of social policy as they pursue 'research in both the historical geography of ideas (with its concern for the spaces of knowledge production and consumption) and the historical geography of social institutions (with its concern for the spaces of control, correction and care)' (Philo 1996, 2–3). A moral-locational analysis was part of the institutional practices in these fields and through this lens of moral topography the city of dark and light took on an almost apocalyptic hue (Driver 1988). Philo (1996) builds on this work to reflect upon the sites and tracts of Reason and Unreason in the nineteenth-century city.

It is also clear that many of the topics which Foucault's historical works introduce are also being urged upon our sets of agenda through the practical politics of those who fight against the limitations and disciplines of medical discourses and practices. Bell (1995) is right to point to the sites of resistance and new configurations of power thrown up by new social movements as crucial in the re-education of political geographers. Dorn and Laws (1994) and Brown (1995) have said much the same of medical geographers. Some of the most exciting work on the geography of AIDS calls upon the methods of cultural studies and ethnography to draw these experiences into the reflections of geographers upon the epidemic (Geltmaker 1992; Brown 1994). Of course, these separations between discourse and practice were never complete, for Foucault's work was both informed by and in turn taken up by several groups in struggle.

I have suggested that there are links between the way we conceive of the history of medical geography and how we might continue to practise medical geography. Most often this relationship is rather loosely figured as the license given by history to some particular account of the essence of the sub-discipline and its wider relations with geography and medicine (Mayer 1990; Paul 1985; among many others). I would propose that medical geography should include within its remit the nature and consequences of medical-geographical ideas and strategies in popular as well as policy circles. It should include within its remit the ways space and place are taken up by medical disciplines and in resistances to those disciplines. It should concern itself with the power relations implicit and explicit in the areas of science, academia, medicine and health; in particular with respect to those ways its own authority requires and reinforces such power relations. To some extent, social theories offer sustained reflections upon such political topics and it would not be difficult to excavate the critical and utopian projects at the heart of structuration theory, symbolic interactionism and humanism as discussed by Jones and Moon (1993) in their review of the relations between medical geography and social theory. In this connection, it is fine to note the number of medical geographers who are engaging seriously with the geographical ideas formulated in Canada and now embedded in the Health for All strategy of the World Health Organization (Taylor 1990). This is not the place to enter into any serious consideration of their work but clearly the operationalization of health indicators (Hayes and Willms 1990), the notion of place-based communities (Coombes 1990; Dorn and Laws 1994) and the utopian idea of total health (Kearns 1993; Mayer and Meade 1994) raise important questions about the relations between discourses and practices.

As medical geographers explore the dispersed medical geographies that emerge when we follow medical geographical ideas out of the sub-disciplinary corral, I believe they might find Foucault's historical tactics of continuing value, and although I have presented three distinct tactics and even illustrated them with separate sets of Foucault's writings, the tactics I have described are certainly interrelated. A concern with what was done as well as said in the name of medicine, returned Foucault to that 'corporeal spatiality' which the much-vaunted objectivity of positivist medical science denied (1973, 199). An interest in the way institutions trained individuals led to an investigation of the way that training was carried out in the wider society both through the state (governmentality) and through individual self-examination (subjectivation). All of this was motivated by a wish to learn from and help those struggling to resist the normalizing powers of the state, medicine and a bourgeois conscience. Critical histories reveal the tangle of strategy and contingency that has produced the apparent naturalness of these present arrangements. Effective histories move us to recognize the legitimacy of challenges to this normality and also serve to further empower resistance. Here is an important set of issues for those tempted out into the wider world of discourses, practices, institutions and subjectivities where instances, analogies and homologies of medical geography may be found.

Acknowledgements

An earlier version of this chapter was prepared for a conference on 'Medical Geography in Historical Perspective' held at the University of Göttingen in June 1996. The conference was funded by the Deutsche Forschungsgemeinschaft (DFG) and organized by Professor Nicolaas Rupke. I am grateful to Professor Rupke and the DFG for the invitation and to the other conference participants for their lively engagement with the claims made in the chapter. I would also like to thank Eilidh Garrett and David Livingstone for advice on an earlier draft. The conclusion owes a great deal to the valuable comments I received from Steve Legg, Miles Ogborn and Chris Philo.

References

Barnett, T. and P. Blaikie (1992) *AIDS in Africa: its Present and Future Impact*. New York: Guilford Press.

Bell, D.J. (1995) 'In Bed with the State: Political Geography and Sexual Politics.' *Geoforum*, 25, 445–52.

Blaikie, P. (1978) 'The Theory of Spatial Diffusion of Innovations: A Spacious Cul-de-Sac.' *Progress in Human Geography*, 2, 268–95.

Blaikie, P., T. Cannon, T. Davis, and B. Wisner (1994) *At Risk: Natural Hazards, People's Vulnerability and Disasters*. London: Routledge.

Blaut, J.M. (1977) 'Two Views of Diffusion.' *Annals of the Association of American Geographers*, 67, 343–9.

Blaut, J.M. (1987) 'Diffusionism: A Uniformitarian Critique.' *Annals of the Association of American Geographers*, 77, 30–47.

Bouchard, D.F., Ed. (1977) *Language, Counter-Memory, Practice: Selected Essays and Interviews*. Oxford: Blackwell.

Brown, M.P. (1994) 'The Work of City Politics: Citizenship Through Employment in the Local Response to AIDS.' *Environment and Planning A*, 26, 873–94.

Brown, M.P. (1995) 'Ironies of Distance: An Ongoing Critique of the Geographies of AIDS.' *Environment and Planning D: Society and Space*, 13, 159–83.

Bunge, W. (1966) *Theoretical Geography*, Second edition. Lund: C.W.K. Gleerup (first edition, 1962).

Burchell, G. (1991) 'Peculiar Interests: Civil Society and Governing "The System of Natural Liberty".' In G. Burchell et al. (1991).

Burchell, G., C. Gordon and P. Miller, Eds (1991) *The Foucault Effect: Studies in Governmentality*. Chicago: University of Chicago Press.

Castel, R. (1991) 'From Dangerousness to Risk.' In G. Burchell et al. (1991).

Cliff, A. and Haggett, P. (1988) *Atlas of Disease Distributions: Analytic Approaches to Epidemiologic Data*. Oxford: Blackwell.

Coombes, Y. (1990) 'Public health: A Community Movement.' In G. Bentham, R. Haynes and I. Langford (Eds) *Fourth International Symposium in Medical Geography: Proceedings*. Norwich: University of East Anglia.

Coronil, F. (1996) 'Beyond Occidentalism: Toward Nonimperial Geohistorical Categories.' *Cultural Anthropology*, 11, 51–87.

Dean, M. (1994) *Critical and Effective Histories: Foucault's Methods and Historical Sociology*. London: Routledge.

Donzelot, J. (1991) 'The Mobilisation of Society.' In G. Burchell et al. (1991).

Dorn, M. and G. Laws, (1994) 'Social Theory, Body Politics and Medical Geography: Extending Kearns's Invitation.' *Professional Geographer*, 46, 106–10.

Draper, H. (1978) *Karl Marx's Theory of Revolution. Vol. 2, The Politics of Social Classes*. New York: Monthly Review Press.

Driver, F. (1988) 'Moral Geographies: Social Science and the Urban Environment in Mid-nineteenth Century England.' *Transactions of the Institute of British Geographers*, NS 13, 275–87.

Driver, F. (1993) *Power and Pauperism: The Workhouse System, 1834–1884*. Cambridge: Cambridge University Press.

Dussel, E. (2000) 'Europe, Modernity and Eurocentrism.' *Nepantla: View from the South*, 1, 465–78.

Farmer, P. (1992) *AIDS and Accusation: Haiti and the Geography of Blame*. Berkeley: University of California Press.

Fee, E. and D.M. Fox, Eds (1992) *AIDS: The Making of a Chronic Disease*. Berkeley: University of California Press.

Foucault, M. (1965) *Madness and Civilisation: A History of Insanity in the Age of Reason*. Translated by R. Howard. New York: Random House (original French edition, 1961).

Foucault, M. (1973) *The Birth of the Clinic: An Archaeology of Medical Perception*. Translated by A.M. Sheridan Smith. New York: Random House (original French edition, 1963).

Foucault, M. (1977) *Discipline and Punish: The Birth of the Prison*. Translated by A. Sheridan. New York: Pantheon Books (original French edition, 1975).

Foucault, M. (1985) *The History of Sexuality, Volume 2: The Use of Pleasure*. Translated by R. Hurley. New York: Random House (original French edition, 1984).

Foucault, M. (1986) *The History of Sexuality, Volume 3: The Care of the Self*. Translated by R. Hurley. New York: Random House (original French edition, 1984).

Foucault, M. (1988a) 'Return of Morality.' In L. Kritzman (1988a).

Foucault, M. (1988b) 'The Concern for Truth.' In L. Kritzman (1988a).

Foucault, M. (1991a) 'Politics and the Study of Discourse.' In G. Burchell et al. (1991).

Foucault, M. (1991b) 'Questions of Method.' In G. Burchell et al. (1991).

Foucault, M. (1991c) 'Governmentality.' In G. Burchell et al. (1991).

Foucault, M. (2004a) *Sécurité, Territoire, Population. Cours au Collège de France, 1977–1978*. Paris: Gallimard.

Foucault, M. (2004b) *Naissance de la biopolitique. Cours au Collège de France, 1978–1979*. Paris: Gallimard.

Geltmaker, T. (1992) 'The Queer Nation Acts Up: Health Care, Politics and Sexual Diversity in the County of Angels.' *Environment and Planning D: Society and Space*, 10, 609–50.

Glacken, C. (1967) *Traces on the Rhodian Shore: Nature and Culture in Western Thought from Ancient Times to the End of the Eighteenth Century*. Berkeley: University of California Press.

Gould, P. (1985) *The Geographer at Work*. London: Routledge.

Gould, P. (1993) *The Slow Plague: A Geography of the AIDS Pandemic*. Oxford: Blackwell.

Haggett, P. (1965) *Locational Analysis in Human Geography*. London: Edward Arnold.

Hartshorne, R. (1939) *The Nature of Geography: a Critical Survey of Current Thought in the Light of the Past*. Lancaster PA: Association of American Geographers.

Hayes, M.V. and S.M. Willms (1990) 'Healthy Community Indicators: The Perils of the Search and the Paucity of the Find.' *Health Promotion International*, 5, 161–6.

Jones, K. and G. Moon (1993) 'Medical Geography: Taking Space Seriously.' *Progress in Human Geography*, 17, 515–24.

Kayal, P.M. (1993) *Bearing Witness: Gay Men's Health Crisis and the Politics of AIDS*. Boulder CO: Westview Press.

Kearns, G. (2001) 'Constructions of the Social.' *Journal of Urban History*, 28, 98–106.

Kearns, G. (2006) 'The Social Shell.' *Historical Geography*, 34, 49–70.

Kearns, R. (1993) 'Place and Health: Towards a Reformed Medical Geography.' *Professional Geographer*, 45, 139–47.

Kearns, R. (1996) 'AIDS and Medical Geography: Embracing the Other.' *Progress in Human Geography*, 123–31.

Kramer, L. (1989) *Reports from the Holocaust: The Making of an AIDS Activist*. New York: St. Martin's Press.

Kritzman, L., Ed. (1988a) *Politics, Philosophy, Culture: Interviews and Other Writings 1977–1984*. London: Routledge, Chapman and Hall.

Kritzman, L. (1988b) 'Introduction.' In idem.

Latour, B. (1993) *We Have Never Been Modern*, Translated by C. Porter. London: Harvester (original French edition, 1991).

Legg, S. (2005) 'Foucault's population Geographies: Classifications, Biopolitics and Governmental Spaces.' *Population, Space and Place*, 11, 137–56.

Livingstone, D. (1992) *The Geographical Tradition: Episodes in the History of a Contested Discipline*. Oxford: Blackwell.

Mayer, J.D. (1990) 'The Centrality of Medical Geography to Human Geography: The Traditions of Geographical and Medical Geographical Thought.' *Norsk Geografisk Tidsskrift*, 44, 175–87.

Mayer, J.D. and M.S. Meade (1994) 'A Reformed Medical Geography Reconsidered.' *Professional Geographer*, 1993, 103–6.

McClintock, A. (1995) *Imperial Leather: Race, Gender and Sexuality in the Colonial Context*. London, Routledge.

Ogborn, M. (1993a) 'Ordering the City: Surveillance, Public Space and the Reform of Urban Policing in England, 1835–1856.' *Political Geography*, 12, 505–21.

Ogborn, M. (1993b) 'Law and Discipline in Nineteenth-century State Formation: The Contagious Diseases Acts of 1864, 1866 and 1869.' *Journal of Historical Sociology*, 6, 28–55.

Oppenheimer, G.D. (1992) 'Causes, Cases and Cohorts: The Role of Epidemiology in the Historical Construction of AIDS.' In E. Fee and D.M. Fox (1992).

Packard, R.M. and P. Epstein (1991) 'Epidemiologists, Social Scientists and the Structure of Medical Research on AIDS in Africa.' *Social Science and Medicine*, 33, 771–94.

Pasquino, P. (1991) 'Theatrum Politicum: The Genealogy of Capital – Police and the State of Prosperity.' In G. Burchell et al. (1991).

Paul, B.K. (1985) 'Approaches to Medical Geography: An Historical Perspective.' *Social Science and Medicine*, 20, 399–409.

Peet, R. (1985) 'The Social Origins of Environmental Determinism.' *Annals of the Association of American Geographers*, 75, 309–33.

Petersen, W., dir. (1995) *Outbreak*, Warner Brothers.

Philo, C. (1992) *The Space Reserved for Insanity: Studies in the Historical Geography of the Mad-business in England and Wales*. Unpublished PhD thesis. Department of Geography, University of Cambridge.

Philo, C. (1995) 'Journey to the Asylum: A Medical-geographical Idea in Historical Context.' *Journal of Historical Geography*, 21, 148–68.

Philo, C. (1996) 'Enlightenment and the Geographies of Unreason.' Unpublished paper prepared for the 'Geography and Enlightenment' conference, Edinburgh, July 1966.

Philo, C. (2000) '"The Birth of the Clinic": An Unknown Work of Medical Geography.' *Area*, 32, 11–19.

Philo, C. (2004) *A Geographical History of Institutional Provision for the Insane from Medieval Times to the 1860's in England and Wales: The Space Reserved for Insanity*. Lampeter: Edwin Mellen Press.

Procacci, G. (1991) 'Social Economy and the Government of Poverty.' In G. Burchell et al. (1991).

Reid, E., Ed. (1995) *HIV and AIDS: The Global Inter-Connection*. West Hartford CT: Kumarian Press.

Said, E. (1978) *Orientalism*. London: Routledge.

Shilts, R. (1987) *And the Band Played On: Politics, People and the AIDS Epidemic.* New York: St. Martin's Press.

Soper, K. (1995) 'Forget Foucault?' *New Formations*, 25, 21–7.

Stoddart, D., Ed. (1981) *Geography, Ideology and Social Concern.* Oxford: Blackwell.

Stoler, A.L. (1995) *Race and the Education of Desire: Foucault's 'History of Sexuality' and the Colonial Order of Things.* Durham, NC: Duke University Press.

Szreter, S.R. (1996) *Fertility, Class and Gender in Britain, 1860–1940.* Cambridge: Cambridge University Press.

Taylor, S.M. (1990) 'Geographic Perspectives on National Health Challenges.' *The Canadian Geographer*, 34, 334–8.

Thomas, W.L., Ed. (1956) *Man's Role in Changing the Face of the Earth.* Chicago: University of Chicago Press.

Watney, S. (1989) 'Missionary Positions: AIDS, "Africa" and Race.' *Critical Quarterly*, 31, 45–78, reprinted in idem (1994) *Practices of Freedom: Selected Writings on HIV/AIDS.* London: Rivers Oram Press.

Wolf, E. (1991) 'A Week on Ward 5A.' In N. McKenzie (Ed.) *The AIDS Reader: Social, Political and Ethical Issues.* New York: Meridian.

Chapter 22

Maps, Race and Foucault: Eugenics and Territorialization Following World War I

Jeremy W. Crampton

In a recent essay, David Harvey suggested that there were 'strong ideas' that geography could profitably explore in order to be more powerfully influential. One of these was geographic knowledge. How is geographic knowledge assembled or produced? In what ways does geographic knowledge become truth? How does what we know affect political or economic outcomes? In order to answer these questions Harvey pointed to a number of 'sites' that produce geographic knowledges, including cartography (see also Livingstone 2003). Harvey wondered why there has been such scant critical attention to the way cartography produces knowledge:

> It is years now since Foucault taught us that knowledge/power/institutions lock together in particular modes of governmentality, yet few have cared to turn that spotlight upon the discipline of Geography itself. (Harvey 2001, 217)

In this chapter I discuss how race and eugenics informed the problem of Europe's boundaries following World War I. Using Foucault's somewhat neglected discussion of state racism, I explore how cartographic knowledge can act as a technology of government and biopolitics through a spatialization of race. In particular I look at the role of eugenic science in the American preparations for peace.

Foucault and Cartography

'I am a cartographer' Foucault once ironically stated (cited in Deleuze 1988, 44), but it seems more likely that he occupied no consistent disciplinary position: 'I think I have in fact been situated in most of the squares on the political checkerboard, one after another and sometimes simultaneously: as an anarchist, leftist, ostentatious or disguised Marxist, nihilist, explicit or secret anti-Marxist, technocrat ... and I must admit that I rather like what they mean' (Foucault 1997, 113).

Yet as this book is designed to elucidate, Foucault was no stranger to the concerns of geographical knowledge. Although Foucault's explicit discussion of cartography is so scant as to be non-existent, his discussion of cartography in a more expansive sense (discourses on territorial partitioning [*quadrillage*], boundary making, and the politics of spatial knowledge) are central to his work on governmentality and biopolitics. That is, Foucault's work suggests that mapping is not just the manufacture or printing of a map at a particular time by a particular person, but that maps are both a product of and intervention in a distributed series of political knowledges.

Foucault's treatment of space is intimately bound up with the activation of geographical knowledges. While these knowledges vary at different times, the nineteenth century saw an explosion of concern in two areas: 'health' (including sexuality) and crime. Why? Foucault's now familiar argument is that modern societies shifted from sovereign rule to one of governmental concern at the level of the population, or biopolitics (Burchell et al. 1991). This fits well with cartography because it also traditionally focuses at the level of the population or group rather than the individual.

Foucault occasionally explicitly links mapping and government, and the need to have a rational plan to manage space. Part of this is knowing and recording where things are, which keeps spaces manageable and secure. He cites the rise of 'new mapping and the closer surveillance of urban space' (Foucault 1988, 142) in the development of policing techniques, and the need for the 'security' of space (Foucault 2004).

His work from the mid-1970s is particularly suggestive, for example on spatial partitioning [*quadrillage*]. Under the rubric of normalization difference can result in 'spatial partitioning and control' (Foucault 1979, 195ff.; 2003a, 43ff.). In Foucault's well-known comparison, whereas the response to the leper was an *expulsion* of the infected, the response to the plague was 'a segmented space, observed at every point' that is, instead of a 'pure community ... that of a disciplined society,' with everybody in their proper place (Foucault 1979, 197–8). With the plague came the realization that the danger was already inside the gates (hence the need for a coterminous and extensive surveillance) and could not be simply expelled. He makes a similar point in a March 1976 lecture that 'the spatial layout of the town' can be a mechanism of discipline and policing. Speaking of nineteenth century 'working-class housing estates', he says that 'one can easily see how the very grid pattern, the very layout, of the estate articulated, in a sort of perpendicular way, the disciplinary mechanisms that controlled the body' (Foucault 2003b, 251). Here the way space is 'geo-coded' through mapping (Pickles 2004; Rose-Redwood 2006) allows space to be better controlled.

In 'Questions on Geography' (chapter 19 of this book) there is a brief reference to maps as instruments of power. In his interview 'Space, Knowledge and Power' he pointed to the important work of the engineers and cartographers at the École des Ponts et Chausées who 'thought out space' (Foucault 1984, 244) such as Charles Minard (Friendly 2002; Robinson 1967). But generally these cartographies remained latent in Foucault's work.

Recent studies have filled in some of the cartographic blanks (Harley 1989; Heffernan 2002; Palsky 2002; Petto 2005; Pickles 2004; Robinson 1982; Sparke 1998). We now know that the nineteenth century was a period of almost ceaseless invention cartographically, a time in fact when many of the maps we still use in modern GIS were invented. What is striking about these maps was that they were directly tied to the governmental concerns about crime and health, that Foucault discusses. Crime mapping emerged in the early nineteenth century along with the police, and as Koch has documented, public health mapping attended not only the plague crises of the seventeenth century, but developed into a rich subdiscipline in the nineteenth century (Koch 2005). These maps and the knowledges they produced were part of the technology of government.

But there is one area that has attracted less attention and that is Foucault's work on the biopolitics of race (Foucault 2003b). Picking up from his rather brief remarks in *The History of Sexuality* (Foucault 1978) Foucault documented a historical shift from a struggle or war between races to a struggle to maintain the purity of race, or what he called 'state racism' (2003b, 82):

> I think that racism is born at the point when the theme of racial purity replaces that of race struggle, and when counterhistory begins to be converted into a biological racism ... [it is used] to preserve the Sovereignty of the State ... by medico-normalizing techniques ... State sovereignty thus becomes the imperative to protect the race. (2003b, 81)

Racism here is not just hating those who are different, but of imposing a 'break (*coupure*) between what must live and what must die', a fragmentation on the continuum of the biological. If life exists as a continually varying diversity, then state racism applies a series of techniques for identifying and ordering groups that are either 'good or inferior' for the health of the population (2003b, 254). These breaks were initially applied between one population and the next, but then also within the population. As with spatial partitions, the problem is that of making these racial orderings from the continuum and of finding or deciding on the basis for identity.

Foucault does not bring out the geographical component of state racism, but as with both crime and health, contemporary writers certainly did. While race maps were not unknown in the nineteenth century (Winlow 2006), the political upheavals of World War I and the subsequent disarray of Europe's political borders provide a particularly powerful instance of how race can be used to 'think out space'. As I discuss below, by choosing race as the criterion (rather than nationality or religion), an apparently intermingled and chaotic space could be revealed as having order. The goal of those involved in redrawing Europe's borders was to identify the distinct races that would occupy racially homogenous territories. Only in this way, they thought, would territories remain stable and peace be guaranteed.

Mapping Race: World War I and the Inquiry

In 1917 the Unites States entered the war and simultaneously began preparing for peace by establishing a secret research group called the Inquiry. Instituted by President Wilson and headquartered at Isaiah Bowman's American Geographical Society (AGS) the Inquiry was charged with determining American policy to be used at the presumptive peace conference (eventually held in Paris from January–

June 1919). The work of the Inquiry offers an instructive instance of Foucault's state racism, in that it understood its mandate as one of geopolitics and race. To redraw the map of post-war Europe the Inquiry sought to isolate both *identity* and *territory*. The peoples or population inside bounded segments of space (regions) should be all alike in the crucial respects. While language had partly been a guide to this since the nineteenth century (Dominian 1917) the ultimate goal was *racial partitioning*. If these territorial units could be identified then this would lead to stable sovereign states across Europe who would be unable to claim extra territories on the basis of racial affiliation of occupants. In other words, not only could distinct natural races be identified, but if their areal extent could be unambiguously determined this would yield viable and peaceful sovereign states.

Bowman's position as Director of the AGS gave him access to many prominent scholars. At its height the Inquiry consisted of about 150 men and women (Gelfand 1963) who worked from September 1917 to December 1918 writing reports and compiling data. In December 1918 they sailed with President Wilson to France to take part in the peace conference itself, where they would staff the territorial commissions. This massive effort has received very little attention from scholars, particularly in terms of its spatializations of race. Yet not only did they have a hand in an important pre-war speech (the Fourteen Points) which was seized upon as the basis for peace by many nations across war-torn Europe (and even the Germans themselves) but they produced many of the key policy positions for the negotiations.

The Inquiry's tasks came at a time when the understanding of geopolitical borders was in transition from older, colonial-based strategic boundaries, to a more modern one based on the qualities and attributes of human populations (Brigham 1919). Those who advocated a strategic approach emphasized that boundaries should be defensible. Good boundaries would run along mountain tops, rivers, or other topographical features (Holdich 1916). While other factors (such as race) could have a role, if a strategic boundary could be found it was preferable. Some advocates of this view, such as Mackinder, also emphasized the balance of power. In 1915 Mackinder argued that there would be little 'ideal map-making' at the peace conference, but it would be rather a case of clipping Germany's power (Wilkinson et al. 1915). In the Inquiry, the strongest advocate of this was Douglas Johnson, a geomorphologist at Columbia University and a Harvard colleague of William Morris Davis. Johnson was responsible for the border between Italy and Yugoslavia that caused something of a crisis at the peace conference. Writing about this border in December 1919 he claimed that:

> ... the natural or geographic frontier lay on a high mountain ridge forming the backbone of Istria and located close to its eastern shore; that all economic relations of the people west of that divide may lie most naturally with the Italian side of the mountain; and, hence, that it may be wisest to push the international boundary away from the racial boundary and on up the slope of the mountain. (Johnson 1919, 516)

Unfortunately the Yugoslavs disagreed, and, with Bowman's support, successfully resisted this border line, causing the Italians to pull out of the conference for a month.

Figure 22.1 Conflicting ethnic claims in the Balkans
Source: NARA RG256 Entry 52 Folder A1–V, Map T-12

The border was not actually resolved until several years later (Sluga 2001) and is still a sensitive frontier.

A second position, the scientific rationalists or neoliberals, rejected geopolitics and held that peaceful coexistence was achieved by following the 'will' of the people through national self-determination. This was the view of President Wilson. In other words geography was not so much defensible features or the *realpolitik* projection of political power, but of populations. As Foucault observed, these populations could be known and identified in their territorial extent (Foucault 2004). There were many in the Inquiry with this view, including Isaiah Bowman, although Sidney Mezes, its rather weak Director, had his doubts.

About eight months after it had been constituted the Inquiry set out its primary task:

> Make a racial map of Europe, Asiatic Turkey, etc., showing boundaries and mixed and doubtful zones.

> On basis of [the above] draw racial boundary lines where possible, i.e. when authorities agree; when they disagree select those *we* had best follow; when *these* disagree map the zone of their disagreement; study density and distribution of peoples in these zones.[1]

The 'authorities' alluded to above were maps produced by various countries depicting the populations of Europe. As Inquiry member Walter Lippmann wrote to Wilson's advisor E.M. 'Colonel' House:

> We made eight maps showing the distribution of nationalities; four of these maps were copied from neutral European or American authorities, while the other four represented the moderate patriotic claims of the Serbs, Bulgars, Greeks, and Albanians respectively. We then put all eight maps together and coloured up the areas which nobody disputed. This left almost all the Balkans in dispute. We then made a second map using only the neutral authorities. This time the areas in dispute were really narrowed down to a place where they were manageable. It proved pretty conclusively that 60–80% of the territory in the Balkans in dispute was put in dispute by propaganda.[2]

As Figure 22.1 shows, what the Inquiry found is that there was very little space in the Balkans that was not disputed one way or another, which forced them to drop from consideration the contesting claims and revert to what they considered 'neutral' European and American maps. These 'neutral' European maps were contrasted with the 'propaganda' of the Serbs, Bulgarians, Greeks and Albanians, allowing the latter to be easily dismissed.

Mezes expressed his discomfort with this scheme. Writing to Lippmann, who was then in Europe, Mezes observed that in the Balkans 'I cannot get away from the thought that topography with its economic and strategic implications, is much more

---

1   'A Preliminary Survey' Inquiry Doc. 893, NARA, undated but probably late July 1918 and not November 1917 as indicated in FRUS (FRUS 1942-7: Vol. I(20)). Emphasis added.
2   Lippmann to House June 7, 1918, NARA/House.

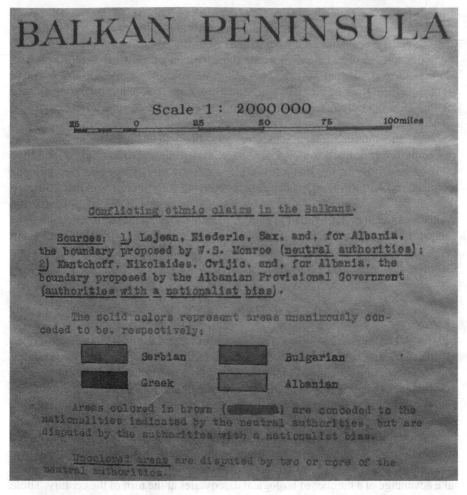

Figure 22.2   Detail of Figure 22.1

important than racial distribution ... it seems to me we must come back to the solid basis indicated by topography.'[3]

Clearly the Inquiry had no wish to impose arbitrary lines across Europe but rather ones that were scientifically justified. President Wilson had made only vague pronouncements favouring 'self-determination' of European countries, but by the end of 1917 the Inquiry had produced a 'Preliminary Survey'. Wilson and House used this for the January 1918 Fourteen Points speech in which Wilson spoke of 'rectifying' borders 'along clearly recognizable lines of nationality' (Wilson 1966–1994). The Inquiry now had the task of identifying these 'clearly recognizable' lines.

---

3   Mezes to Lippmann, November 16, 1918, NARA/Lippmann.

This meant the collection of an impressive set of reports, documents, fieldwork, maps and statistics for European, Middle Eastern and African countries. While many existing maps were purchased or extracted from the AGS library, the Inquiry also made hundreds more of its own under the direction of Bowman's old professor, Mark Jefferson, as Chief Cartographer (Martin 1968). Most of the third floor and all of the fourth floor of the AGS's headquarters were devoted to this research, and the President honoured them with a visit in October 1918 (American Geographical Society 1919).

Thus, the question for the Inquiry was simultaneously one of *knowledge*, especially territorial and spatial knowledge, and second a *rationality* or reasoned basis on which to deploy that knowledge. Both the knowledge and the rationale were centred on race. Perhaps the most interesting information was that collected through fieldwork, either that done by its own members (Douglas Johnson and Walter Lippmann both travelled to Europe and interviewed many representatives of countries with disputed territory) or by hiring certain specialists. It is the relationship of the Inquiry to one of these, Charles Davenport and his Eugenic Record Office that illustrates the operation of state racism.

Charles Davenport and the Eugenicists

Charles B. Davenport (1866–1944) was a leading eugenicist who has been described as 'the chief American advocate of eugenics' (Allen 1986, 225). The ideas behind eugenics were first articulated by Francis Galton in the 1880s. They centred on the idea that racial qualities could be improved through control of breeding, and that the causes of many of society's ills were due to hereditary defects. Thus alcoholism, lack of social integration and rebelliousness, and even criminality were due to inherited birth defects that could be eliminated in a rational breeding program, including forced sterilization. For eugenicists, the solution to these social problems lay in biological corrections, rather than social corrections (Marks 1995). Eugenicists were able to take advantage of a number of biological discoveries at the turn of the century. These included the discovery of the ABO blood types, and the rediscovery of Mendel's laws of inheritance that showed that a new organism's genes were inherited half from the paternal and half from the maternal ancestry.

Mendel had proposed these laws based on work he had conducted with plants. Mendel had inferred that genes are present in two ways in organisms; its genetic constitution or genotype and the physical characteristics of these genes expressed in the phenotype (Marks 1995). Mendel also inferred that there were 'units' of inheritance, and it was these units that eugenicists latched on to. If those units that caused social ills could be eliminated through selective breeding (including forced sterilization and the rational choice of marriage mates, or what Davenport called 'falling in love intelligently' (Marks 1995, 81)) then this would improve the 'race'.

Additionally, the new Hardy-Weinberg law showed that these Mendelian laws of inheritance applied at the level of the population. That is, all things being equal, the genetic diversity (the gene pool) of a population will be passed on to the next generation. However, the gene pool could be profoundly altered by changes in the gene flow, for example through intermarriage or by selective breeding; and control of both of these were policies advocated by eugenicists. Immigration laws passed during the 1920s were heavily influenced by a number of eugenicists both inside and outside of government.

Davenport's laboratory, the Eugenics Record Office, was established in 1910 at Cold Spring Harbor and continued until the end of 1939, when war in Europe left the lab in a politically untenable position given its close ties to German eugenicists. During its heyday, however, the ERO received funding from major institutions, such as the Carnegie Institution of Washington (CIW), and philanthropists such as John D. Rockefeller Jr. Davenport's most significant early benefactor was Mrs. E.H. Harriman, the widow of a railroad magnate. Harriman had inherited an estate estimated to be worth approximately $70 million (Allen 1986, 234), and Davenport was able to convince her to establish the ERO at least partly because her daughter was a student at Cold Spring Harbor in 1906. All told, between 1910 and 1918 Mrs. Harriman contributed upwards of half a million dollars to Davenport's eugenic enterprise (Kevles 1985, 55) (about $5.5 million in buying power today). Harriman funded Davenport until 1917, when the CIW agreed to take over the operating costs of the ERO. Davenport acted as Director, while his friend Harry H. Laughlin (1880–1943) was superintendent. The ERO was a significant force in the eugenics movement, and provided both a point of focus and scientific credibility to the wider eugenics movement in the Unites States and abroad.

In addition to being a research centre the ERO also had the purpose of collecting and centralizing data on human heredity. It pursued this goal through its *Trait Book*, a complete listing of traits that might be found in families that were collected by ERO caseworkers and recorded on 3 X 5 cards (over 1 million by 1939) (Allen 1986). These fieldworkers were trained in summer courses led by Davenport and Laughlin, and had trained 258 workers (mostly women) by 1924 (Bix 1997). Davenport was able to draw on the services of these fieldworkers in his work for the American Inquiry in 1918, particularly Mary T. Scudder. The data collection was meant to be scientific and objective, but was based on subjective impressions and 'community reactions', which were a 'euphemism for "common gossip"' (Allen 1986, 243). Visiting committees sent by an increasingly worried Carnegie Institute twice slammed the ERO for relying on these impressionistic records, eventually recommending in 1935 that the ERO was wasting resources and be wound down.

But this was still far in the future during the 1910s and 1920s. The Carnegie provided extensive funding for the ERO, and eugenics attracted few critics (such as anthropologist Franz Boas). Davenport argued that debilitating weaknesses were Mendelian traits that could be diagnosed. One of the major traits was the extremely general 'feeblemindedness', which was inherited: 'it follows that two parents who are feeble-minded shall have only feeble-minded children and this is what is empirically found' (Davenport 1921, 393). Feeblemindedness could thus be ascribed to a large number of 'abnormalities' that were detrimental to the state. It is important to realize that Davenport and Laughlin were heavily involved in political activity. The ERO sought to have the Census Bureau collect eugenic data in 1930, they advocated a pan-American eugenics society, they drew up forced sterilization laws (passed in 35 states by 1935 it resulted in 35,878 sterilizations or castrations by 1940, Black 2003), and they pushed to have American overseas consulates perform eugenic tests on prospective immigrants (Allen 1986, 249–50). As discussed below, much of this activity during the 1910s and 1920s was focused on passing restrictionist immigration laws such as the quota-based 1924 law which restricted immigration from southeastern Europe.

For Laughlin, immigrants from these areas – along with Jews – were degrading the Nordic stock of America, and special vigilance was required to prevent 'race-crossing' or inter-marriage between southeastern Europeans and white Americans. Laughlin and Davenport often argued that such inter-marriages watered down the good qualities of the one stock by the degeneracy of the other, and it explains their special concern with people from this region (Davenport and Steggerda 1929). In fact, as geneticists and breeders know the intermixing of genes is actually *healthy* for the genetic pool of the population, promoting 'hybrid vigour'. Laughlin and Davenport's policies, insofar as they are based on biological reasoning, are therefore completely opposite from actuality, and their ardour to create 'pure' races would actually result in very weak populations genetically.

Nevertheless, these views were not unusual at the time. In 1916 one of the best-known works on eugenics, *The Passing of the Great Race*, by Madison Grant, was published. It went through numerous editions and argued that there were three distinct races in Europe: the superior 'Nordic' race (with phenotypical characteristics of blond hair and high brows) and the lesser Mediterranean and Alpine races (Grant 1932). In order to prevent the eclipse of this Nordic race, Grant forcefully advocated sterilization:

> This is a practical, merciful and inevitable solution of the whole problem and can be applied to an ever widening circle of social discards, beginning always with the criminal, the diseased and the insane and extending gradually to types which may be called weaklings rather than defectives and perhaps ultimately to worthless race types. (Grant 1932, 51)

Grant's work was highly influential and he was close to both Laughlin and Congressman Johnson, as well as a councillor of the American Geographic Society from 1913–1935 (Wright 1952). Indeed, Grant's maps of the 'Nordic race' were published in the *Geographical Review* (Grant 1916) edited by Isaiah Bowman. Both Grant and Laughlin provided the government with race data (Laughlin even travelled to Europe with Secretary of Labor James Davis as a Special Immigration Agent in 1923 to help prepare for the 1924 law, see Black 2003).

In this political activity we find the concerns of governmentality underlined by Foucault; abnormality, deviancy, moral imbecility, marriageability, patriotism, and so on that affect the quality of the 'race'. State racism is the official manifestation of eugenic concerns. While there have long been struggles between races (of anti-Semitism since the Middle Ages for example) Foucault suggests that what underpins modern government is racial purity. It is here then that Foucault is particularly valuable in his analysis of a seemingly unrelated set of knowledges that might otherwise be dismissed as some long-ago racism.

State racism is essentially the view of the eugenicists; not so much racial *elimination*, but the *purification and protection* of the worthy stock from the unworthy. Eugenics can be understood in this light as a technology of government that focuses on the health of the population. While the Nazi program is perhaps one of the most obvious examples – a 'paroxysmal point' Foucault calls it in several places (2003b, 259, 260; see also Foucault 1978, 149) – this racism is found in all or nearly all modern states and was not a product of a single monstrous state. As Black has argued, the Nazi state in fact enjoyed a rich relationship with American scientists (Black 2003). Certainly the ERO had major connections to the German race-hygiene (*Rassenhygiene* or eugenic) movement, and its *Eugenical News* often translated German articles, provided positive reviews of German research, and after 1933 (when the Nazis came to power) lauded their model sterilization laws, and even helped to distribute the German eugenic film *Erbkrank* (*The Hereditarily Diseased*) in America in 1937 (Black 2003). And on his release from prison, no less than Adolf Hitler wrote a fan letter to Grant (Black 2003). Eventually these German connections came to so embarrass the CIW that the ERO was closed on the last day of 1939.

Race-based Mapping and the Inquiry

During the war, Davenport held the rank of Major in the Sanitary Corps of Office of the Surgeon General. Davenport was officer in charge of a Division of Anthropology in the Department of Medical Records. Its duties were to make biometric measurements of recruits and to 'assist the War Department in all questions about racial dimensions and differences' (Anonymous 1918, 112). His appointment was arranged by the Committee on Anthropology of the National Academy of Sciences (NAS). Davenport's connection to the NAS and the National Research Council (as its Vice-Chairman of the Committee on Anthropology) made him an obvious candidate to assist the Inquiry.

On February 5, 1918 Isaiah Bowman wrote to the Chairman of the National Research Council (NRC). The NRC had been formed by Presidential Executive Order in 1916 by members of the NAS to coordinate wartime research. It was first chaired by the astrophysicist George E. Hale (1916–18) and then Berkeley palaeontologist John C. Merriam (1918–19) (Cochrane 1978; Kevles 1968). Bowman himself would be chair in the early 1930s (Smith 2002). During the war the NRC actively participated in the war effort, and 'brought about an unprecedented and fruitful collaboration of university and industrial scientists with the military' (Kevles 1968, 431).

Bowman wanted to know if its Committee on Anthropology could perform 'an intensive anthropological study of European peoples'.[4] By this Bowman meant the territories occupied by the distinct peoples of Europe:

> The specific work which we had in mind for the Committee to do was the preparation of reports on each of the Ethnographic units of Europe. It is not so important at this time to know the internal characteristics of each group as to the know *the location of boundary lines between the different groups*. A great deal of critical work is required for the determination of these lines...the more closely the studies are made to apply to the political questions of the time the more valuable they will be.[5]

Hale asked Davenport in the Anthropology Committee to help.[6] Davenport immediately recruited both Madison Grant and Frederick Hoffman (a medical statistician and anthropometrist).

Bowman was desperate for a good 'ethnographer' that is, someone who could study ethnic or racial groups and derive 'ethnographic units' of territory. Part of the solution would involve fertility, because it would help determine just how much 'ethnographic penetration' had been and would be made in the disputed areas.[7] If the ethnic and racial makeup of these areas could be known along with the likely population growth rates, this factor could be used to allocate territory. Wilson's emphasis on 'clearly recognizable lines of nationality' (Wilson 1966–94, vol. 45, 459) raised the issue of what constituted a viable self-determining sovereign state. As advocates of a neoliberal scientific rationality, the Inquiry understood that racially based borders would bring conflict and territorial claims in Europe to an end – an 'ethnographic cartography' (Noyes 1994; Palsky 2002). And the Inquiry would resolve territorial disputes using this ethnographic cartography all the way through to the Peace Conference itself.

The NRC anticipated that race-based mapping would be tricky, not because *racism* might be a factor, but rather because some scholars might not have the required objectivity. Hale warned Bowman that the 'personal and race' prejudices of the Smithsonian's Alěs Hrdlička were affected by his 'nationality' (although Hrdlička's family had emigrated to America when he was thirteen, Montagu 1944).[8] As with the Inquiry, the NRC were leery of others' racial prejudices while being blind to their own. Hale also invited William Ripley, author of *The Races of Europe*, a book as racist as Grant's own, to help in the data collection.[9] Bowman sent a brief

---

4   Bowman to Hale, February 5, 1918, NARA/Hale.
5   Bowman to Hale, February 13, 1918, NARA/Hale. Emphasis added.
6   Hale to Bowman, February 14, 1918, NARA/Hale.
7   Bowman to Davenport, June 12, 1918, NARA/Davenport.
8   Hale to Bowman, February 25, 1918, NARA/Hale.
9   Ripley declined the invitation, despairingly pointing out that 'during the last few days it does appear as if our entire civilization hangs trembling in balance. The possibility of pan-Germanism over running the continent is too appalling to leave room for anything except contemplation of salvage work.' Ripley to Hale, March 27, 1918, NARA/Hale. Ripley is referring to the German spring offensive ('Kaiserschlact'), which made the deepest territorial

letter to Hale complaining about the invitation, but notably *not* because of Ripley's racial views but rather that Hale had mentioned Colonel House's name (the Inquiry supposedly operated under conditions of secrecy).

Nevertheless, the NRC did commission Hrdlička to undertake ethnographic research on a strip of territory from the Baltic to the Aegean. For Bowman, 'the vital ethnographic problems of Europe are very largely concentrated in this strip', it was the very 'storm centre'.[10] It included Poland's eastern border (the Chelm or Kolm region), a disputed area between Romania and Bulgaria, Albania's borders, Macedonia, parts of the Eastern front, Lithuania and the Baltic provinces.[11]

The focus was squarely on eastern and southeastern Europe. Bowman reiterated that they sought the 'ethnic character' of these areas and that it 'should be related to religion and the local sense of nationality and affiliation so as to make the end result contribute to the political purposes of the Inquiry'. Bowman also wanted extensive immigration and emigration data for each area.[12] The Inquiry's political purpose of course was to prepare for the Peace Conference by grounding their position in detailed ethnographic reports for each disputed area. In Paris, these reports were distilled into an important policy document known as the 'Black Book' which was then used in the Territorial Commissions (Crampton 2006).

If the Inquiry and the NRC saw fit to enlist Davenport and the ERO it was not so much due to an inadvertent racism but a fundamental assumption that people could be categorized into a small number of distinct natural races occupying identifiable segments of territory. If virulent racists like Madison Grant could be on the Council of the AGS for decades and publish in the *Geographical Review*, it was because race was an accepted explanatory variable.

Davenport well understood what Bowman wanted; writing to an assistant he repeated Bowman words: he needed 'not only the racial character in the narrow sense of the population of this district but also of the prevailing religion, language and so much of the history as will throw some light upon the local sense of nationalist and affiliation'.[13] This 'narrow sense' for Davenport meant the biological race of the inhabitants as observable through skin colour, face and hair characteristics and head-shape. Tremendous effort was expended in imposing a racially distinct order on the chaotic and messy populations of Europe. But race would always tell; Davenport approvingly quoted Reclus 'the Teutonic invaders have remained what they were 700 years ago, aliens in the land (Reclus '94)'.[14]

Ethnographic cartographers had long sought the holy grail of a clear and undisputed map of European races. Grant (1916), Davenport (1911), Dominian (1917), Laughlin

---

gains by either side since 1914. By July and August however the Allies had regained much of this ground.

10 Bowman to Hrdlička, March 18, 1918 and August 31, 1918, NARA/Hrdlička.
11 Bowman to Davenport, February 19, 1918, APS/Davenport.
12 Bowman to Davenport, February 19, 1918, APS/Davenport.
13 Davenport to Mabel Earle, February 25, 1918, APS/Davenport.
14 Davenport to Bowman, March 1918, APS/Davenport.

(United States Congress 1924) and Bowman (1921) were just a few of the scholars who spatialized race with ethnographic maps. As the Serbian geographer Jovan Cvijić put it in the *Geographical Review*, maps allowed you to identify the dominant 'zones of civilization' (Cvijic 1918) even in the messy Balkans.

By mid-March Davenport had sent in his first report on the Baltic Provinces ('The Finnish Peoples') and further reports in August and November 1918. The initial report provided detailed summaries on various Baltic peoples, the Finns, the Germans and the Lithuanians. For example the latter were identified as having Aryan qualities 'features fine, the very fair hair, blue eyes and delicate skin distinguish them from Poles and Russians … tall … Ripley says the pure [head] type approximates quite closely to the Anglo-Saxon model, that is dolichocephalic'. The Letts, meanwhile, 'are a purer blond type than the Lithuanian…[one authority] says they have elongated faces, noses long and straight, mouths small; in general the women are rather pretty, have blonde hair and blue eyes.'[15]

But the Inquiry's focus on southeastern Europe would play into Davenport's hands. Using the Inquiry/NRC relationship as cover, Davenport soon started focusing on foreign nationals in the USA, rather than in Europe. These included Jews, Albanians, Bohemians, Bulgarians, Yugoslavs, Polish and Greeks. Davenport asked Bowman to help hire his former fieldworker and co-author Mary Scudder to 'locate the leading men who represent the different Central European races and … to list the societies, churches, newspapers and special libraries of such races'.[16]

Davenport had two purposes in redirecting the focus of the Inquiry. Both of these related to Davenport's concerns over people from southeastern Europe. First, he saw an opportunity to exploit these organizations for purposes of influencing opinion in Europe. He wrote to Bowman that the recent upsurge of immigrants from southeastern Europe 'could be exerted on their blood relatives "at home"' and that the way to do this would be to compile a list of foreign nationals in the United States 'who are absolutely loyal and reliable, who are capable of effective propaganda work, who could undertake missions to their native countries … loyal men or women, good talkers and organizers'. This should be done secretly, and Davenport warned that it would be expensive, but he at least was prepared to face the cost: 'even though it costs $100,000,000 to send 10,000 men to these countries it were well worth the expense'.[17] In effect Davenport was proposing a massive fifth column of propagandists, or missionaries as he called them:

> It is urged that we send an army forthwith to Russia. Let us send first a small body of missionaries to insure that we shall receive a welcome when we come. After the missionaries have explained matters we may send our generals, munitions and, later, soldiers to fight with the Russians and not against them. If our army should go to Russia now, while it is under the influence of German lies, the coming of our troops would be

---

15 NARA Inquiry Document 110.
16 Davenport to Bowman, April 17, 1918, NARA/Davenport.
17 NARA/Inquiry Document 110.

the signal of armed resistance to us on the part of those who may readily be made to be our friends.[18]

It is difficult to credit this as a serious proposal. Yet Davenport also wrote to Sidney Mezes (Director of the Inquiry) with the same idea. While Davenport was only too ready to denounce the peoples of southeastern Europe as 'slovenly' and 'more given to crimes of larceny, kidnapping, assault, murder, rape' and (his bete noir) 'sex-immorality' (quoted in Kevles 1985, 41) he was not afraid to exploit those in America who expressed friendship towards the United States. 'During the past few weeks my assistant [Scudder] and I have been cataloguing the organizations which these people have formed in New York City.'[19] Davenport suggested focusing on people from Russia, the Baltic provinces and the Balkans.

Thus, Scudder's reports to Davenport included assessments of the 'loyalty' of the Jews or Slavs. In one interview for example, Scudder met with an Anthony Tanaskovich, who was business manager of the *Jugo-Slav World*: 'he is a tall man with a somewhat longer face than some Slavs. There is almost a tinge of red in his brown hair. He is a nice clean looking man ... he told me Frank Zotti [editor of a rival publication] ... is a paid Austrian spy.' In her interview with Dr. Nikolas Papantonopulos, a dentist in Chicago, she describes him as 'of medium height and has a large, long head and a smooth, somewhat yellow tinged skin, dark brown eyes and hair, a large nose, and thin ears that stand out from his head like wings'. Many of the interviews include either nationalistic claims to disputed territory – Mr. Angelinoff, a Bulgarian claimed that 'Macedonia is overwhelmingly Bulgarian' or ethnic slurs against other peoples – Angelinoff calls the Serbians 'very devils in sheep's clothing'. Mr. Bagdziunas, a Lithuanian, is quoted as saying that 'the men who especially tried in every way to avoid the draft were first the Greeks, then the Italians, and then the Lithuanians'. 'He hates a Jew,' Scudder writes, 'he says that they lack patriotism and loyalty and that is the reason they have no country ... he declares that as a rule Lithuanians are a calm and even tempered people ... strongly Socialistic ... there are about 80,000 Lithuanians' in Chicago.[20]

These extensive reports are hardly scientific documents and would be impossible to substantiate. As with the 3 X 5 cards the ERO collected, the data were a mix of subjective impressions and gossip. Certainly their use for eugenical purposes would be severely limited. It is doubtful if the Inquiry made much of them. But Davenport had a second reason for collecting data under the imprimatur of the National Research Council and the Inquiry: immigration reform.

---

18 Ibid.
19 Davenport to Mezes, May 6, 1918, NARA/Davenport.
20 NARA Inquiry Doc. 110.

Race and Immigration Restriction

During the 1910s and 1920s Davenport, the ERO, and the indeed the Carnegie Institute worked to a larger agenda to restrict immigration into the United States. Laughlin, Davenport, Merriam at the Carnegie, and Congressman Albert Johnson, the chair of the House Committee on Immigration and Naturalization, assembled data to show the genetic inferiority of people from central and southern Europe. This region was often ranked as having a significantly 'degenerate' stock compared to western Europe (Sluga 2002). In the 1896 words of the former Superintendent of the Census and immigration restrictionist Francis Walker they were 'beaten men from beaten races' (Hannah 2000, 109; Walker 1896). These political activities helped pass the quota-based immigration act of 1924 (the Johnson-Reed Act) that turned the clock back to the 1890 census and restricted immigration of southeastern Europeans.

A look at the immigration picture of the early twentieth century illustrates this concern. The numbers of immigrants entering the United States not only rose dramatically, but underwent a shift in origins (Table 22.1).

In 1890 the proportion of foreign born nationals from eastern Europe was only 5.5%, but by 1910 it stood at 22% and by 1920 had risen to over 26%. Including 'Mediterraneans' one of Madison Grant's 'lesser' races, over 40% of the foreign born in America were from southeastern Europe by 1930. In response, immigration restrictionists moved to influence public policy, first by requiring new immigrants to pass a literacy test, and then implementing immigration laws in 1921 and 1924. The Harvard geographer Robert DeCourcy Ward (President of the AAG in 1917), and a longtime eugenicist evoked a 'symbolics of blood' (Foucault 1978, 148), proclaiming that 'the real, fundamental, lasting reasons for [the 1921 law's] continuance is biological' (Ward 1922b) and not socio-economic. Ward was by profession a climatologist who taught both J.K. Wright and Mark Jefferson, and was

Table 22.1   **Immigrants into the United States from Southeast and Eastern Europe**

Foreign born population, United States					
Year	Total	SE. Europe	%	E. Europe	%
1850	2233602	9672	0.43	1520	0.07
1860	4138697	32312	0.78	10586	0.26
1870	5567229	93824	1.69	63408	1.14
1880	6679943	248620	3.72	182371	2.73
1890	9249547	728851	7.88	512464	5.54
1900	10341276	1674648	16.19	1134680	10.97
1910	13515886	4500932	33.3	2956783	21.88
1920	13920692	5670927	40.74	3731327	26.8
1930	14204149	5918982	41.67	3785890	26.65

Source: http://www.census.gov/population/www/documentation/twps0029/twps0029.html

Chairman of Boston's Immigration Restriction League. 'We have, of late years, not been getting the best of Europe' he complained (Ward 1922a, 316).

At the Carnegie, Merriam solicited the help of Madison Grant on what he called 'the problem of race migration following the present war'. Complimenting Grant, Merriam continued 'your knowledge of the immigration situation from many sides gives you an excellent position for considering the subject, not merely for the use of our own country, but also for suggestion of the international policy which should be followed by the allies'.[21] The first eugenic immigration law was passed in 1917 which debarred Asians. But for Laughlin, testifying before Congress in 1924, this was far from sufficient. A whole series of 'filterings' were necessary, of which the examination at Ellis Island would not be the first. These filters, ideally carried out in the home country, would run a battery of tests on the applicant's intelligence and moral qualities, as well as a full physical examination and work up of 'family stock' (heredity). Calling for 'refined biological standards' for admission to the United States, Laughlin complained that 'while many superior immigrants of many races have recently come to the United States, along with them there has been an unduly large percentage of dross' (United States Congress 1924, 1263). The Chair of the House Committee on Immigration and Naturalization, the eugenicist Albert Johnson, could only agree.

If the 1924 law was the first to impose strict quotas in an attempt to control immigration, it was not the last. Other quota laws, including a National Origins Law of 1929, lasted until well into the 1960s. In 2006 the American Congress again considered immigration law amid concerns that an estimated 11–12 million undocumented (or illegal) immigrants posed a threat to jobs and national security. Despite the fact that a significant number of these immigrants were from Europe, or had overstayed their visas, debate characterized them as 'Mexican'. Some Senators called for mass deportation or a moratorium on immigration. In the face of this, America saw some of the largest mass rallies in favour of immigrant rights for many decades. The terms of the debate were remarkably similar to those of the 1910s and 1920s, substituting 'Mexicans' for southeastern Europeans.

Conclusion: Cartographic Spatializations of Race

Today it is common to assert that geography and identity cannot be equated. Mol and Law claim further that '[i]t is no longer assumed that geography and identity map onto one another. And the resulting complexity – self, other, here, there – defies the cartographic imagination' (Mol and Law 2005, 637). Speaking of the geography of racial distributions the anthropologist Jonathan Marks states that '[w]e don't know how many there are, where to draw the boundaries between them, or what those boundaries and the people or places they enclose would represent' (Marks 1995, 275). And yet if life is a continuously varying diversity, this has not stopped attempts

---

21   Merriam to Grant, August 8, 1918, NARA/Merriam.

at such spatial divisions, nor their representation on maps. Today the biological discussion of race concentrates on those last remaining bits of the genetic code that seem to vary spatially (the gene pool residual). The case of 'racialized medicine' is a good example where race is once again supposedly a useful factor.

It is of course possible to identify race if you want to. Race as a category has been around since the eighteenth century: '[p]rior to that time, and even into the nineteenth century, human variation was always interpreted as varying in local terms' (Marks 2005, n.p.). Large-scale 'para-continental' groupings were new, and usefully allowed people to think of themselves as civilized versus others who not only were uncivilized, but were a threat. Foucault calls this other the 'barbarian' who exists in a relation to the civilized: '[t]he barbarian is always the man who stalks the frontiers of States, the man who stumbles into the city walls' (Foucault 2003b, 195). This is the other who threatens the purity of the race. But the idea of race as natural, large-scale differences is arbitrary. As Marks adds, '[t]he development of the concept of race can profitably be seen as an expression of this so-called "bio-power" – constituting an authoritative, scientific answer to the basic question, "What kinds of people are there?"' (Marks 2005, n.p.). It is in this sense then that racism did not emerge from nationalism, but from the cut (*coupure*) into the continuously varying diversity of human variation (see Elden 2002). Maps provided the spatial imaginary to do this, using race-based data.

We have seen how the Inquiry took as its predicate the derivation of scientific racial boundaries across Europe; boundaries that were assumed to reflect an underlying racial partitioning that could be discerned on maps. The Inquiry knew well that this was not a simple reduction of identity to space. But they assumed that territory and its rightful populations could be discerned if you looked hard enough and assembled the right data. Once 'propaganda' ('politics') had been removed from the equation, a clear track could be cut through the morass of competing claims. As the Serbian geographer Jovan Cvijic put it in the pages of the *Geographical Review*, his ethnic fieldwork allowed him to discern 'natural barriers' in Europe which picked out 'zones of civilization'. While Cvijic recognized that with trade, migration and communication, people could be 'dove-tailed', he argued that each ethnicity had left 'a deeper impress than others' which would allow such zones to be identified (Cvijic 1918, 470).

We have seen too that these racial motivations were in fact mainstream during this period. Davenport and the ERO were happy to work with the Inquiry as part of their larger scheme to pursue their eugenic principles of biological discrimination. Foucault's remarks in *Discipline and Punish* are still descriptive: 'a meticulous tactical partitioning in which individual differentiations were the constricting effects of a power that multiplied, articulated and subdivided itself' (Foucault 1977, 198). Hence the need for all those index cards and the focus on problematic populations in southeastern Europe – Bowman's 'storm centre'.

How should we understand the turn to Davenport and the ERO, of Grant's position on the AGS Council, or of AAG president Robert Ward's immigration writings? It is at least necessary to know the intellectual history of race-based mapping in geography. There are two reasons for this. One is that the results of the decisions at Versailles, and particularly the idea that race could be unambiguously spatialized, would later haunt the twentieth century. The Balkan conflict of 1991–1995 is only one example. Additionally, we see today, if only in transmuted form, much of the same discourse applied to immigration questions. The binary division of us-and-them still has many racial overtones.

But perhaps most importantly this episode illustrates that it was not just the work of certain racist men, nor equally just the result of some spirit of the times. If Foucault is right that biopolitics is characteristic of modern societies (and it is certainly a sweeping claim) then there is also relevance for us today. Mapping race is not just something that was an experiment carried out at a certain time and place. Rather, it is part of an ongoing series of geographical knowledges that allow the biopolitics of the population to be known en mass in its territories.

Archives Consulted

National Archives (NARA), Record Group 256:
Entry 1 (Correspondence). Folders Davenport, Hale, Hrdlicka, Merriam, Scudder.
Entry 51 (Cartographical Records).
Microform Series M1107, Roll 9, Inquiry Document 110.
American Philosophical Society (APS), Davenport Papers. Folders Bowman, Earle, Hoffman, Mezes, Scudder.

Acknowledgements

Many thanks to the archivists who assisted me in finding the material cited here. My thanks also to Krystyn R. Moon for census data assistance, and to Stuart Elden, Kara C. Hoover, and Matthew Farish for suggestions for improvement. Any errors are the responsibility of the author.

References

Allen, Garland (1986) 'The Eugenics Record Office at Cold Spring Harbor, 1910–1940: An Essay in Institutional History.' *Osiris 2*, 2, 225–64.
American Geographical Society (1919) 'The American Geographical Society's Contribution to the Peace Conference.' *Geographical Review*, 7(1), 1–10.
Anonymous (1918) 'Science Notes and News.' *Science*, 48(1231), 112.
Bix, Amy Sue (1997) 'Experiences and Voices of Eugenics Field-Workers: "Women's Work" in Biology.' *Social Studies of Science*, 27, 625–68.
Black, Edwin (2003) *War against the Weak. Eugenics and America's Campaign to Create a Master Race*. New York and London: Four Walls Eight Windows.

Bowman, Isaiah (1921) *The New World; Problems in Political Geography*. Yonkers-on-Hudson, NY: World Book Company.

Brigham, Albert Perry (1919) 'Principles in the Determination of Boundaries.' *Geographical Review*, 7(4), 201–19.

Burchell, Graham, Colin Gordon and Peter Miller, Eds (1991) *The Foucault Effect: Studies in Governmentality: With Two Lectures by and an Interview with Michel Foucault*. Chicago: University of Chicago Press.

Cochrane, Rexmond C. (1978) *The National Academy of Sciences the First Hundred Years 1863–1963*. Washington, DC: National Academy of Sciences.

Crampton, Jeremy W. (2006) 'The Cartographic Calculation of Space: Race Mapping and the Balkans at the Paris Peace Conference of 1919.' *Social and Cultural Geographies*, 7(5), 731–52.

Cvijic, Jovan (1918) 'The Zones of Civilization of the Balkan Peninsula.' *Geographical Review*, 5(6), 470–82.

Davenport, Charles B. (1911) *Heredity in Relation to Eugenics*. New York: Henry Holt.

Davenport, Charles B. (1921) 'Research in Eugenics.' *Science*, 54(1400), 391–7.

Davenport, Charles B. and M. Steggerda (1929) *Race Crossing in Jamaica*. Washington, DC: Carnegie Institute of Washington.

Deleuze, Gilles (1988) *Foucault*. Minneapolis: University of Minnesota Press.

Dominian, Leon (1917) *The Frontiers of Language and Nationality in Europe*. New York: Henry Holt for the American Geographical Society.

Elden, Stuart (2002) 'The War of Races and the Constitution of the State: Foucault's *Il Faut Defendre La Société*.' *boundary 2*, 29, 125–51.

Foucault, Michel (1977) *Discipline and Punish: The Birth of the Prison*. New York: Pantheon Books.

Foucault, Michel (1978) *The History of Sexuality*. New York: Pantheon Books.

Foucault, Michel (1979) *Discipline and Punish: The Birth of the Prison*. New York: Vintage Books.

Foucault, Michel (1984) 'Space, Knowledge, and Power.' In Paul Rabinow (Ed.) *The Foucault Reader*. New York: Pantheon. 239–56.

Foucault, Michel (1988) 'The Dangerous Individual.' In Lawrence D. Kritzman (Ed.) *Politics, Philosophy, Culture: Interviews and Other Writings of Michel Foucault, 1977–1984*. New York: Routledge. 125–51.

Foucault, Michel (1997) 'Polemics, Politics, and Problematizations.' In Paul Rabinow (Ed.) *Michel Foucault Ethics*. New York: The New Press. 111–19.

Foucault, Michel (2003a) *Abnormal*. New York: Picador.

Foucault, Michel (2003b) *Society Must Be Defended: Lectures at the Collège De France, 1975–76*. New York: Picador.

Foucault, Michel (2004) *Sécurité, Territoire, Population, Cours Au Collège De France 1977–78*. Paris: Gallimard-Seuil.

Friendly, Michael (2002) 'Visions and Re-Visions of Charles Joseph Minard.' *Journal of Educational and Behavioral Statistics: A Quarterly Publication Sponsored by* the American Educational Research Association and the American Statistical Association, 27(1), 31 (22 pages).

Gelfand, Lawrence Emerson (1963) *The Inquiry; American Preparations for Peace, 1917–1919*. New Haven: Yale University Press.

Grant, Madison (1916) 'The Passing of the Great Race.' *Geographical Review*, 2(5), 354–60.

Grant, Madison (1932) *The Passing of the Great Race or the Racial Basis of European History*. New York: Scribner's Sons.

Hannah, Matthew (2000) *Governmentality and the Mastery of Territory in Nineteenth-Century America*. Cambridge: Cambridge University Press.

Harley, J.B. (1989) 'Deconstructing the Map.' *Cartographica*, 26(2), 1–20.

Harvey, David (2001) 'Cartographic Identities: Geographical Knowledges under Globalization.' In David Harvey (Ed.) *Spaces of Capital: Towards a Critical Geography*. New York: Routledge. 208–33.

Heffernan, M. (2002) 'The Politics of the Map in the Early Twentieth Century.' *Cartography and Geographic Information Science*, 29(3), 207–26.

Holdich, Thomas H. (1916) 'Geographical Problems in Boundary Making.' *The Geographical Journal*, 47(6), 421–36.

Johnson, Douglas Wilson (1919) 'A Geographer at the Front and at the Peace Conference.' *Natural History*, XIX(6), 511–17.

Kevles, Daniel J. (1968) 'George Ellery Hale, the First World War, and the Advancement of Science in America.' *Isis*, 59(4), 427–37.

Kevles, Daniel J. (1985) *In the Name of Eugenics. Genetics and the Uses of Human Heredity*. Berkeley and Los Angeles: University of California Press.

Koch, Tom (2005) *Cartographies of Disease: Maps, Mapping and Medicine*. Redlands, CA: ESRI Press.

Livingstone, David N. (2003) *Putting Science in Its Place: Geographies of Scientific Knowledge*. Chicago: University of Chicago Press.

Marks, Jonathan (1995) *Human Biodiversity*. Hawthorne, NY: Aldine de Gruyter.

Marks, Jonathan (2005) 'The Realities of Races.' Online. http://raceandgenomics. ssrc.org/Marks.

Martin, Geoffrey J. (1968) *Mark Jefferson, Geographer*. Ypsilanti, MI: Eastern Michigan University Press.

Mol, Annemarie and John Law (2005) 'Boundary Variations: An Introduction.' *Environment & Planning D: Society and Space*, 23, 637–42.

Montagu, M.F. Ashley (1944) 'Aleš Hrdlička, 1869–1943.' *American Anthropologist*, 46, 113–17.

Noyes, J.K. (1994) 'The Natives in Their Places: "Ethnographic Cartography" and the Representation of Autonomous Spaces in Ovamboland, German South West Africa.' *History and Anthropology*, 8(1–4), 237–64.

Palsky, Gilles (2002) 'Emmanuel De Martonne and the Ethnographical Cartography.' *Imago Mundi*, 54, 111–19.

Petto, Christine M. (2005) 'From L'etat, C'est Moi to L'etat, C'est L'etat: Mapping in Early Modern France.' *Cartographica*, 40(3), 53–78.

Pickles, John (2004) *A History of Spaces. Cartographic Reason, Mapping and the Geo-Coded World*. London: Routledge.

Robinson, Arthur H. (1967) 'The Thematic Maps of Charles Joseph Minard.' *Imago Mundi*, 21, 95–108.

Robinson, Arthur H. (1982) *Early Thematic Mapping in the History of Cartography*. Chicago: University of Chicago Press.

Rose-Redwood, Reuben S. (2006) 'Governmentality, Geography, and the Geo-Coded World'. *Progress in Human Geography*, 30(4), 469–86.

Sluga, Glenda (2001) *The Problem of Trieste and the Italo-Yugoslav Border: Difference, Identity, and Sovereignty in Twentieth-Century Europe*. Albany: State University of New York Press.

Sluga, Glenda (2002) 'Narrating Difference and Defining the Nation in Late Nineteenth and Early Twentieth Century "Western" Europe.' *European Review of History*, 9(2), 183–97.

Smith, N. (2002) 'The American Century: Consensus and Coercion in the Projection of American Power.' *Progress in Human Geography*, 26(2), 282–4.

Sparke, Matthew (1998) 'A Map That Roared and an Original Atlas: Canada, Cartography, and the Narration of Nation.' *Annals of the Association of American Geographers*, 88(3), 463–95.

United States Congress (1924) *Europe as an Emigrant-Exporting Continent and the United States as an Immigrant-Receiving Nation. Statement of Dr. Harry H. Laughlin*. Washington, DC: Government Printing Office.

Walker, Frances Amasa (1896) 'Restriction of Immigration.' *The Atlantic Monthly*, 77(464), 822–9.

Ward, Robert DeCourcy (1922a) 'Some Thoughts on Immigration Restriction.' *The Scientific Monthly*, 15(4), 313–19.

Ward, Robert DeCourcy (1922b) 'What Next in Immigration Legislation?' *The Scientific Monthly*, 15(6), 561–9.

Wilkinson, Spenser, H. Mackinder and L.W. Lyde (1915) 'Types of Political Frontiers: Discussion.' *The Geographical Journal*, 45(2), 139–45.

Wilson, Woodrow (1966–1994) *The Papers of Woodrow Wilson*. Edited by Arthur Stanley Link. Princeton, NJ: Princeton University Press.

Winlow, H. (2006) 'Mapping Moral Geographies: W.Z. Ripley's Races of Europe and the United States.' *Annals of the Association of American Geographers*, 96(1), 119–41.

Wright, J.K. (1952) *Geography in the Making. The American Geographical Society 1851–1951*. New York: American Geographical Society.

Chapter 23

Beyond the Panopticon?
Foucault and Surveillance Studies

David Murakami Wood

... I have heard the key
Turn in the door once and turn once only
We think of the key, each in his prison
Thinking of the key, each confirms a prison. (T.S. Elliot, *The Waste Land*, lines 411–15)

I was saying simply this: perhaps everything is not as simple as one believes. (Foucault 1994c, 629)

Introduction

Surveillance Studies is a transdisciplinary field that draws from sociology, psychology, organization studies, science and technology studies, information science, criminology, law, political science and geography. It emerged through combination of the mainstream liberal sociological approach of Rule (1973) via Giddens (1985) which, following Zuboff (1998) and Gary Marx (1988), was combined with Foucault, in particular *Discipline and Punish* (1977), and its reading of Bentham's *Panopticon*. There are of course other theoretical approaches: Marxism; Weber; Machiavelli; anarchism and situationalism, all offer useful avenues. But a relatively smooth story can be told of the movement from Foucault's panopticism to a 'surveillance society' (Lyon 1993; 1994), via a 'new surveillance' (Marx 1988; 2003) of computerized and increasingly automated 'social sorting' (Gandy 1993; Lyon 2001; Lyon 2002a) based on 'categorical suspicion' (Marx 1988; Norris and Armstrong 1999).

This is a simplification, but how much this story can be sustained by the foundations provided by *Discipline and Punish*, and the different places assigned to the Panopticon, panopticism, this book and Foucault, are the matters to be discussed here. This chapter considers the central arguments in *Discipline and Punish*, outlines some common surveillance-related critiques before discussing developments beyond criticism and interpretation. It concludes by suggesting ways forward largely through a combination of Deleuze and Actor-Network Theory.

Discipline and Punish

Foucault is clear about his aims: the final sentence of the book summarizes that it 'must serve as a historical background to various studies of the power of normalization and the formation of knowledge in modern society' (308). This is a history of knowledge not material things, even though spatio-temporally specific materialities may be inscribed with or represent this knowledge. In the first pages, following the horrific description of the public torture and death of Damiens, and Faucher's rules for young prisoners, Foucault sets out to:

> ... regard punishment as a complex social function ... regard punishment as a political tactic ... make the technology of power the very principle both of the humanization of the penal system and of the knowledge of man ... [and] study the metamorphosis of punitive methods on the basis of a political technology of the body in which might be read a common history of power relations and object relations. (23–4)

This is a deliberate attempt to understand these developments free of moral judgements – because as a writer, Foucault too sees all of us, himself included, caught in the same webs of power, that constitute those things we regard as objective truth or the 'facts as they are': there is no free classical subject. In opposition to those who conflate Foucault and Orwell or Arendt, and both the positive and negative view of liberals and the Frankfurt school respectively, the modern subject is seen as an intrinsic part of what we call progress – it is the result of the process of civilization that began with the enlightenment. His project then is to expose that supposedly objective fact to scrutiny and make it reveal its composition and history – its genealogy.

Why do so many apparently find this problematic? Bruno Latour (2005) remarks that the Francophone and Anglophone academic worlds are divided not just by language but by underlying assumptions about communication: French writers assume a semiotic education that understands the text as having an existence *in itself*, and a relationship with everything outside. One has to conduct a double reading of the book as text and as interactive: Mottier (2001) thus describes them as 'book-bombs' or toolkits for thought and action.

What then does the book actually argue? It tells of sovereign monarchical power (*arche*) with its capricious and limited but directed, spectacular and often lethal impact on bodies, transformed by knowledge of both the individual human and humans together. These rational, scientific, and humanist reforms changed conceptions of justice and the place of the body in this schema. Part One thus deals with torture and the spectacle of public execution as 'power that not only did not hesitate to exert itself directly on bodies, but was exalted and strengthened by its visible manifestations' (Foucault 1977, 57).

Part Two covers the development of enlightened understanding of criminality and the humanization of the penality in the eighteenth century. This begins with attempts to formulate punishments that match or mirror directly the crime of the offender through signing on the body. As a development from the symbolic act of public execution, this remained 'a lesson for all', however these 'fair' systems of punishment were rapidly replaced by 'the law of detention for every offence of any importance, except those requiring the death penalty' (116).

Part Three introduces new elements, in particular: the concepts of the soldier and general militarization; and the emerging mechanical science of the body, exemplified in La Mettrie's *L'Homme-machine* (136). Both these knowledges were productive of the notion of the body as 'docile', trainable through repetitive disciplinary practices. This is the origin of the modern subject: the malleable, improvable person. Foucault's argument is that 'in the course of the seventeenth and eighteenth centuries, the disciplines became general formulas of domination' (137).

Crucial here is not just the body itself, but the spatial and temporal distribution and regulation of the body: time was divided into smaller units to allow for total control of activity, likewise space was constructed so as to enclose but also to partition. Bringing these concepts together is the idea of productive ordering: the classification and arrangement of all kinds of properties and entities into '*tableaux vivants*', to maximize their usefulness (148). Things had to be intelligible to be manageable, and manageable to be productive. This management of the space-time of all entities changed our conception of time itself, and produced the idea of 'progress'; the 'geneses' of individuals was paralleled by the 'evolution' of society, a development that maintains its hold.

This training of malleable subjects in a progressive social order is accomplished through two main 'instruments'. The first is hierarchical observation exemplified by the military camp, a 'diagram of a power that acts by means of general visibility' (171). The second is normalizing judgement found within institutions such as schools and factories. The visibility effected through hierarchical observation is not enough; a 'micro-penality of time' (178) has to be operationalized, acting in gaps in law, enforced and corrective through both punishment and reward. This makes explicit the boundary between normal and abnormal – it is not repression but the generation of a shared sameness – but also, through examination, categorizes people according to worth within that normal community.

Already, it can be seen that the Panopticon is not the only point of the book. It is not even the only exemplar of panopticism. This has been pointed out most recently by Norris (2003) and Elden (2003), both of whom emphasize the importance of plague. For the control of the town of Vincennes faced with epidemic is where Foucault begins elaborating panopticism. The town becomes a camp, a blockaded space where normal rules are suspended to fight an outside evil:

> This enclosed, segmented space, observed at every point, in which the individuals are inserted in a fixed place, in which the slightest movements are supervised, in which all events are recorded, in which an uninterrupted work of writing links the centre and the periphery, in which power is exercised without division, according to a continuous hierarchical figure, in which each individual is constantly located, examined and distributed among the living beings, the sick and the dead – all this constitutes a compact model of the disciplinary mechanism. (197)

Plague represents those things the ordering gaze seeks to overcome: abnormality, disorder, chaos, license, whose previously dominant conception was that of the leper: excluded and rejected. Instead the disordered are to be ordered.

Foucault is clear that, 'Bentham's *Panopticon* is the architectural figure of this composition' (200): simply one aspect. Elden (2003) argues that it is 'the culmination of technologies of power rather than their beginning' (245–6), however, like the camp or plague town, the Panopticon represents *both* the summation of power/ knowledge in the text thus far, *and* another transformation. By making the inmates totally visible to an assumed gaze, the Panopticon acts 'to induce in the inmate a state of conscious and permanent visibility that assures the automatic functioning of power'. It is 'light' power as opposed to the 'heavy' power of the monarch's dungeon, because the prisoner, by watching himself, 'becomes the principle of his own subjection' (203). The transformation occurs because unlike both camp and plague-town, the Panopticon is a permanent structure. It is normal not exceptional. It is therefore conceptually generalizable: a disciplinary *dispositif* rather than a disciplinary blockade. It is not that the Panopticon actually is built everywhere – it is a utopia – but that it is the purest expression of the trajectory from exclusion and blockade towards generalized discipline. The principles of this discipline are: the transformation of the disciplines emerging from monasticism and the army, etc. from negative to positive and productive, that is 'to become attached to some of the great essential functions: factory production, the transmission of knowledge, the diffusion of aptitudes and skills, the war-machine' (211); 'the swarming of disciplinary mechanisms' (211) or their tendency to diffuse outward from those institutions into society as more flexible forms; and finally, state control of these mechanisms, most of which had previously been religious, through organizations like the police.

Here Foucault makes explicit connections to marxism and liberalism, arguing that disciplines are essentially ordering techniques that have a 'very close relationship' with production and the division of labour, indeed 'each makes the other possible and necessary; each provides a model for the other' (221). At the same time the normality constituted by disciplinary mechanisms was bourgeois, and this class's advances in democracy and legal equality have to be understood alongside panopticism, 'the other, dark side of these processes' (222). Foucault asks why disciplinary advances are not celebrated as are other historical developments, such as record-keeping or mining or the Panopticon compared to the steam engine and microscope. It has not only because it seems 'inglorious' and alien to our internalized heroic story of progress. This remains important: as Roy Boyne remarks, 'any deep critique of surveillance as a principle would have to imply a critique of social democracy and social welfare simultaneously, and may help explain the relative calm with which contemporary development of surveillance powers has been received' (292).

Ultimately, in the early nineteenth century, this resulted in a distributed 'carceral archipelago', 'a multiple network of diverse elements' (307), exemplified by the 1840 opening of Mettray children's prison, 'the disciplinary form at its most extreme, the model in which are concentrated all the coercive technologies of behaviour' (293). But this is another culmination/transformation, with the simultaneous emergence of psychology, psychiatry and scientific medicine, which sought to reach inside the body. It is *not* the Panopticon, but the combination of institutions and an era when 'medicine, psychology, education, public assistance, "social work" assume an ever greater share of the powers of supervision and assessment' (306) that produces 'an art of punishing more or less our own' (296). In the 'carceral', it is not that everywhere becomes a prison, but that 'prison and its role as link are losing something of their purpose' (306). The issue then is 'the steep rise in the use of these mechanisms of normalization and the wide-ranging powers which, through the proliferation of new disciplines, they bring with them' (306).

Critiques

There are many critiques of *Discipline and Punish* on the edges of the concerns of this chapter, on visibility (see: Jay 1993; Bogard 1996; Gordon 2002; Norris 2003; Yar 2003; Simon 2004); and power (see: Herbert 1996; Porter 1996; Alford 2000). Many of the latter fall back on structuralist Marxism – although Coleman's (2003) reading is more subtle – and ignore Foucault's argument that we cannot take 'the facts as they are' as objective, a critique that has been more fully developed in Actor-Network Theory (Latour 1993; 2005).

There are also many arguments about how much real prisons resemble the Panopticon. Ignatieff (1989), Lyon (1993) and Boyne (2000) all note that Bentham's original plans for the Panopticon were not implemented. Alford (2000) goes further, arguing that Foucault's theory is invalidated because nothing panoptic can be seen in contemporary US prisons, where only the entrances and exits are controlled and prisoners merely counted because 'one inmate is exactly like another' (129). He argues that 'If you have to look, you have already ceded a measure of power, the power not to look and not to care' (2000, 127). But this criticism is flawed for two related reasons. First, once again, it assumes that Foucault was writing a conventional history of prisons, rather than a genealogy of modern punishment. Second, Foucault had described exactly this type of power earlier in the book: the 'heavy' power of the monarch's dungeon. The fact that societies contain mixtures of modes of ordering, that modernism remains an incomplete or failed project (c.f.: Bauman 1991a; Latour 1993), or that the death penalty still exists, does not invalidate a genealogy of the modern subject. Alford (2000) later notes that prisoners live in a *pre-modern* style, so why criticize Foucault's model of *modern* subjectivity for being inadequate? McCorkhill (2003) claims that there are examples of panoptic prisons: women's prisons, which tend to emphasize moral reform. However she overestimates the degree to which their psychological approaches contradict Foucault: where she is correct in arguing, after Bartky (1988) that Foucault neglected gender (see also: Koskela 2000; 2003), her evidence seems to support the post-carceral model that emerges at the end of the book.

Many question Foucault's interpretation of Bentham's *Panopticon* (1791), for example, Lyon (1993), Hannah (1997) and Boyne (2000), and there are other non-

Foucauldian readings (see: Himmelfarb 1965; Jacobs 1977; Ignatieff 1989; Markus 1993; Kaschadt 2002). However, it is quite common now to see Bentham entirely through Foucault or through surveillance studies' reading of Foucault, for example, Dubbeld (2003) implies almost in passing that Bentham himself proposed the idea of the disciplinary gaze, and that Foucault merely documented the generalization of such surveillance, thus rereading Foucault back onto Bentham.

But perhaps Foucault is partly to blame. Lyon (1993) claims that 'Foucault did, despite himself' see the Panopticon 'in a "totalizing" way' (675), and Simon (2004) argues that Foucault's 'stark' and 'jarring' description makes the Panopticon 'prone to iconic simplification' (3). However one should not forget that the Panopticon is put forward as a diagram rather than either material object or summative theory. Foucault warned against the foregrounding conducted by many contemporary scholars as far back as 1978:

> Concerning the reduction of my analyses to the simple figure which is the metaphor of the panopticon, I believe that here one can respond on two levels. One can say: compare what they attribute to me to what I have said; and here, it is easy to show that the analyses of power that I have carried out cannot be reduced to this figure alone, not even in the book in which they have gone to look, that is to say, *Surveillir et Punir*. In fact, if I show that the panopticon was a utopia, a type of pure form elaborated at the end of the Eighteenth Century to furnish the most convenient formulation of a constant exercise of instant and total power, if then I had shown the birth, the formulation of this utopia, its raison d'etre, it is also true that I had directly shown that it concerned precisely a utopia which never functioned as it was described and that all the history of the prison – its reality – consisted precisely of always having passed this model by. (Foucault 1994c, 628)

I will concentrate for the remainder of this section on arguments about the development of surveillance in the book.[1] As with the prison, many of those tracing the history of surveillance make qualified acceptance of Foucauldian ideas. For example, Torpey (2000) argues that, whilst 'suggestive' (16), 'Foucault's considerations of these matters lack any precise discussion of the techniques of identification that have played a crucial role in the development of modern territorial states resting on the distinctions between citizens/nationals and aliens' (5). This is true. One of the biggest flaws in Foucault's attempt at the 'grand sweep' of history is his Franco-centrism – with the exception of Bentham, there is little in the book beyond Francophone nations, and little awareness of the issues around border and boundaries, indeed the act of categorization, and its political technologies, is underplayed. Despite the important section on tables, Foucault if anything overly privileges the conventionally spatial. The criticism that Foucault missed the importance of the file, 'papers', borders and the definition and measurement of the citizen is more important than those of Alford (2000 – supra) as these figures are as vital in the generation of the modern subject as the Panopticon. This is one of several places where actor-network theorists who

---

1   Foucauldian historical geography is considered by Gerry Kearns (chapter 20 in this book).

specifically build on Foucault (see below), in this instance, John Law (1994), would help fill gaps. However it is not enough to say 'Foucault missed *X*': there must be consideration of why *X* affects subjectivity and power.

The one seemingly universally accepted 'fact' about Foucault within surveillance studies is that he did not deal with the impact of contemporary, and in particular digital, technologies. Lyon (1993) says that, 'Curiously enough, Foucault himself seems to have made no comments about the relevance of panoptic discipline to the ways that administrative power has been enlarged and enhanced by computers especially since the 1960s' (659); Haggerty and Ericson (2000), Mottier (2001), and Morgan (2004) echo this claim. We shall return to this issue below.

Torpey also criticizes Foucault's formulation of the carceral as 'a nightmarish, dystopic, even absurd vision' (16). Lyon (1993) questions Foucault's rhetoric here too, but doesn't note that here at least this appears to refer specifically still to the end of the eighteenth century and the beginning of the nineteenth. This criticism is only partially valid in the context of the whole work, because it ignores both changes in power, and the fundamental lack of moral judgement as to purpose. Foucault cannot be held to argue that even the carceral is an entirely negative development, an ethically worse mode of ordering than the cruelty of public torture, or that 'surveillance society' is a completely intolerable and totalitarian repression. Instead discipline and the advances of the 'enlightenment' are inextricably interlinked, but that makes neither less real. One has to return to the book as text and this rhetoric must be seen to serve a semiotic purpose, if only as a warning sign: dystopia is vital in this context.

But as Deleuze (1992), Staples (1994), Lianos (2003), Norris (2003), and Elden (2003) all point out, this is a historically bounded project. It seems very strange therefore, for those studying surveillance almost two centuries after 1840, to take panopticism as a theoretical base for any new sociotechnological development, as if nothing had changed in terms of power/knowledge, rather than to follow Foucault in his method, and trace the inextricably interlinked historical evolution of punitive technologies, and power and object relations. This is not to argue that Foucault was entirely right, but criticisms that Foucault neglects the role of modern technologies should be a simple statement of fact, and a call to further theoretical development. On the contemporary it is, at best, suggestive.

However this is complicated by the fact that, even if it is ambiguous in the book, elsewhere Foucault is clear about the fact he *did* regard panopticism as a description of surveillance in contemporary society. Elden argues that he does 'designate the disciplinary society under the general rubric of panopticism' (2003, 247) largely due to the notion of 'police' as a 'general set of rules and regulations for the government of society' (247). In a presentation in 1973, he said:

The Panopticon is the utopian vision of a society and a kind of power which is, fundamentally, the society which we know today, a vision which has been effectively realized. This type of power of power can perfectly well be called panopticism. We live in a society where panopticism rules. (Foucault 1994b, 594)

This seems to be fundamentally different from the way in which the book ends, with the carceral and the move to professionalized control by specialist knowledges. But, of course, this is outside the text.

Developments

Beyond Panopticism

Lyon (1993) asks whether the panoptic can be generalized or applied to contemporary societies. Elden (2003) argues that 'rather than the Panopticon being the model for the disciplinary, surveillance society, the surveillance society is, taken to its extreme, exemplified by the Panopticon'. Here we examine attempts to take the Panopticon and panopticism beyond its specific occurrence in the book.

Lyon is cautious, arguing 'electronic surveillance does exhibit panoptic qualities in certain settings' (674), especially consumer surveillance. Hannah (1997) answers by claiming that Foucault 'never precisely spelled out the ways in which the panoptic logic of visibility had to change in order to operate effectively in an environment where its subjects did not suffer continuous confinement' (344). He is thus interested in the imperfect panopticism for the 'already normal' and wants to compare the lives of the free to the institutionalized, however despite the imperfection of an urban Panopticon, '*complete unity of the authoritative subject is unnecessary for the reasonably successful enforcement of normality in today's society*' (353).

Simon (2004) contends that this imperfection actually undermines the whole project because it is not the already normal that one actually wants to see and that 'the individuals one hopes to detect are the very individuals that have the best chance of evading detection' (8). Enclosure in the Panopticon is thus vital and this simply does not apply in streets. Norris and Armstrong (1999) and McCahill (2001) argue that CCTV cameras, despite their many important effects, do not create an urban Panopticon, however Koskela (2000) disagrees: the contemporary city *is* a power-space and 'through surveillance cameras the panoptic technology of power has been electronically extended, making our cities like enormous panopticons' (243). Fundamentally for Koskela, Foucault and his interpreters emphasize the spatialization of power rather than the effect of power on space because Foucault actually had a rather vague, architectural, conceptualization of space.

However the new chosen 'site' of the Panopticon *par excellence* is neither the city nor the workplace,[2] but the 'panoptic sort' (Gandy 1993) of the database. Poster (1990; 1996) defined databases as an electronic 'superpanopticon'. Placeless and in-between, databases do not formally confine in any way, but they are, as Poster argues, 'performative machines, engines producing retrievable identities'. It does not simply work on the docile body, but it creates entirely new electronic 'subjects'. But why consider databases as panoptic? Why not the subsequent concept of the

---

2   There is no space here for even a cursory consideration of the massive literature on workplace surveillance.

'carceral'? Why not prior diagrams like 'the table' or the 'examination'? Bauman (1991b) has argued that databases are not disciplinary devices at all but merely confirm credibility. However, the database, for all that it is undoubtedly another tool of hierarchical organization and normalization, is not panoptical. It comes from the technological stream that Torpey criticizes Foucault for ignoring, the 'file'. If Foucault had continued his genealogical historical account into the twentieth century, it seems unlikely he would have described databases as superpanoptic, rather he would have treated the 'database' as a particular political technology, a diagram, a mode of ordering, of its own space/time of power/knowledge.

Beyond Discipline and Punish

There are two broad ways of moving beyond the book: through thematic or methodological trajectories suggested within the book; or through Foucault's other work. I argue that the most productive are those that emerge from Deleuze's work.

First, there are those that emphasize the movement towards a consumer society and take seriously Bauman's dualism of seduction and repression, for example Staples (1994) argues that this breaks down boundaries and makes things porous, and also becomes linked with entertainment too and a sense of pleasurable surveillance (Lyon 1993; Weibel, 2002). According to Elmer's version of this 'consumers are not exclusively disciplined – they are both *rewarded*, with a preset familiar world of images and commodities, and *punished* by having to work at finding different and unfamiliar commodities if they attempt to opt-out' (245). Finally, drawing on Foucault's suggestions about the role of the bourgeoisie in discipline in the book and elsewhere, Hunt has for several years looked at consumption in terms of discipline and regulation (1996; 1999) and argues that 'moral regulation projects have been key vehicles for the articulation of the politics of the middle classes' (2004, 563).

A third approach focuses on technological development, exemplified by Graham (1998) on automation; Norris et al. (1998) and Graham and Wood (2003) on algorithmic surveillance; Jones (2000) on 'digital rule'; Introna and Wood (2004) on face-recognition; and Staples (1994) and Nellis (2005) on the electronic tagging of offenders. Drawing on Poster (1990), Marx (1988) and of course Lyon, they argue that new technologies of various kinds have enabled new forms and qualities of surveillance, but their main theoretical wellspring is the genealogical thought-experiment of Gilles Deleuze (1992) who argues that there was a change from disciplinary societies that accelerated after the second world war with the disciplinary institutions, now in 'generalized crisis' (3), being replaced by *societies of control*. Whereas discipline is analogical and moulding, control is digital and modulating. The control *dispositif* is the *code*, and it is the numerical language of control that is made of codes that mark access to information, or reject it. Instead of dealing with the mass/individual pair, individuals have become '*dividuals*', and masses, samples, data, markets, or '*banks*'. These dividuals exist both as the old physical body of the modern subject, but as with Poster's conclusion, as multiple subjects in databases. The key difference between Poster and Deleuze is that the former is trying to hammer this square peg into the round hole of panopticism, whereas by thinking genealogically, Deleuze is able to envisage a new round peg, a new *dispositif*. Code has attracted much attention from the humanities and social sciences, however in a great deal of the writing, code appears as just a kind of metaphor, something akin to poetry, or free floating items which infiltrate themselves into daily life through a rather unspecific set of processes. There are notably exceptions, in Thrift and French (2002), the work of American legal scholar, Lessig (2000) and the explorations of critical media theorists like Galloway (2004), who argues for the notion of the 'protocol' of the distributed network of the Internet as a more accurate diagram: the seemingly contradictory combination of enabling restrictions, utter vertical control but total horizontal liberty that increasingly defines our societies and what we are. Galloway takes us from Foucault's sovereign state and modern subject, through our information society, the society of control ruled by protocol, and beyond to a potential 'bioinformatic' future.

Staples (1994) shares Deleuze's historical analysis but instead of focusing on databases and dividuals, argues that the 'new economy of discipline' moves beyond the walls as 'generalized surveillance and control' (649) but still focuses on the body. Lyon too has tried to bring attention to the body even within the technological turn, drawing full attention to the way in which most new techniques are concerned with bodily traces, or biometrics, which require new kinds of examination (Lyon 2001; 2002b). For both Staples (1994) (and also Deleuze), this 'no longer requires the delimitation of space through architecture' (653), but space can and must still be divided into smaller segments.

The third major approach is those that concentrate on the visual and mass media, firstly through Mathiesen's (1997) concept of the viewer society and the synopticon (the many watching the few), as a discipline of consciousness. According to Simon (2004), the synoptic may solve the conundrum that 'panopticism, as a totalizing system, fails without an equally sophisticated cultural apparatus for reminding citizens that they are being watched' (14). But, as we saw, the Panopticon included this element of theatre from the beginning in Bentham and Foucault. According to Elmer (2003) many of these accounts are biased towards spectatorship and fail to see how synopticism and panopticism work together.

The second visual approach is through the Baudrillardian concepts of simulation and the hyperreal. Baudrillard (1983, 1987), Bogard (1996) and Pecora (2002) argue that simulation and surveillance increasingly are linked and that moving beyond to simulation and anticipatory surveillance means that there is no (or less) need for a Panopticon. Hope (2005) however believes such simulation to be limited as 'although simulation can encourage social order and self-policing, it cannot hold individuals accountable for acts committed' (362). Graham (1998) criticizes Bogard for lacking empirical analysis of sociospatial relations. Graham analyses the twin development of surveillance and simulation through contemporary information technologies, tracing four key developments: networks; the power and capacity of computers; the movement to visualization and simulation; and advances in 'georeferencing' systems such as GIS, GPS and virtual/intelligent environments; the key being increasing simultaneity, such that 'the gap between virtual control and real control disappears' (Bogard 1996, 9). This combination of surveillance and simulation (and indeed stimulation) was analyzed much earlier by Shearing and Stenning (1985) who argue quite convincingly that the theme park of Disney World should replace the Panopticon.

Elden (2003) and Ventura et al. (2005) argue that Foucault's later work on governmentality provide stimulating avenues. Some, like Hunt (2004), attempt to combine this with marxist perspectives, other like Stenson (2005) argue for a realist governmentality, based on the understanding of 'biopolitics and the struggle for sovereign control' (280; c.f.: Foucault 1979). Garland (2001) has also pointed in this direction.[3] This understanding is often complemented by an awareness of the changes in capitalism in the 1970s and neo-liberal deregulation, coupled with the concept of risk society (Beck 1992; Douglas 1996). Stenson (2005) argues that vague threats are categorized as risks which mandates new mapping of populations aimed at pre-empting or containing those threats (Rose, 2000).

Risk society also provides a context for the renewal of interest in self-surveillance. McGillivray (2005) claims that too much attention has been paid to the earlier Foucault and concepts of docility and disciplinary society, and that later writings, e.g.: *The Care of the Self* (Foucault 1990) offers more interesting work on self-actualization and reflexive subjects. Armstrong (1995) traces the rise of surveillance medicine (replacing hospital medicine which succeeded bedside medicine), which transforms everyone into a medical subject, and increases the responsibility of patients to look after themselves: 'the ultimate triumph of Surveillance Medicine would be its internalization by all the population' (400). The keys are pathologization and vigilance, and this trajectory has been extended by Vaz and Bruno (2003) who argue that there can be 'no neat line distinguishing power from care' (273) and that self-surveillance 'constitutes a subject that judges and condemns his or her own acts, intentions, desires and pleasures according to "truths" that are historically produced' (279). This they agree makes it difficult to question care and surveillance and also opens up possibilities of exclusion.

Beyond Foucault

However, should we 'forget Foucault' (Baudrillard 1987) altogether? Lyon (2003) argues that 'it is not clear that [models like the Panopticon] are entirely helpful ways of understanding surveillance today' (4) and that the less glamorous technologies of social sorting technologies and networked surveillance need more attention in themselves.

This networking has proved one area where moving entirely beyond Foucault has been suggested. One way is via the theories of Foucault's contemporaries Deleuze and Guattari (1987) on *rhizomes*, networks that send up shoots from anywhere, and

---

3   Garland has come under a stinging neo-Foucauldian attack from Voruz (2005) for confusing genealogy with simple history.

assemblages, heterogeneous objects brought together and working as an entity, giving greater permanence to flows. This is largely via Haggerty and Ericson (2000). Without offering any detailed critique they dismiss many surveillance studies as works which 'offer more and more examples of total or creeping surveillance, while providing little that is theoretically novel' (607). They claim that in late modernity 'we are witnessing a rhizomatic levelling of the hierarchy of surveillance, such that groups which were previously exempt from routine surveillance are now increasingly being monitored' (606). Echoing the Panoptic city argument, they rather tendentiously claim that this results in the progressive 'disappearance of disappearance', with the anonymity previously afforded by the city increasingly difficult. Sean Hier (2003) further 'probes' this conceptualization, looking at this decentralizing, perhaps even democratizing, interpretation of surveillance through an examination of welfare regimes.

In contrast to the democratizing possibilities of levelling hierarchies, Graham (1998) argues that 'the worry is that future surveillant-simulation techniques will embed subjective normative assumptions about disciplining within cybernetic computerized systems of inclusion and exclusion, where even opportunities for human discretion are removed?' (499). Boyne (2000) claims that 'few would argue this' (299), and Graham too dismisses totalizing impact scenarios. But Lianos (2001; 2003; c.f.: Lianos and Douglas 2000) strongly argues that the efficient servicing of consumer demand, society and sociality has moved towards a new situation of 'unintended control', whereby social interactions increasingly operate in socio-technical environments within which negotiation is not possible. Lyon had in fact noted the possibility that efficiency might unintentionally lead to social control as far back as 1993, but Lianos's spatial formulation of 'Automated Socio-technical Environments' (ASTEs) which has much in common with Thrift and French's (2002) description of the 'automatic production of space', is a productive concept, and has been taken forward by Norris (2003), Graham and Wood (2003) and Murakami Wood and Graham (2006).

Finally, there is another emerging strand of post-Foucauldian theory, which is both richer, more practical and less obscure than the surveillant assemblage, and which avoids the technological fetishism to which the 'automatic construction of space' arguments are prone despite their best intentions. 'Actor-Network Theory' (ANT) (Latour 2005) is the only comprehensive attempt to develop a post-Foucauldian understanding of power, arguing that society is always what results from the complex iterations between human, inhuman and nonhuman (actor-networks or collectives) rather than being a given thing or a pre-condition, or indeed exclusively 'human'. Boyne (2000) and Simon (2004) mention Latour, but despite the spread of ANT through the social sciences, my own collaborative work (Donaldson and Wood 2004; Murakami Wood and Graham 2006), and one paper by Ball (2002) appear to be the only detailed consideration of ANT for surveillance. Ball argues for an interactional approach, that recognizes that any form of surveillance implies that humans and non humans are arranged in a relation. She argues against reifying technological architectures of surveillance, which separates the social and the technical, and theorizes four common elements of surveillance practice: the *re-presentation* of material elements through technology; the *meaning* (or 'socially constructed interpretations of data, and subjects, categories and cultures within each surveillance domain'; the *manipulation* of relations within the actor-network; and finally the *intermediation*, which sustains the actor-network. My work has been concerned with how surveillant practices simultaneously reconstruct boundaries and knowledge, recasting Foucault's abnormal as 'strange materialities' that evolve because of the ability of subjects of surveillance to object to categorical work (Donaldson and Wood 2004) and in turn recasting surveillance as not simply about people, but about defining the relationship of all sorts of actants to boundaries: it is the determination of particular spaces and relationships to those spaces through categorization, boundary maintenance (in terms of both space and identity), observation and enforcement (Murakami Wood and Graham 2006).

Conclusion

Simon (2004) argues that 'Surveillance studies has gone further than Foucault in demonstating how information collected from individualized persons is organized, and manipulated to alter, manage or even control the life-chances of those persons.' This is particularly true in the case of Lyon, Norris, Graham and others who have made a sustained critical and empirical effort to do this.

However, many scholars of surveillance assume too much. Sometimes the problems result from shallow theoretical eclecticism. Sometimes the Panopticon is used simply as a badge of identity: this can be acceptable, however not when the Panopticon, standing for *Discipline and Punish* or for Foucault, is used as a straw man to be knocked down in order for the author(s) to set up their own approach. The Panopticon remains a useful figure, however every new technology is not the Panopticon recreated, nor does panopticism describe every situation. Foucault did not think so and provided multiple diagrams of power/knowledge, and importantly as Staples (1994) remarks, nothing vanishes: 'these new applications should be seen as capilliary extensions of disciplinary power that invest, colonize, and link-up, pre-existing forms', and the traces of all these forms remain (Hunt 2004), and may be reinvigorated, adapted, or persist in spatio-temporally specific ways. Certainly Boyne (2000) is correct to ask to what extent are we still 'pre-panoptical', following Bauman (1991a) and Latour (1993) in questioning the success of the modern project, and Agamben's (1998) work seems to suggest a return of the camp and the pre-panoptic exclusionary model. Surveillance is a mode of social ordering (Donaldson and Wood 2004), and concerns enforced categorization above all else: what is in and out, or what is to be considered social, remains a powerful categorization, that must be exposed to scrutiny.

We remain 'at the beginning of something' (Deleuze 1992, 7) but, as Foucault showed, we are always also the sum of things, thus the first priority remains taking seriously Foucault's genealogical method for contemporary (and future) subjectivity.

But rather than a hagiographic approach, we need a creative relationship with Foucault, challenging and disrupting existing forms of thought (Armstrong 1995).

Some, like Poster, have produced admirably productive failures in this area, but I would argue that Simon (2004) is correct in calling for a return to Deleuze's critique of panopticism, which, in its seven pages, extends *Discipline and Punish* beyond most other work in the field. However if *code*, or *protocol* is to be the new dispositif, it needs more complex theorizing and investigation. If one is to find a new language to discuss the relationship of life and technology, then one must move away, as both media and surveillance studies seem unwilling to do, from the vague poetics of Deleuze and Guattari, still struggling to accommodate Marxism because of some mistakenly moral conception of the project of academia trapped in the very prison of modernity that Foucault identified, to something that is capable of producing a genealogy of the present (and perhaps the future). In this area Actor-Network Theory remains the brightest hope for post-Foucauldian studies of surveillance because it is the only approach to combine methodological advances from simple genealogy with a continued refusal to allow moral assumptions to predetermine analysis.

References

Agamben, G. (1998) *Homo Sacer: Sovereign Power and Bare Life*. Stanford: Stanford University Press.
Alford, C.F. (2000) 'Would it Matter if Everything Foucault said about Prison were Wrong? *Discipline and Punish* after Twenty Years.' *Theory and Society*, 29, 125–46.
Armstrong, D. (1995) 'The Rise of Surveillance Medicine.' *Sociology of Health & Illness*, 17(3), 393–404.
Ball, K. (2002) 'Elements of Surveillance: A New Framework and Future Directions.' *Information Communication and Society*, 5(4), 573–90.
Bartky, S.L. (1988) 'Foucault, Feminity and the Modernisation of Patriarchal Power.' In I. Diamond and L. Quinby (Eds) *Feminism and Foucault*. Boston: Northeastern University Press, 61–86.
Baudrillard, J. (1983) *Simulations*. New York: Semiotext(e).
Baudrillard, J. (1987) *Forget Foucault!* New York: Semiotext(e).
Bauman, Z. (1991a) *Modernism and Ambivalence*. Cambridge: Polity Press.
Bauman, Z. (1991b) *Intimations of Postmodernity*. London: Routledge.
Beck, U. (1992) *Risk Society*. Cambridge: Polity Press.
Bentham, J. (1791) *Panopticon; or the Inspection-House*. 2 vols. London: T. Payne.
Bogard, W. (1996) *The Simulation of Surveillance*. Cambridge: Cambridge University Press.
Boyne, R. (2000) 'Post-Panopticism.' *Economy and Society*, 29(2), 285–307.
Coleman, R. (2003) 'Images from a Neoliberal City: The State, Surveillance and Social Control.' *Critical Criminology*, 12, 21–42.

Dandeker, C. (1990) *Surveillance, Power and Modernity*. Cambridge: Polity Press.

Deleuze, G. (1992) 'Postscript on the Societies of Control.' Translated by M. Joughin. *October*, 59, 3–7.

Deleuze, G. and F. Guattari (1987) *A Thousand Plateaus: Capitalism and Schizophrenia*. Translated by B. Masumi. Minneapolis: University of Minnesota Press.

Donaldson, A. and D. Wood (2004) 'Surveilling Strange Materialities: The Evolving Geographies of FMD Biosecurity in the UK.' *Environment and Planning D: Society and Space*, 22(3), 373–91.

Douglas, M. (1996) *Purity and Danger: An Analysis of Concepts of Pollution and Taboo*. Harmondsworth: Penguin.

Dubbeld, L. (2003) 'Observing Bodies. Camera Surveillance and the Significance of the Body.' *Ethics and Information Technology*, 5, 151–62.

Elden, S. (2003) 'Plague, Panopticon, Police.' *Surveillance & Society*, 1(3), 240–53.

Elmer, G. (2003) 'A Diagram of Panoptic Surveillance.' *New Media and Society*, 5(2), 231–47.

Foucault, M. (1977) *Discipline and Punish: The Birth of the Prison*. Harmondsworth: Penguin.

Foucault, M. (1979) *The History of Sexuality Vol.1: Introduction*. London: Penguin.

Foucault, M. (1990) *The Care of the Self (The History of Sexuality Vol. 3)*. Translated by R. Hurley. Harmondsworth: Penguin.

Foucault, M. (1994a[1973]) 'La societe punitive.' Lecture to Collège de France. Reprinted in *Dits et Ecrits 1954–1988, Vol 2*. Paris: Editions Gallimard, 456–70.

Foucault, M. (1994b[1973]) 'La verite et les formes juridique.' Lecture and roundtable discussion at the Pontificacal Catholic University, Rio de Janiero, 21–25 May. Reprinted in *Dits et Ecrits 1954–1988, Vol 2*. Paris: Editions Gallimard, 538–646.

Foucault, M. (1994c[1978]) 'Precisions sur le pouvoir: reponses a certaines critiques.' Interview with P. Pasquino, February in *Aut-Aut*, 167–8, 3–11. Reprinted in *Dits et Ecrits 1954–1988, Vol 2*. Paris: Editions Gallimard, 625–35.

Galloway, A.H (2004) *Protocol: How Control Exists After Decentralization*. Cambridge, MA: MIT Press.

Gandy, O.H. (1993) *The Panoptic Sort: A Political Economy of Personal Information*. Boulder, CO: Westview Press.

Giddens, A. (1985) *The Nation State and Violence*. Cambridge: Polity Press.

Garland, D. (2001) *The Culture of Control: Crime and Social Order in Contemporary Society*. Oxford: Oxford University Press.

Gordon, N. (2002) 'On Visibility and Power: An Arendtian Corrective of Foucault.' *Human Studies*, 25, 125–45.

Graham, S. (1998) 'Spaces of Surveillant Stimulation: New Technologies, Digital Representations, and Material Geographies.' *Environment and Planning D: Society and Space*, 16, 483–504.

Graham, S. and D. Wood (2003) 'Digitising Surveillance: Categorisation, Space, Inequality.' *Critical Social Policy*, 23(2), 227–48.

Haggerty, K. and R. Ericson (2000) 'The Surveillant Assemblage.' *British Journal of Sociology*, 51(4), 605–22.

Hannah, M. (1997) 'Imperfect Panopticism: Envisioning the Construction of Normal Lives.' In G. Benko and U. Strohmayer (Eds) *Space and Social Theory*. Oxford: Blackwell, 344–59.

Herbert, S. (1996) 'The Geopolitics of the Police: Foucault, Disciplinary Power and the Tactics of the Los Angeles Police Department.' *Political Geography*, 15(1), 47–57.

Hier, S. (2003) 'Probing the Surveillant Assemblage: On the Dialectics of Surveillance Practices as Processes of Social Control.' *Surveillance & Society*, 1(3), 399–411.

Himmelfarb, G. (1965) 'The Haunted House of Jeremy Bentham.' In R. Herr and H. Parker (Eds) *Ideas in History: Essays Presented to Louis Gottschalk by his Former Students*. Durham, NC: Duke University Press, 199–238.

Hope, A. (2005) 'Panopticism, Play and the Resistance of Surveillance: Case Studies of the Observation of Student Internet Use in UK schools.' *British Journal of Sociology of Education*, 26(3), 359–73.

Hunt, A. (1996) *Governance of the Consuming Passions: A History of Sumptuary Law.* Basingstoke: Palgrave Macmillan.

Hunt, A. (1999) *Governing Morals: A Social History of Moral Regulation.* Cambridge: Cambridge University Press.

Hunt, A. (2004) 'Getting Marx and Foucault into Bed Together!' *Journal of Law and Society*, 31(4), 592–609.

Ignatieff, M. (1989) *A Just Measure of Pain: the Penitentiary in the Industrial Revolution 1750–1850*. Harmondsworth: Penguin.

Introna, L. and D. Wood (2004) 'Picturing Algorithmic Surveillance: The Politics of Facial Recognition Systems.' *Surveillance and Society*, 2(2/3), 177–198. http://www.surveillance-and-society.org/articles2(2)/algorithmic.pdf.

Jacobs, J.B. (1977) *Stateville: Penitentiary in Mass Society*. Chicago: University of Chicago Press.

Jay, M. (1993) *Downcast Eyes: the Denigration of Vision in Twentieth-Century French Thought*. Berkeley: University of California Press.

Jones, R. (2000) 'Digital Rule: Punishment, Control And Technology.' *Punishment and Society*, 2(1), 5–22.

Kaschadt, K. (2002) 'Jeremy Bentham – The Penitentiary Prison or Inspection House.' In T.Y. Levin et al., 114–19.

Koskela, H. (2000) '"The Gaze Without Eyes": Video-surveillance and the Changing Nature of Urban Space.' *Progress in Human Geography*, 24(2), 243–65.

Koskela, H. (2003) '"Cam Era" – The Contemporary Urban Panopticon.' *Surveillance & Society*, 1(3), 292–313.

Latour, B. (1993) *We Have Never Been Modern*. Translated by C. Porter. Cambridge, MA: Harvard University Press.

Latour, B. (2005) *Reassembling the Social: An Introduction to Actor-Network-Theory*. Oxford: Oxford University Press.

Law, J. (1994) *Organizing Modernity*. Oxford: Blackwell.

Lessig, L. (2000) *Code – and Other Laws of Cyberspace*. New York: Basic Books.

Levin, T.Y., U. Frohne and P. Weibel (2002) *CTRL [SPACE]: Rhetorics of Surveillance from Bentham to Big Brother*. Karlsruhe: ZKM/Cambridge, MA: MIT Press.

Lianos, M. (2001) *Le Nouveau Contrôle Social: toile institutionnelle, normativité et lien social*. Paris: L'Harmattan-Logiques Sociales.

Lianos, M. (2003) 'After Foucault.' Translated by D. Wood. *Surveillance & Society*, 1(3), 412–30.

Lianos, M. and M. Douglas (2000), 'Dangerization and the End of Deviance: The Institutional Environment.' *British Journal of Criminology*, 40, 261–78.

Lyon, D. (1993) 'An Electronic Panopticon? A Sociological Critique of Surveillance Theory.' *The Sociological Review*, 653–78.

Lyon, D. (1994) *The Electronic Eye: The Rise of the Surveillance Society*. Cambridge: Polity Press/Blackwell.

Lyon, D. (2001) *Surveillance Society: Monitoring Everyday Life*. Buckingham: Open University Press.

Lyon, D. (2002a) 'Editorial. Surveillance Studies: Understanding Visibility, Mobility and the Phenetic Fix.' *Surveillance & Society*, 1(1), 1–7.

Lyon, D., Ed. (2002b) *Surveillance as Social Sorting: Privacy, Risk and Automated Discrimination*. London: Routledge.

Lyon, D. (2003) *Surveillance after September 11th*. Cambridge: Polity Press.

McCahill, M. (2001) *The Surveillance Web: The Rise of Visual Surveillance in an English City*. Devon: Willan Press.

McCorkhill, J.A. (2003) 'Surveillance and the Gendering of Punishment.' *Journal of Contemporary Ethnography*, 32(1), 41–76.

McGillivray, D. (2005) 'Fitter, Happier, More Productive: Governing Working Bodies through Wellness,' *Culture and Organisation*, 11(2), 125–38.

Markus, T.A. (1993) *Buildings and Power: Freedom and Control in the Origin of Modern Building Types*. London: Routledge.

Marx, G.T. (1988) *Undercover*. Berkeley: University of California Press.

Marx, G.T. (2003) 'What's New about the "New Surveillance"? Classifying for Change and Continuity.' *Surveillance & Society*, 1(1), 9–29.

Mathiesen, T. (1997) 'The Viewer Society: Michel Foucault's "Panopticon" Revisited.' *Theoretical Criminology*, 1(2), 215–33.

Morgan, M.J. (2004) 'The Garrison State Revisited: Civil-military Implication of Terrorism and Security,' *Contemporary Politics*, 10(1), 5–19.

Mottier, V. (2001) 'Foucault Revisited: Recent Assessments of the Legacy.' *Acta Sociologica*, 44, 329–36.

Murakami Wood, D. and S. Graham (2006) 'Permeable Boundaries in the Software-sorted Society: Surveillance and the Differentiation of Mobility.' In M. Sheller and J. Urry (Eds) *Mobile Technologies of the City*. London: Routledge, 177–91.

Nellis. M. (2005) 'Out of this World: The Advent of the Satellite Tracking of Offenders in England and Wales.' *The Howard Journal*, 44(2), 125–50.

Norris, C. (2003) 'From Personal to Digital: CCTV, the Panopticon and the Technological Mediation of Suspicion and Social Control.' In D. Lyon (Ed.), 249–81.

Norris, C. and G. Armstrong (1999) *The Maximum Surveillance Society: The Rise of CCTV*. Oxford: Berg.

Norris, C., J. Moran, and G. Armstrong (1998) 'Algorithmic Surveillance: The Future of Automated Visual Surveillance.' In C. Norris, J. Moran and G. Armstrong (Eds) *Surveillance, Closed Circuit Television and Social Control*. Aldershot: Ashgate, 255–75.

Pecora, V.P. (2002) 'The Culture of Surveillance.' *Qualitative Sociology*, 25(3), 345–58.

Porter, S. (1996) 'Contra-Foucault: Soldiers, Nurses and Power.' *Sociology*, 30(1), 59–78.

Poster, M. (1990) *The Mode of Information: Poststructuralism and Social Context*. Cambridge: Polity Press.

Poster, M. (1996) 'Databases as Discourse; or, Electronic Interpellations.' In D. Lyon and E. Zureik (Eds) *Computers, Surveillance and Privacy*. Minneapolis: University of Minnesota Press, 175–92.

Rose, N. (2000) 'The Biology of Culpability: Pathological Identity and Crime Control in a Biological Culture.' *Theoretical Criminology*, 4, 1, 5–34.

Rule, J.B. (1973) *Private Lives, Public Surveillance: Social Control in the Information Age*. London: Allen Lane.

Shearing, C.D. and P.C. Stenning (1985) 'From the Panopticon to Disney World: The Development of Discipline.' In A. Doob and E. Greenspan (Eds) *Perspectives in Criminal Law*. Toronto: Canada Law Books, 335–49.

Short, E. and J. Ditton (1998) 'Seen and Now Heard: Talking to the Targets of Open Street CCTV.' *British Journal of Criminology*, 38(3), 404–28.

Simon, B. (2004) 'The Return of Panopticism: Supervision, Subjection and the New Surveillance.' *Surveillance & Society*, 3(1), 1–20.

Simon, M. and S. Feeley (1994) 'Actuarial Justice: Power/Knowledge in Contemporary Criminal Justice.' In D. Nelken (Ed.) *The Future of Criminology*. London: Sage, 173–201.

Staples, W.G. (1994) 'Small Acts of Cunning: Disciplinary Practices in Contemporary Life.' *The Sociological Quarterly*, 35(4), 645–64.

Stenson, K. (2005) 'Sovereignty, Biopolitics and the Local Government of Crime in Britain.' *Theoretical Criminology*, 9(3), 265–87.

Thrift, N. and S. French (2002) 'The Automatic Production of Space.' *Transactions of the Institute of British Geographers*, 27(4), 309–35.

Torpey, J. (2000) *The Invention of the Passport: Surveillance. Citizenship and the State*. Cambridge: Cambridge University Press.

Vaz, P. and F. Bruno (2003) 'Types of Self-Surveillance: From Abnormality to Individuals "At Risk".' *Surveillance & Society*, 1(3), 272–91.

Ventura, H.E., J.M. Miller and M. Deflem (2005) 'Governmentality and the War on Terror: FBI Project Carnivore and the Diffusion of Disciplinary Power.' *Critical Criminology*, 13, 55–70.

Voruz, V. (2005) 'The Politics of *The Culture of Control*: Undoing Genealogy.' *Economy and Society*, 34(1), 154–72.

Weibel, P. (2002) 'Pleasure and the Panoptic Principle.' In T.Y. Levin et al. (Eds), 206–23.

Yar, M. (2003) 'Panoptic Power and the Pathologisation of Vision: Critical Reflections on the Foucauldian Thesis.' *Surveillance & Society*, 1(3), 254–71.

Zuboff, S. (1988) *In the Age of the Smart Machine*. New York: Basic Books.

Chapter 24
Beyond the European Province: Foucault and Postcolonialism

Stephen Legg

Introduction

The colonization of most of the free world between the 16th and 21st centuries has brought not only territorial but also epistemic and historiographical violence and domination. The end of formal occupation has not signalled the withdrawal of colonial categories, procedures and technologies of rule, nor has it beheaded Europe as the sovereign subject in deference to which many postcolonial[1] histories and geographies are constructed (Chakrabarty 2000). Whilst Michel Foucault has provided many of the tools that are necessary to unpick the power-knowledge relationships of post-Enlightenment Europe, especially in their spatial groundedness, his silence on the colonial construction of European modernity and the mutual constitution of 'metropole' and 'periphery' is astounding.

This chapter will begin by examining the haunting presence of colonialism in Foucault's writings and will then explore how geographers have tried to commune with our discipline's colonial past and postcolonial present. The use of Foucault in the work of Edward Said and the Subaltern Studies Group will be investigated to suggest a movement towards an analysis of the lived and the governmental that chimes with much existing geographical research into the postcolonial.

The path I tread here is only one of the many routes through a field of study that could span, at least, Alexander the Great to George W. Bush and Tony Blair, and every country on earth whether as a colonized, colonizing, or indirectly influenced nation. Postcolonial forces operate at every scale, from trans-national flows of capital or bodies, global imaginary geographies, national stereotypes, urban re-mappings, to domestic routines and individual psychology. Postcolonial theory itself is a complex mix of theorists, including Homi Bhabha, Jaques Derrida, Franz Fanon and Gayatri Chakravorty Spivak. Moreover, Foucault has been used to analyze

---

1   I use the term 'postcolonial' here to refer to the interaction between colonized and colonizing populations following initial contact, although this need not have been face to face, such as in the mediated contact of trade networks. The term thus encompasses the experiences of both groups during and after the period of formal rule, if there was one. See Gandhi (1998, 3–4) for a discussion of the term.

postcolonial relations throughout the world, including Latin America (Trigo 2002, Outtes 2003), Africa (Mbembe 2001), ex-settler colonies (Clayton 2000; Dean and Hindess 1998; Henry 2002) and South Asia. The predominance of the latter in postcolonial theory may be a problem in itself, globalizing the experiences of a few colonies into the universal experience of the colonized. Such tendencies can be countered by a continuing commitment to studying the particular and specific instances of colonization and postcolonial experience within globally structuring systems of postcolonial rule.

The Absent Presence of Colonialism in Foucault

Peter A. Jackson (2003) has summarized the many critiques of Foucault that claim that the 'difference' he theorizes is that of 'complexity', difference *within* a society, rather than 'multiplicity', differences *between* societies. In his mostly local or national scale of study this is true, a fact compounded by his focusing on Europe in general, and France in particular. There are enough passing references to show that Foucault was aware of the importance of the colonial world, yet the significance of these traces of colonialism is much debated. In 1989 Uta Liebman Schaub suggested that the non-West operated as a counter-discourse or subtext that affected Foucault's mode of thought; the unspoken ground from which he attacked Western thought. Schaub (1989, 308) even suggested that Foucault, like many of his contemporaries, was influenced by eastern philosophy. However, critical commentary has focused more on how Europe and its colonies were mutually constitutive, and whether this was acknowledged in Foucault's writings. These constitutions can be separately considered, rhetorically if not historically, as practical, epistemic, and disciplinary.

A Practically Constitutive Outside

> A whole series of colonial models was brought back to the West, and the result was that the West could practice something resembling colonization, or an internal colonialism, on itself. (Foucault 1975–76 [2003], 103)

In a 1976 lecture Foucault admitted that the techniques and weapons Europe transported to its colonies had a 'boomerang' effect on the institutions, apparatuses and techniques of power in the West (see above). However, this is one of his few acknowledgements that the compendium of power techniques he assembled regarding Europe had extra-European origins (for further brief comments see Foucault 1972, 210; Foucault 1977, 29, 314; Foucault 1980, 17, 77, and the quotation below from Foucault 1961).[2] In a summary of postcolonial research, Timothy Mitchell showed that the panopticon itself, along with school monitoring, population government and

---

2   The ongoing translation of Foucault's lecture courses promises to add much, however, to postcolonial readings of his work. See references in *Psychiatric Power* (Foucault [1973–74] 2006, chapter four), and, especially *Security, Territory, Population* (Foucault forthcoming-b);

its cultural analysis, British liberalism's imagination, English literature curriculums and colonial medicine all had some of their many origins in the colonies (Mitchell 2000, 3). Driver and Gilbert (1998) have also shown how the material landscape of London was, in various ways, an intensely imperial space. These examples are beyond the more obviously 'colonial' techniques of slavery, shipping, and plantations that impacted back on Europe. All of these imperial techniques were topographically re-inscribed in Europe and often failed to reveal their travels and complicity in consolidating the effects of territorial expansion. Despite his brilliance at thinking 'power-in-spacing', Gayatri Chakravorty Spivak (1988 [2000], 1449–50) justly claims that Foucault's analysis actually produced a miniature version of colonialism, one that replayed the management of space and peripheral populations through the screen allegories of doctors, prisons, and the insane.

While Edward Said's eventual rejection of Foucault concerned his broader philosophy, he also criticized Foucault's Eurocentrism and tendency to universalize from French case studies (Said 1984a, 10). The ethnocentrism of this work clashed with Said's belief that discipline was used to administer, study and reconstruct, then to occupy, rule and exploit, almost all of the world (Said 1984b, 227). To Said, Foucault's carceral system was strikingly like the Orientalism he described. The systems were, of course, linked by networks of discursive and practical connections (Lester 1998). But beyond the humanitarian debates sparked by colonialism or the commodities and images consumed in Europe, there were also more fundamental processes of mutual constitution. Colonial environments threatened an intermixing of races, genders and classes that demanded reinforced distinctions of race, sexuality, culture and class (Mitchell 2000, 5). These thematics found their way back to the metropole and relayed a symbolic and material reworking of the European Self.

An Epistemologically Constitutive Outside

> Within the universality of Occidental *ratio* there is to be found the dividing line that is the Orient: the Orient that one imagines to be the origin, the vertiginous point at which nostalgia and the promises of return originate; the Orient that is presented to the expansionist rationality of the Occident but that remains eternally inaccessible because it always remains the limit. (Foucault 1961, iv, translated in Schaub 1989, 308)

Pre-dating Said's (1978) *Orientalism* by 17 years, Foucault acknowledged in a previously un-translated passage (although see Foucault 2005, xxx) the formative role of an imagined Orient on European collective memory (see above). While Said famously drew out this imagination, Ann Laura Stoler (1995) has done much to examine how imperial notions of race and sexuality constituted the European bourgeoisie. Drawing on Foucault's histories of sexuality (1979, 1986a, 1986b) and the *Society Must be Defended* lecture courses (1975–76 [2003]), Stoler showed that discourses of sex were on a 'circuitous imperial route' and that bourgeois identity

---

here Europe itself is portrayed as a post (Holy Roman) imperial space, while the constitutive nature of the colonial economy is explicitly addressed.

was itself racially coded. Within the complex routings by which biopower sought to regulate national populations, sex became a state target while race discourses became the effect, taking up and re-moulding older forms of racism. While Mitchell (2000, 13) warns that this represents a double overlooking of Empire, negating the colonial origins of 18th–19th century racisms, Stoler acknowledged the paradoxical nature of a colonial biopolitical state that claimed to augment life, yet administered the right to kill. It was the role of race to decide who would live and die, the administration of what Achille Mbembe (2003) has termed 'necropolitics'. This racialized politics of classification was taken up in Stoler's (2002) later consideration of the normalizing activities of the state in the colonies themselves. Racism was here shown to thrive upon lines of unclear difference, combining pseudo-scientific symbolics of blood with cultural contagion theory.

As such, Stoler (2002, 142) showed that though Eurocentric, Foucault was not blind to race and its potential imperial connections. She also showed that, given Foucault's two years spent in Tunisia (1966–68), this Eurocentrism remains intriguing, as does the lack of study of the *Archaeology of Knowledge* (Foucault, 1972) that he wrote on the basis of his lectures there. Robert Young (2001, 395–397) has written of Foucault's experiences and interest in political struggles at this time, but also how he used his distance from home to critically and ethnographically consider France and the West. As against *Madness and Civilization* (Foucault 1967), Foucault (1972) argued against the Other's separated and silenced existence. Homi Bhabha (1992 [2000], 130) has similarly claimed that within Foucault's 'massive forgetting' there is a metaleptic presence of postcolonialism. In *The Order of Things* (Foucault 1970, 369) anthropology emerges to confront the universalist claims of history, marking it out as the product of a European homeland. Historicist claims are thus exposed as dependent upon the technologies of colonialism, establishing anthropology as the counter-discourse of modernity.

However, such interpretations read much into the silences and cracks of Foucault's writings. This corpus, Mitchell Dean (1986 [1994], 289) has suggested, saw Foucault pull back from the challenge of deconstructing the 'West' as a critical ethnographer and re-colonize his radical insights within an analysis of western modernity that, Mitchell (2000, 16) argues, reproduced the spatialization of modernity. The historical time-scheme of colonizing Europe captured the histories of overseas and returned them to the ordering, historicist logic of the colonial core. Undoing this process, and bias in Foucault's writings, is not just a task of re-writing history, but of pursuing discourses, and disciplines, that though complicit with colonial states in the past, preserve the potential to mobilize counter-discourses of modernity.

A Discipline Constituted Outside

Felix Driver (1992) used Foucault's writings to excavate a colonial history of the geographical discipline that paid attention to its institutional, rather than philosophical or scientific, genealogy. He suggested a thoroughly Foucauldian reading that would pay attention to the various types of powers at play within the rise of geography as a discipline and the internal contradictions and resistances it came across in the consolidatory age of *Geography Militant* (Driver 1999). Stressing the spatiality of the discipline, Daniel Clayton (2001/02) has emphasized the need to trace these resistances in the colonial margins, as well as the imperial metropole.

Derek Gregory (1998) further mapped out the imaginary geographies by which geography as a discipline had imposed its Eurocentric worldview on the territories it surveyed. As with the sovereign Europe Foucault analyzed, the discipline of geography has been one of 'constitutive exclusions and erasures' (Gregory 1998, 72), viewing certain things and ignoring others through representational 'geo-graphs'. For example, the geo-graph of 'absolutizing time and space' established Europe as the sovereign centre, but also divided the periphery into those more or less deserving of rights and along axes of alterity, forming a structured yet unstable hierarchy of difference. Other modalities concerned exhibiting the other, normalizing the subject and abstracting culture and nature, which all contributed to the view of the world presented by the geographical discipline to its students and author audiences through its home institutions.

While the implications of geography within the colonial past is increasingly clear, the colonial present requires constant attention. Jennifer Robinson (2003) has focused attention on how to bring about postcolonial geographical practice. Robinson links Chakrabarty's assertion of Europe as the historical core to the geographical practices that put it there and to the universalizing tendencies of some post-1960s geographical theory. To undermine the epistemic violence of these traditions Robinson suggests: we acknowledge location, and the limits to analysis it poses; that we reincorporate area and development studies in innovative formations; that we engage with regional scholarship that disrupts dominant locations; and that we transform the conditions for the production and circulation of knowledge, regarding publication, sources and readership. These processes must, of course, take place within active research. Geographical research along these lines has been framed within readings of Foucault following Said's influential interpretation.

Said: The Presence of Foucault

There is a certain irony in the discrepancy between the Foucault that Said propounded in his earlier theoretical writings, and the afterlife of Foucault's analytical categories that were taken up in colonial discourse analysis and postcolonial studies more broadly. While Said initially stressed the worldliness of texts and the materiality of discourse, the various studies that claimed his lineage were often focused on an individual text or the relationships between separate texts, rather than their historical and geographical contingency. Yet, while Said was an early champion of Foucault, it is also the case that he (1993 [2004], 214) rejected Foucault for his political quietism, while also claiming that he had got all he needed from Foucault by the publication of *Discipline and Punish* (Foucault 1977). From this point onwards the distance between Said's humanism and Foucault's anti-humanism became more pronounced.

Despite this, Said moved in the 1990s towards a geographically grounded form of analysis which has more in common with Foucault's post-1978 lectures and writings on government than his earlier linked, but distinct, work on the materiality of discourse. This trajectory, and the positioning of geographical research within it, will now be traced.

The Materiality and Discontinuity of Discourse

In 1972, in the first edition of the journal *boundary 2*, Edward Said advocated the use of Michel Foucault (Said 1972, the article was re-written and published in Said, 1975, 277–343).[3] Against later criticism of Said's approach being atemporal and textual, he emphasized four particular elements of Foucault's work. 'Reversability' supplanted the search for origins, development, or authors with the primacy of discourse and verbal usage. 'Discontinuity' undermined the idea of unlimited, silent, and continuous discourses in favour of the discontinuous practicalities that cross, juxtapose and ignore each other. This emphasis on difference, Said suggested, could be extended to include the differences not just within, but between societies, privileging histor*ies* over History (referencing Foucault 1961). As such, the idea of discourse from Foucault (1970, 1972) was one of dispersal and fragmentation that saw any seriality as an internal order within dispersal. The third Foucauldian method was that of 'specificity' which saw the boundaries of individual discourses policed by what is deemed wrong or forbidden, while the final method was that of locating 'exteriority', the transcendental homelessness of subjectivities incompatible with a discursive norm, whether deemed mad, dangerous or, like the Marquis de Sade, a subject of total desire.

However, it was the idea of discourse presented in 1978's *Orientalism* which had a longer lasting effect, one which Young (2001, 386) claims is dissimilar to that of *The Archaeology of Knowledge*. *Orientalism* depicted the dichotomization and essentialization of Europe's worldwide geopolitical imagination. The discourse of orientalism could be traced in academic disciplines, a broader ontological and epistemological division between East and West, and finally in the institutions that governed the Orient. While flitting between different writings and institutions, Said focused on certain texts without attendant study of their environments of production. The emphasis on texts written from other texts led to an analysis of stereotypes that were posed as mis-representations, marking a move from a Foucauldian discourse analysis to a more Gramscian investigation of ideological representations. Timothy Brennan (2000) has, indeed, asserted that *Orientalism* is not Foucauldian due to its humanist specializations, sweeping syntheses, aesthetic indulgence and totalizing appetites. The sprawling debate from this tension is summarized in Ashcroft and

---

3   Against this, I can find no reference to Said in Foucault's writings. This is despite a brief correspondence following the publication of *Orientalism* (Salusinszky 1987, 136) and a meeting in 1979 in Foucault's flat, where Said noticed his *Beginnings* (1975) on the bookshelf (Said 2000a).

Ahluwalia (1999, 76–80), but within this argument the significance is perhaps that without Gramsci's notion of hegemonic power relations, Said felt that Foucault alone lacked political bite.

The Spiderless Web

In 1984 Said marked the beginning of his formal distancing from Foucault. While still favouring Foucault's political view of language and his geopolitical interest in the control of territory, he launched two critiques based around notions of agency and power. Firstly, he questioned Foucault's lack of interest in explaining why people or things were distributed as they were (Said 1984b, 220). Without immediacy or intentionality the historical evolutions of power Foucault suggests would have no drive. As Alison Blunt (1994, 54) has suggested, contra Foucault, it *does* matter who is writing; their conditions of authorship, gendered identity, or perception of audience must play a part. Similarly, Alan Lester's emphasis on trans-imperial networks of discursive connections maintains a focus on the agency of individuals exercised in facilitating flows and constructing networks (Lester 2002, 29). Said later referred to the tension between the anonymity of discourse and the will to power of particular egos as an 'almost terrifying stalemate' (Said 1984a, 6) and forcefully rejected the notion that he suggested there was no voice to answer back against resistance (Said 2002, 1).

Said's criticism of agency fed into the later comments on Foucault's supposedly passive and sterile view of power, which, he claimed, failed to consider why power was gained and held on to. The existence of class struggle, imperialist war, and resistance show us that power does remain with rulers, monopolies and states: as Said (1984b, 221) put it, you cannot have the web without the spider. As such, Foucault failed to consider the intentionality and effort of history, refused to imagine a future rather than analyze the present, and failed to consider the space of existence beyond the power of the present (Said 1984b, 245–7).

This critical position was maintained throughout Said's later writings. In his 1984 obituary article for Foucault, Said respectfully emphasized Foucault's influence and his entangling of power and resistance, yet still decried the pessimism and determinism of his later work (Said 1984a, 3, 6). Said's (1986) article on 'Foucault's imagination of power' stands as his most vociferous rejection of Foucault's account of the supposedly unremitting and unstoppable expansion of power. As against Noam Chomsky's insurgent consideration of what could vanquish power relations, and his utopian postulations of what cannot be imagined, Foucault was claimed to only imagine what one could do with power if one had it, and what one could imagine if one had power. As such, Foucault's imagination, unlike Gramsci's, was thought to be with power, rather than against it. Paul Bové (1986 [2001]) approved of Said's rejection of Foucault, warning of the 'immoral consequences' of the latter's system, which prevented a recognition of resistance, denied the imagination of alternative orders and explained all social phenomenon by the structure of power. Said's wariness of Foucault's emphasis on assimilation and acculturation was re-emphasized in a 1986 interview (Salusinszky 1987, 137) and was unchanged by 1993 when Foucault was portrayed as scribing the victory of power (Said, Beezer and Osbourne 1993 [2004], 214).

Said acknowledged that his *Culture and Imperialism* (1993) was written against the negative effects of Foucault in the book to which it was the sequel, *Orientalism* (Said in Said et al. 1993 [2004]). Against the impression of an orientalism that continued to grow without contestation, a wider geographical scope and an emphasis on the contestation of territory allowed Said to examine people's counter-will as framed by Raymond Williams's cultural reading of Gramsci. In the 20 years since his *boundary 2* article, the Foucault of reversibility, discontinuity, specificity and exteriority was lost amongst the more abstract Foucault of power-knowledge relations. This bias fails to do justice to the relevance and utility of Foucault's earlier and later writings on archaeology, discourse and governmentality that are undergoing a current re-assessment beyond Said's dismissal.

Travelling with Foucault

Said (1984b, 227) famously argued that theories travel, each having points of origin, a distance that is traversed, conditions that are confronted, and transformations that occur along the way. Said took Foucault both to America, institutionally, and to the Orient, theoretically. Between the two, Foucault's writings seeped into the emergent field of postcolonial studies and were incredibly influential. But theories also travel through time. As has been shown above, Foucault has travelled to places he never envisaged, confronted conditions he didn't expect, and has been over time, in cases, transformed beyond recognition. Ashcroft and Ahluwalia (1999, 82) admitted that Said only took what he needed from Foucault (also see Gregory 2004b), resulting in an ambivalent privileging of authors and literature which *itself* contracted the scope for resistance. Indeed, it was Said's *lack* of a Foucauldian approach, rather than its presence, which decreased his attention on the non-representational spaces of the everyday in which the subaltern vocabulary of resistance is often located (see Smith 1994, 494). As such, the field of colonial discourse analysis, which played such a key role in establishing postcolonial studies, bore a bias towards the colonial mindset and its representation in textual accounts (see the emphasis on literary sources in, for instance, Ashcroft, Griffiths and Tiffin 1989; Behdad 1994; Lowe 1991; Slemon 1989; Spurr 1993; Suleri 1992).

Driver (1992, 33) suggested that both Foucault's *Discipline and Punish* and Said's *Orientalism* were similarly misread, downplaying the heterogeneity of modern discourses, the controversies and resistances they contain, and the specificity of discursive regimes. However, Young (2001, 407) suggested that it is Said's misrepresentation of Foucault that lays his work open to such misreadings. Young showed how Said came to interpret Foucault as dealing with textuality, estranging the Orientalist discourse from its material circumstances and welding it to representations. The effect of this reading, Young (2001, 389) argued, can be traced through to the common criticisms of colonial discourse analysis. He categorized these as follows:

- Historicity: the generalization from a few literary texts that tend to be de-historicized and un-situated in non-discursive texts.
- Textuality: the treatment of texts as historical documents, without accompanying materialist historical inquiry or political understanding.
- Representation: if all truth is representation, what was mis-represented? How can the subaltern speak?
- Homogeneity and determinism: notions of discourse that override historical and geographical difference and problematize how people become subjects in such discourses.

Young argued that an analysis more loyal to *The Archaeology of Knowledge* would negate many of the criticisms outlined above. The archaeological model of discourse eschews a disembodied study of intertexts, of representations and interpretation, in favour of studying the practical emergence of knowledge at the interface of language and the material world. Discourse analysis should, therefore, be situated at the contact zone of materiality, bodies, objects and practices. As the network which links together statements, objects and subjects, discourses must be fragmented and heterogeneous, yet are unified by particular rules that operate on all individuals. However, these rules lead to multiplicity, not uniformity, of choice and action (as was still asserted in Foucault 1979, 100).

As such, Young argued that Foucault's conception of discourse is actually antithetical to postcolonial theories that posit a subjective voice of the colonized against an objective, colonizing discourse (also see Brennan 2000). Rather, discourses are unstable and cause the proliferation of subaltern discourses, whether as speaking from outside colonial discourses or mounting counter-discourses in direct confrontation (also see Terdiman 1985). Thus, a Foucauldian colonial discourse analysis would not be so vulnerable to the four criticisms outlined above, focused as it would be around using discourse to study colonial practice in successive administrative regimes (for such a place bound approach see Chatterjee 1995, 24). This brings colonial discourse analysis closer to work both on colonial governmentality and a material geographical analysis.

Re-materializing Postcolonial Geography

Most geographers will take Young's arguments as reaffirmation, rather than revelation. Although not always referencing Foucault directly, but often in Foucauldian terminology, there is an entrenched tradition within the discipline that argues for a material grounding of postcolonial analyses (see Clayton 2004). Neil Smith (1994), in his review of *Culture and Imperialism*, showed that Said's newfound commitment to resistance was constrained within his textual reading of discourse, thus presenting the struggle for decolonization as a literary affair. Jane

Jacobs (1996, x) attempted to reorient the spatial emphasis in colonial discourse analysis from metaphor to 'real' geographies. While not actually dismissing textual representations as unreal, Jacobs traced imperial remains not just *in*, but also *through* and *about* space. It was at the contact zone of materiality and practices that Jacobs sought out the 'promiscuous geographies of dwelling in place' that activated imperial pasts in postcolonial presents. While Clive Barnett (1997) reassured those who feared a 'descent into discourse', Driver and Gilbert (1998, 14) repeated worries about the textual nature of postcolonial cultural geographical work and argued for an appreciation of the imperial inheritance in different types of urban space, whether architectural, spectacular or lived.

Reading Foucault's work on the political function of discourses, Alan Lester (1998, 2001, 2002) has been at the forefront of empirical research into not just the material practicalities of colonial rule but also the networking functions of international colonial discourses. His attention to the various sites in which power and knowledge were intertwined has led to a sophisticated understanding of grounded imperial power, with all the tensions and contestations that this involved. James Sidaway (2000, also see Sidaway, Bunnel and Yeoh 2003) repeated calls for a movement beyond discourse and representations to material practices, actual spaces and real politics, although these are all very much central to a Foucauldian understanding of discourse itself. More in line with Foucault's writings, Cole Harris (2004) has recently argued for an examination of the physical dispossession of the colonized rather than their misrepresentation.

Accompanying these calls for a more material approach, Cheryl McEwan (2003) has criticized the postcolonial tendency to separate discourses from lived experience, its failure to propose solutions, and its privileging of theory and culture over political and ethical responsibilities. In response, she suggested re-materializing postcolonialism, exploring the lived nature of postcoloniality, and advocated tactics for linking the textual with macro-issues. Conjoining the political-economic, the ethical, and the material should create opportunities in the present for, as Jacobs (2001) insisted, postcolonial study has a contemporary effect. Derek Gregory (2004a) has recently demonstrated the capacity of Foucauldian history and cultural geography to disrupt any complacency about the colonial past. In a series of accounts regarding the colonial historico-geographical present in Palestine, Afghanistan and Iraq, Gregory has traced the violent, physical and material manifestations of imaginary geographies bred through decades of colonial administration. These discourses are filled with the intentional voices of perpetrators, commentators and victims, and are scarred with the searing potential of counter-discourses to erupt in the space between the contradictory statements of neo-colonial discourses.

What is most surprising about Said's work after his rejection of Foucault is not only how much he retains his geographical emphasis, but the degree to which this emphasis becomes not just imaginary but also governmental. Corollaries develop not just with Young's Foucauldian colonial discourse analysis but also with a colonial application of Foucault's (1978 [2001]; 1979) later writings on governmentality and biopower. While *Orientalism* had acknowledged institutions of administration as the third facet of orientalist discourse, Said (1984b, 219) later expressed his interest in Foucault's (1980, 77) writing on Geography; the control of territories, their demarcation and the study of armies, campaigns and territories (also see Gregory 1995). Here he also expressed the need to go beyond a purely linguistic discourse not just in the *Orientalism* tripartite of philology, ontology and institution, but also to the colonial bureaucracy and its virtual power of life and death over the Orient.

This movement was continued in *Culture and Imperialism*, despite his stubborn textualism (however, for some instances of Said grounding texts in material context see Gregory 1995, 453). Interest was expressed in the 'actual geographic underpinnings' beneath social space and the ways in which geographical projections make possible the construction of knowledge (Said 1993, 93). Physical transformations were noted, ranging from ecological imperialism and urban reconstruction down to the micro-physics of organizing everyday interaction (1993, 132). But the geographical element was also essential to anti-imperialism, at first through imagining the recovery of loss, and later the recovery of territory (1993, 271). This was part of Said's ongoing rethinking of the 'struggle over geography' (Said in Said et al. 1994, 21), which was affirmed in his later comments on memory and geography (Said 2000b). Here orientalism itself was stressed to be about the mapping, conquest and annexation of densely inhabited, lived-in places, as part of an unending struggle over territory and memory.

By the late 1990s Said was advocating a form of geographical research that explored the diverse range of governmental tactics used to order space and the various different forms of memory production that negotiated this space. Such writings cannot be considered outside of his committed involvement with the Palestinian cause, which did not always feature in his theoretical work (see Gregory 1995; Said 2000b). The Subaltern Studies Group (SSG) also produced theoretically sophisticated material that remained oriented around the present. Said (1988) had praised the SSG under their editor Ranajit Guha, for their innovative archival work and for searching out non-elite histories not only in elite writings but also in mundane, everyday texts. He later acknowledged this level of research as, perhaps, more important than his preferred level of representations:

> Now there is of course a subcultural tradition, for example, as Guha and others have shown, a whole range of colonial writing which is not artistic but is administrative, is investigative, is reportorial, has to do with conditions on the ground, has to do with interactions depending on the native informant. All that exists, there is no question of that. I was trying to adumbrate, perhaps a less important, but to my way of thinking, a larger picture of a certain kind of stability. (Said 2002, 7)

Subaltern Studies: From Gramsci to Governmentality

Ranajit Guha (1982) established the Subaltern Studies publication series in an attempt to grant credit and autonomy to the peasant classes of India as a politicized, active section of the population; the non-elite. While the Gramscian notion of the subaltern would later be extended from the military or class concept to that of race, sexuality, caste or language, the emphasis remained on detailing the existence of action that could not be teleologized into a colonial, nationalist, or Marxist narrative. In over 20 years the literature by Subaltern Studies authors has converged with certain postcolonial themes, with an increasing use of Said but a decline from heavily Marxist origins to a 'spirit of Marx' (Chaturvedi 2000, vii) in later work.

The Spirit of Foucault

Partha Chatterjee has consistently worked to bring the SSG in line with Foucault's and Said's writings. While his initial contribution (Chatterjee 1983) dealt with the transition from feudalism to capitalism and Marx's theories on property, this was presented as an analysis of 'modes of power' and ended with an avocation of Foucault's capillary and embodied understanding of power relations. However, marking the qualified application of western theories to India that would characterize the SSGs work, Chatterjee asserted that modern power in the 'Third World' was combined with older modes of control and different state formations to those in Europe (for a reaffirmation of this view see Chatterjee 1995, 8).

Having first read Said in 1980 (Chatterjee 1992, 194), Chatterjee (1984) applied his theories to India in claiming that nationalists operated within orientalist discourses and with orientalist stereotypes *themselves*. As such, the representational structure of nationalist thinking corresponded at times to the structure of power it tried to repudiate. David Arnold's work on the Madras police force applied Foucault's (1977) work to India, looking at the removal of social intermediaries, the surveillance and discipline of the force itself, and political criticism of the police as anti-national during the non-cooperation movement (Arnold 1984). Later work on anti-plague measures showed that attempts to initiate mass state intervention between the 1890s and 1930s was met with a hostile response, not passivity or docility (Arnold, 1987). This reaction was against the latent claim for increased power over the body, as also expressed in dictates on widow immolation, whipping and medicine. Arnold's (1994) later work also included an investigation of colonial prisons as lived spaces of resistance but also as abstract spaces for the collection of knowledge about Indian bodies.

This usage of Foucault was, I would suggest, forestalled and redirected by a shift that took place in the mid-1980s. This marked a turn to 'discourse' as it was increasingly being defined by postcolonial studies, rather than being akin to Foucault's original notion. The rupture was triggered by a debate over the epistemological validity of the subaltern as an autonomous subject of history. Spivak (1985, 338) argued that the attempt to discover or establish a peasant or subaltern consciousness was positivistic, denoting a single, underlying consciousness. In the place of this romantic quest should be, she claimed, a charting of the subaltern-effect, the knotting of strands, whether political, economic, historical, or linguistic, that gave the effect of the operating subject. The fact that a strategically essentialist concept of the subject might be necessary to tie this knot was accepted as a valid risk for the political interest of the SSG project. This argument was affirmed by Rosalind O'Hanlon (1988) who criticized the retention of a humanist subject alongside the growing use of anti-humanist, post-structuralist theory. In 1988 Guha's retirement signalled the increase of post-modern theory within the group and a turn to the discursive construction of the subaltern(-effect).[4]

However, the 'discourse' used here was as much influenced by Spivak's readings of Derrida than that of Foucault. Spivak (1985, 330) had defined the SSG project as being about confrontation and change, but this was a change in sign-systems that classified, for example, crime as insurgency. These were 'discursive displacements' that charted people or events as political signifiers. As such, the SSG was claimed to examine the 'socius' as a sign-chain in which action marked a breaking of this chain. However, in this approach all attempts at displacement must be failures due to the breadth of colonial organization and the failure of the Indian bourgeoisie to politicize the peasantry. The focus from the fourth *Subaltern Studies* volume (1985) thus shifted to analyzing the difference of the subaltern that emerged within elite discourses (Prakash 1994). Chatterjee (1986), for example, showed how the agency of the common people was appropriated by the nationalist elite, leaving them as silenced fragments of a strengthening nation (Chatterjee 1993). This historiographical move *did* produce an innovative reading of sources for subaltern traces and stereotypes, yet the end result that was sought was one of failure. The textualism and political pessimism that resulted from such an approach has recently been challenged, but this has been within an understanding that subaltern studies be framed as a form of postcolonial criticism.

Gyan Prakash (1990) situated subaltern studies as a post-foundational history. He claimed it had overcome the depictions of India in orientalist texts as passive and separate, and in nationalist texts as autonomous and essential. He also criticized the essentialist notions of anthropology and area studies, along with the structural explanations of Marxist and social historians, much to the ire of O'Hanlon and Washbrook (1992). Against these traditions, and inline with Said's call to reject, not reverse, colonial categories, the SSGs charting of multiple and changing subject positions was claimed to be fully post-foundational, and postcolonial (Prakash 1994).

The SSG has come under constant and sustained attack, from within India and without (Chaturvedi 2000). Perhaps one of the most provocative critiques came from Sumit Sarkar (1996 [2000]), a former contributor to the series and member of the editorial team. Sarkar mourned the decline in the study of underprivileged groups and the attendant increase in studying the power-knowledge relationships of colonialism, which often inserted religious community as the consciousness of the non-West. Sarkar criticized Chatterjee for depriving both the masses *and* the

---

4   This shift can also be attributed to various personal factors. For instance, many of the SSG members acquired familial and institutional commitments that precluded long research trips to the archive in favour of textual analysis, while the previous approach had already occupied some contributors for a decade (Dipesh Chakrabarty, personal communication).

intelligentsia of agency, the latter of whom were just subjects within a derivate discourse of European nationalism and orientalism (for comments on Chatterjee's pessimistic view of the fate of women in the nationalist movement see Legg 2003). While reviewers had explained any essentialism within the SSG as residual Marxism, Sarkar stressed the ability of socio-economic analysis to fracture essential notions of identity. However, the Subaltern Studies authors have increasingly been returning to Foucault's work, especially that on government, to seek new ways of framing and searching for subaltern agency. Again, this return to the material and biopolitical has been pre-empted by a seam of postcolonial geographical research.

Spaces of Biopower

Apart from the theoretical calls to re-materialize, geographers have specialized in empirical research that has reinforced postcolonial development and elaboration of Foucault's theories. For instance, Jonathan Crush (1994) combined theories of panopticism with those of capitalist work-regimes to analyze South African mine compounds. Here architecture was used to increase visibility throughout the delimited space, although cultural forms of resistance proliferated in response through, for example, the production of liquor, hyper-masculine behaviour or the smuggling of banned medicines. James Duncan (2002) has, similarly, examined the attempted production of abstract space and bodies in Ceylonese coffee plantations. However, the workers engaged not only in resistance through insubordination or desertion, but also through exploiting the cracks in abstract space; minimalizing output, feigning sickness, and forging networks of counter-surveillance to indicate when the colonial gaze is untrained on the workers. Jennifer Robinson (2000) also focused on the embodied gaze, in the case of housing managers in 1930s South Africa. Moving away from the masculine vocabulary of many accounts of panopticism, Robinson showed that the surveying gaze took the form of friendly, female enquiry, forging links over racial boundaries. Indeed, in non-institutional cases the form of power seemed more liberal, ruling from a distance and through the powers of freedom.

Foucault's writings on governmentality have proven appealing to geographers for a variety of reasons. Firstly, they present an analytical programme for investigating modern regimes of government (Foucault 1978 [2001]). This may be through the individual categories of episteme, identity, visuality, techne and ethos (Dean 1999; Rose 1996), or through looking across these categories for evidence of regime change (Legg 2006b; Watts 2003). Secondly, the literature refers to a mode of power that has overcome, though retains features of, the power regimes of sovereignty and discipline with that of regulatory government. Regulation involves gathering information about people and territories, calculating and classifying this knowledge, and exerting power from a distance to normalize and stabilize a specific population.

The first task is what increasingly attracted the attention of Said, the geographies of which have been investigated by Matthew Hannah. In the 1870s the United States government sought to increase its knowledge concerning the Sioux Native American population through a social cycle of control concerning observation, judgement and enforcement (Hannah 1993). Attempts to fix the Sioux in one place only increased governmental awareness of how little information they had about these people and how problematic census taking would be. The census was one of the main means of establishing power-knowledge grids over opaque territories. Hannah's (2000) study of the extension of population assessments across the United States illustrates how closely the European colonizing nations shared techniques with internally colonizing postcolonial states.

In the case of British Columbia, Daniel Clayton (2000) has examined the processes of cultural interaction, modes of representation and local power relations during Western encounters with the natives between the 1770s and 1840s. Clayton examines just how Foucault's Eurocentric ideas can map onto peripheral areas through a genealogical tracing of relations through three phases of encounter structured by relations of science, profit, and imperial geopolitics. Following Clayton's work, Cole Harris (2004) has shown how natives were allocated reservation spaces, thus allowing development and reorganization outside these areas. While initial dispossession rested on the physical violence of the state as encouraged by capitalist interests, the legitimation of the scheme was cultural while the actual management of the dispossessed was disciplinary, combining the full spectrum of governmental tactics. Bruce Braun (2000) has also used the Canadian context to draw out the links between the physical sciences and the governmentality of the Victorian state.

While at times physically violent or overbearingly disciplinary, colonial and postcolonial states also sought to govern, which was the eventual outcome of many of the processes outlined above. Robinson (1997) has shown that apartheid in South Africa lasted so long because it manipulated populations through 'locations' that segregated different sub-groups who could be governed through their representatives. These biopolitical manipulations sought to normalize populations in terms of their behaviour while keeping them in visible and controllable places. However, the identity assumptions of biopolitical regimes in colonial contexts often fit neither into Foucault's assumptions about modern liberalism, or the genocidal extremes of the Nazi or Stalinist state. Rather, as Gregory (1998, 85–86) suggested, colonized people were often treated as the objects, not subjects, of rule in systems less individualizing than those of Europe (also see Chatterjee 1995, 8, and Vaughan 1991). This led to calculations that often prioritized cost and political threat over welfare, although such calculations were perfect material for critiques not just of colonial violence or intrusion, but of their active mismanagement (Legg 2006a; 2007).

As Stoler argued, sexual politics were central to the colonial state and marked the hub of 'biopower', the dovetailing of discipline and government. Exploring these intersections, Mike Kesby (1999) has used Foucault's writings on sexuality to explore corporeal demarcations of patriarchal space in rural Zimbabwe that influenced who the colonial authorities negotiated with and how. Philip Howell (2004a) has also argued that Foucault can be used in the colonies in terms of his work on biopower, normalization and spatial ordering. All these elements come together in his investigation of the regulation of prostitution in colonial Hong Kong. Here he makes clear that the European models based on self-disciplining subjects were not applicable, and gave way to the racial objectification and geographical segregation of a reluctantly expansive state (also see Howell 2004b). These themes of discipline, biopolitics, and government have informed a range of work by authors associated with the SSG and others working on South Asia.

Subaltern Negotiations of Governmental Spaces

David Arnold consolidated his work on colonial biopolitics with his *Colonizing the Body* (Arnold 1993), which explored the expansion of European medical practices, their cautious reception by indigenous populations, and how they were signified as representing more than simple health practice. David Scott (1995) has investigated 'colonial governmentality' as theory and practice in Ceylon/Sri Lanka. Scott stressed the need to examine the targets of rule, how they are conceived and the means used to conduct them through space, while simultaneously considered the effects of race and religion on these European developed technologies of control.

The most thorough application to date of the colonial governmentality approach has been provided by Gyan Prakash (1999). Prakash analyzed scientific structures and regulations as 'civilizing' strategies that targeted the population, yet in the process opened up a sphere of political activity in which nationalists could challenge the government. These processes were traced across a variety of geographical scales, from the institutions of the museum and Asiatic Society to the body, civic works and the imagination of the nation itself. Satish Deshpande (2000) has also adapted Foucault's work to the Indian nation, analyzing aspirational Hindu communalism as a heterotopia that attempts to mediate the utopic and the real.

The scope of practices within the framework of governmentality proportionally increases the scope across which one can look for resistance. This can operate from the level of societal or economic processes to the level of local technologies and bodies. Spivak (2000) has bridged the international and corporeal in suggesting that the 'new subaltern' is positioned by organizations like the World Bank or multi-national corporations as intellectual property whether in terms of agri- or herbi-cultural knowledge. Dipesh Chakrabarty and Partha Chatterjee have, however, looked instead to how governmental categories are lived and negotiated by subaltern populations.

Chakrabarty (2002), in his book *Habitations of Modernity: Essays in the Wake of Subaltern Studies*, has investigated the governmental roots of modern ethnicity. Noting how the notions of race explicated by Foucault and Stoler tend to be viewed in India as external, Chakrabarty traces the links between internal views of community and caste and the processes of ethnicity and government. The governmentality work is used to examine the structuring of the colonial Indian political imagination and the founding of categories that outlived the administration and contained the seeds of ethnic violence.

Chatterjee (2004) has produced a sophisticated account of the negotiation of population politics by the governed themselves. Here, politics is located not just as the outcome of the universal ideals of civic nationalism, but also as the cultural uptake of the categories mobilized by governmental rationalities. Against his earlier pessimism, Chatterjee holds up hope against governmental technologies merely being instruments of class rule in a global capitalist order. He claims that '(b)y seeking to find real ethical spaces for their operation in heterogeneous time, the incipient resistances to that order may succeed in inventing new terms of political justice' (Chatterjee 2004, 23). The argument is that most people in India today have tenuous rights and are not part of the elite civil society. This is despite still being within the government's reach through policies that target the 'political society' of the subaltern. Chatterjee suggested these tactics emerged in the 1980s, despite hinting at their colonial origins in an earlier paper (Chatterjee 2001, 175). Within this space, population groups can claim the rights of a community and a voice that arises from the *violation* of property laws and civic regulations that are so central to governmental order. Mediators are employed to bargain with the state for concessions that *are* delivered due to the sub-population's rights, not as citizens, but through their existence as living beings.

Although Chatterjee does not use these terms, I would suggest the subaltern he targets is one that precociously straddles the positions of *zoe* (the simple fact of living) and of *bios* (normalized behaviour and individual rights). Georgio Agamben (1998) has drawn on Foucault's writings to trace the genealogy of *homo sacer*, the subject so stripped of rights that he (*sic*, in Agamben's gendered language) can be sacrificed without penalty; s/he is bare life. Agamben traces the states of exception in which homo sacer have been produced, from ancient Rome to Auschwitz, which Derek Gregory (2004a) extends to Palestine, Iraq and Afghanistan. However, in going on to claim the camp as the *nomos* of modernity, surely Agamben conforms to the pessimism and determinism of which Foucault has been criticized? What other reactions could there be to the state of exception? What if the subjects so paraded there are re-embraced, their exposition demanding the restitution of rights in a state of reception? Chatterjee sees hope in the politics of objectification. The Indian Emergency of the 1970s represented an exceptional biopolitical stripping of the urban poor, denying them the right to biologically reproduce through sterilization. However, the demolitions and deaths at Delhi's Turkman Gate, Chatterjee (2004, 135) reminds us, led to a nationwide outcry, juridical protection for the poor, and contributed to the downfall of Indira Gandhi's government.

In a cross-disciplinary collaboration, Corbridge, Williams, Srivastava and Véron (2005) have brought detail to the politics Chatterjee describes, while carrying his hope against objectification through to an empirical study. They do this through explaining in detail how the rural subaltern see, and negotiate, the state. Taking Foucault's assertion that governmental techniques *make* the state as much as they are deployed by it, Corbridge et al. demonstrate how marginal populations meet the state, whether embodied in administrators or the policy initiatives of 'political society'. Development policies in the 1990s increasingly came to stress 'participation' as a means of conducting conduct and facilitating self-help that drew the state into new forms of personal contacts with its population. Here it had to negotiate local power networks, misunderstandings, authority figures, corruption, feedback and mobilized resistance from local mediators. The case studies show that most people actually experience a limited and capricious state and demand greater assurances and information before engaging with the policies it suggested. This approach rightly posits resistance and agency as central to governmental rationalities that must forge spaces of connection between the central state and marginal populations whilst remaining sensitive to the culture and politics of the locale. It is within such governmental negotiations of the economic, biopolitical and the social that current research is applying Foucauldian theory to the historically conditioned yet urgently contemporary moments of the postcolonial.

Conclusions

Current trends in postcolonial research, both within and without the geographical discipline, are pushing scale-sensitive examinations of material places that open up spaces to consider the activities of the subjectivized and the subaltern. At the non-representational level of the lived it is possible to trace discourses as Foucault described them; as the material and corporeal production of knowledge and practice. As Said suggested in his later work, and his political activism throughout his life, this necessitates an examination of postcolonial work on the ground as well as in imaginary geographies. While his turn to resistance remained locked at the representational level, the Subaltern Studies literature struggled to locate this resistance on the ground, while simultaneously looking at the discursive production of the oppressed. Foucault's (1975–76 [2003]) *Society Must be Defended* lectures ended with a discussion of biopolitics after dwelling on race, but actually began with lectures on subjugated knowledges and the power of memory. As he urged towards the end of his life, no doubt in reaction to accusations of his political pessimism, resistance and local configuration had to be acknowledged in all power relations. It is at this level of realization and mobilization that geographical research on the postcolonial has excelled. If, as Chakrabarty suggests, Europe remains the sovereign subject of much postcolonial history, historiographical regicide must be worked towards through a combination of the tactics described above: a sensitive and cosmopolitan scholarly practice; a geography that is attuned to material as well as textual power relations; research of compatible yet different modes of power at a variety of scales; and an awareness of the agency and resistance of the individuals that may be the target government, capitalist, nationalist or communal regimes, but are never wholly constituted by them.

Acknowledgements

This chapter was written as part of a Junior Research Fellowship at Homerton College, Cambridge. Many thanks for their support and comments go to Dan Clayton, Stuart Corbridge, Jim Duncan, Derek Gregory, Gerry Kearns, Phil Howell, Miles Ogborn, and Si Reid-Henry. Thanks also go to Gyan Prakash and Dipesh Chakrabarty for their email correspondence. All mistakes are, of course, my own.

References

Agamben, G. (1998) *Homo Sacer: Sovereign Power and Bare Life*. Stanford, California: Stanford University Press.

Arnold, D. (1984) 'Bureaucratic Recruitment and Subordination in Colonial India: The Madras Constabulary, 1859–1947.' In R. Guha (Ed.) *Subaltern Studies IV*. Delhi: Oxford University Press. 1–53.

Arnold, D. (1987) 'Touching the Body: Perspective on the Indian Plague.' In R. Guha (Ed.) *Subaltern Studies V*. Delhi: Oxford University Press. 55–90.

Arnold, D. (1993) *Colonizing the Body: State Medicine and Epidemic Disease in Nineteenth-century India*. Berkeley: University of California Press.

Arnold, D. (1994) 'The Colonial Prison: Power, Knowledge and Penology in Nineteenth-century India.' In D. Arnold and D. Hardinman (Eds) *Subaltern Studies VIII*. Delhi: Oxford University Press. 148–87.

Ashcroft, B. and P. Ahluwalia (1999) *Edward Said: The Paradox of Identity*. London; New York: Routledge.

Ashcroft, B., G. Griffiths and H. Tiffin (1989)*The Empire Writes Back: Theory and Practice in Post-colonial Literatures*. London: Routledge.

Barnett, C. (1997) '"Sing Along with the Common People": Politics, Postcolonialism, and Other Figures.' *Environment and Planning D-Society & Space*, 15(2), 137–54.

Behdad, A. (1994) *Belated Travellers: Orientalism in the Age of Colonial Dissolution*. Cork: Cork University Press.

Bhabha, H. (1992 [2000]) 'Postcolonial Criticism.' In D. Brydon (Ed.) *Postcolonialism: Critical Concepts in Literary and Cultural Studies*. London: Routledge. 105–33.

Blunt, A. (1994) 'Mapping Authorship and Authority: Reading Mark Kingsley's Landscape Descriptions.' In A. Blunt and G. Rose (Eds) *Writing Women and Space: Colonial and Postcolonial Geographies*. New York: Guilford Press. 51–72.

Bové, P.A. (1986 [2001]) 'Intellectuals at War: Michel Foucault and the Analytics of Power.' In P. Williams (Ed.) *Edward Said, Volume 1*. London; Thousand Oaks, California; New Delhi: SAGE. 41–62.

Braun, B. (2000) 'Producing Vertical Territory: Geology and Governmentality in Late Victorian Canada.' *Ecumene*, 7(1), 7–46.

Brennan, T. (2000) 'The Illusion of Travelling Theory: Orientalism as Travelling Theory.' *Critical Inquiry*, 26, 558–83.

Chakrabarty, D. (2000) *Provincializing Europe: Postcolonial Thought and Historical Difference*. Princeton, NJ; Oxford: Princeton University Press.

Chakrabarty, D. (2002) *Habitations of Modernity: Essays in the Wake of Subaltern Studies*. Chicago; London: University of Chicago Press.

Chatterjee, P. (1983) 'More on Modes of Power and the Peasantry.' In R. Guha (Ed.) *Subaltern Studies II*. Delhi: Oxford University Press. 311–49.

Chatterjee, P. (1984) 'Gandhi and the Critique of Civil Society.' In R. Guha (Ed.) *Subaltern Studies III*. Delhi: Oxford University Press. 153–95.

Chatterjee, P. (1986) *Nationalist Thought and the Colonial World: A Derivative Discourse?* London: Zed for the United Nations University.

Chatterjee, P. (1992) 'Their Own Words? An Essay for Edward Said.' In M. Sprinker (Ed.) *Edward Said: A Critical Reader*. Oxford: Blackwell. 194–220.

Chatterjee, P. (1993) *The Nation and its Fragments: Colonial and Postcolonial Histories*. Princeton, NJ: Princeton University Press.

Chatterjee, P. (1995) 'The Disciplines in Colonial Bengal'. In P. Chatterjee (Ed.) *Texts of Power: Emerging Disciplines in Colonial Bengal*. Minneapolis; London: University of Minnesota Press. 1–29.

Chatterjee, P. (2001) 'On Civil and Political Society in Postcolonial Democracies.' In S. Kaviraj and S. Khilnani (Eds) *Civil Society: History and Possibilities*. Cambridge: Cambridge University Press.

Chatterjee, P. (2004) *The Politics of the Governed: Reflections on Popular Politics in Most of the World*. New York: Columbia University Press.

Chaturvedi, V. (2000) *Mapping Subaltern Studies and the Postcolonial*. London; New York: Verso.

Clayton, D. (2000) *Islands of Truth: The Imperial Fashioning of Vancouver Island*. Vancouver: University of British Columbia Press.

Clayton, D. (2001/02) 'Absence, Memory, and Geography.' *BC Studies*, 132, 65–79.

Clayton, D. (2004) 'Imperial Geographies.' In J. Duncan, N. Johnson and R. Schein (Eds) *Companion to Cultural Geography*. Oxford: Blackwell. 449–68.

Corbridge, S., G. Williams, M. Srivastava and R. Véron (2005) *Seeing the State: Governance and Governmentality in Rural India*. Cambridge: Cambridge University Press.

Crush, J. (1994) 'Scripting the Compound – Power and Space in the South-African Mining-Industry.' *Environment and Planning D-Society & Space*, 12(3), 301–24.

Dean, M. (1986 [1994]) 'Foucault's Obsession with Western Modernity.' In B. Smart (Ed.) *Michel Foucault: Critical Assessments*. London; New York: Routledge. 285–99.

Dean, M. (1999) *Governmentality: Power and Rule in Modern Society*. London: SAGE.

Dean, M. and B. Hindess (1998) *Governing Australia: Studies in Contemporary Rationalities of Government*. Cambridge: Cambridge University Press.

Deshpande, S. (2000) 'Hegemonic Spatial Strategies: The National-space and Hindu Communalism in Twentieth-century India.' In P. Chatterjee and P. Jeganathan (Eds) *Subaltern Studies XI*. London: Hurst. 167–211.

Driver, F. (1992) 'Geography's Empire – Histories of Geographical Knowledge.' *Environment and Planning D-Society & Space*, 10(1), 23–40.

Driver, F. (1999) *Geography Militant: Cultures of Exploration in the Age of Empire*. Oxford: Blackwell.

Driver, F. and D. Gilbert (1998) 'Heart of Empire? Landscape, Space and Performance in Imperial London.' *Environment and Planning D-Society & Space*, 16(1), 11–28.

Duncan, J.S. (2002) 'Embodying Colonialism? Domination and Resistance in Nineteenth-century Ceylonese Coffee Plantations.' *Journal of Historical Geography*, 28(3), 317–38.

Foucault, M. (1961) *Folie et déraison: histoire de la folie à l'âge classique*. Paris: Plon.

Foucault, M. (1967) *Madness and Civilization: A History of Insanity in the Age of Reason*. London; Sydney: Tavistock.

Foucault, M. (1970) *The Order of Things: An Archaeology of the Human Sciences*. London: Tavistock.

Foucault, M. (1972) *The Archaeology of Knowledge*. London: Tavistock.

Foucault, M. (1973-74 [2006]) *Psychiatric Power: Lectures at the Collège de France, 1973–1974*. London: Palgrave Macmillan.

Foucault, M. (1975-76 [2003]) *Society Must be Defended: Lectures at the Collège de France 1975–1976*. London: Penguin.

Foucault, M. (1977) *Discipline and Punish: The Birth of the Prison*. Harmondsworth: Penguin.

Foucault, M. (1978 [2001]) 'Governmentality.' In J. D. Faubion (Ed.) *Essential Works of Foucault, 1954–1984: Power*. Vol. 3. London: Penguin. 201–22.

Foucault, M. (1978 [Forthcoming]) *Security, Territory, Population: Lectures at the Collège de France 1978*. Basingstoke; New York: Palgrave Macmillan.

Foucault, M. (1979) *The History of Sexuality Volume 1: An Introduction*. London: Allen Lane.

Foucault, M. (1980) *Power/Knowledge: Selected Interviews and Other Writings, 1972–1977*. Brighton: Harvester Press.

Foucault, M. (1986a) *The History of Sexuality Volume 2: The Use of Pleasure*. London: Viking.

Foucault, M. (1986b) *The History of Sexuality Volume 3: Care of the Self*. London: Allen Lane.

Foucault, M. (2005) *History of Madness*. London: Routledge.

Gandhi, L. (1998) *Postcolonial Theory: A Critical Introduction*. Edinburgh: Edinburgh University Press.

Gregory, D. (1995) 'Imaginative Geographies.' *Progress in Human Geography*, 19(4), 447–85.

Gregory, D. (1998) 'Power, Knowledge and Geography – The Hettner Lecture in Human Geography.' *Geographische Zeitschrift*, 86(2), 70–93.

Gregory, D. (2004a) *The Colonial Present*. Oxford: Blackwell.

Gregory, D. (2004b) 'The Lightning of Possible Storms.' *Antipode*, 36(5), 798–808.

Guha, R. (1982) 'On Some Aspects of the Historiography of Colonial India.' In R. Guha (Ed.) *Subaltern Studies I*. Delhi: Oxford University Press. 1–8.

Hannah, M. (1993) 'Space and Social-Control in the Administration of the Oglala Lakota (Sioux), 1871–1879.' *Journal of Historical Geography*, 19(4), 412–32.

Hannah, M. (2000) *Governmentality and the Mastery of Territory in Nineteenth-century America*. Cambridge: Cambridge University Press.

Harris, C. (2004) 'How did Colonialism Dispossess? Comments from an Edge of Empire.' *Annals of the Association of American Geographers*, 94(1), 165–82.

Henry, M.G. (2002) *The Disciplining Spectacle: Power, Performance, and Place in Twentieth-century Auckland*. Unpublished Doctoral Thesis, Department of Geography, University of Auckland.

Howell, P. (2004a) 'Race, Space and the Regulation of Prostitution in Colonial Hong Kong: Colonial Discipline/Imperial Governmentality.' *Urban History*, 31(3), 229–48.

Howell, P. (2004b) 'Sexuality, Sovereignty and Space: Law, Government and the Geography of Prostitution in Colonial Gibraltar.' *Social History*, 29(4), 444–64.

Jackson, P.A. (2003) 'Mapping Poststructuralism's Borders: The Case for Poststructuralist Area Studies.' *SOJOURN*, April.

Jacobs, J. (1996) *Edge of Empire: Postcolonialism and the City*. London; New York: Routledge.

Jacobs, J.M. (2001) 'Symposium: Touching Pasts.' *Antipode*, 33(4), 730–34.

Kesby, M. (1999) 'Locating and Dislocating Gender in Rural Zimbabwe: The Making of Space and the Texturing of Bodies.' *Gender, Place and Culture*, 6(1), 27–47.

Legg, S. (2003) 'Gendered Politics and Nationalised Homes: Women and the Anti-colonial Struggle in Delhi, 1930–47.' *Gender, Place and Culture*, 10(1), 7–27.

Legg, S. (2006a) 'Governmentality, Congestion and Calculation in Colonial Delhi.' *Social and Cultural Geography*, 7(5), 709–29.

Legg, S. (2006b) 'Postcolonial Developmentalities: From the Delhi Improvement Trust to the Delhi Development Authority.' In S. Corbridge, S. Kumar and S. Raju (Eds) *Colonial and Postcolonial Geographies of India*. London: SAGE. 182–204.

Legg, S. (2007) *Spaces of Colonialism: Delhi's Urban Governmentalities*. Oxford: Blackwell.

Lester, A. (1998) *Colonial Discourse and the Colonisation of Queen Adelaide Province, South Africa*. Twickenham: Historical Geography Research Group.

Lester, A. (2001) *Imperial Networks: Creating Identities in Nineteenth Century South Africa and Britain*. London: Routledge.

Lester, A. (2002) '"Constructing Colonial Discourse": Britain, South Africans and Empire in the 19th Century.' In A. Blunt and C. McEwan (Eds) *Postcolonial Geographies*. London; New York: Continuum. 29–45.

Lowe, L. (1991) *Critical Terrains: French and British Orientalisms*. Ithaca; London: Cornell University Press.

Mbembe, A. (2001) *On the Postcolony*. Berkeley; Los Angeles; London: University of California Press.

Mbembe, A. (2003) 'Necropolitics.' *Public Culture*, 15(1), 11–40.

McEwan, C. (2003) 'Material Geographies and Postcolonialism.' *Singapore Journal of Tropical Geography*, 24(3), 340–55.

Mitchell, T. (2000) 'The Stage of Modernity.' In T. Mitchell (Ed.) *Questions of Modernity*. Minneapolis; London: University of Minnesota Press. 1–34.

O'Hanlon, R. (1988) 'Recovering the Subject: Subaltern Studies and Histories of Resistance in Colonial South Asia.' *Modern Asian Studies*, 22, 189–224.

O'Hanlon, R. and D. Washbrook (1992) 'After Orientalism – Culture, Criticism, and Politics in the Third-World.' *Comparative Studies in Society and History*, 34(1), 141–67.

Outtes, J. (2003) 'Disciplining Society Through the City: The Genesis of City Planning in Brazil and Argentina.' *Bulletin of Latin American Research*, 22(2), 137–64.

Prakash, G. (1990) 'Writing Postorientalist Histories of the Third-World Perspectives from Indian Historiography.' *Comparative Studies in Society and History*, 32(2), 383–408.

Prakash, G. (1994) 'Subaltern Studies as Postcolonial Criticism.' *The American Historical Review*, 99(5), 1475–90.

Prakash, G. (1999) *Another Reason: Science and the Imagination of Modern India*. Princeton, NJ; Chichester: Princeton University Press.

Robinson, J. (1997) 'The Geopolitics of South African Cities – States, Citizens, Territory' *Political Geography*, 16(5), 365–86.

Robinson, J. (2000) 'Power as Friendship: Spatiality, Femininity and "Noisy Surveillance".' In R. Paddison, C. Philo, P. Routledge and J.P. Sharp (Eds) *Entanglements of Power: Geographies of Domination & Resistance*. London: Routledge. 67–92.

Robinson, J. (2003) 'Postcolonialising Geography: Tactics and Pitfalls.' *Singapore Journal of Tropical Geography*, 24(3), 273–82.

Rose, N. (1996) 'Governing "Advanced" Liberal Democracies.' In A. Barry, T. Osbourne and N. Rose (Eds) *Foucault and Political Reason: Liberalism, Neo-Liberalism and Rationalities of Government*. London: University College London Press. 37–64.

Said, E. (1972) 'Michel Foucault as an Intellectual Imagination.' *boundary 2*, 1(1), 1–36.

Said, E. (1975) *Beginnings: Intention and Method*. New York: Basic Books.

Said, E. (1978) *Orientalism*. London: Routledge & Kegan Paul.

Said, E. (1984a) 'Michel Foucault, 1927–1984.' *Raritan*, 4(2), 1–11.

Said, E. (1984b) *World, the Text, the Critic*. London: Faber.

Said, E. (1986) 'Foucault and the Imagination of Power.' In D.C. Hoy (Ed.) *Foucault: A Critical Reader*. Oxford: Blackwell.

Said, E. (1988) 'Foreword.' In R. Guha and G.C. Spivak (Eds) *Selected Subaltern Studies*. New York; Oxford: Oxford University Press.

Said, E. (1993) *Culture and Imperialism*. London: Chatto & Windus.

Said, E. (2000a) 'Diary.' *London Review of Books*, 22(11), 42–3.

Said, E. (2000b) 'Invention, Memory, and Place (Geography, Palestine).' *Critical Inquiry*, 26(2), 175–92.

Said, E. (2002) 'In Conversation with Neeladri Bhattacharya, Suvir Kaul, and Ania Loomba.' In D.T. Goldberg and A. Quayson (Eds) *Relocating Postcolonialism*. Oxford: Blackwell. 1–14.

Said, E., A. Beezer and P. Osbourne (1993 [2004]) 'Orientalism and After.' In G. Viswanathan (Ed.) *Power, Politics and Culture: Interviews with Edward W. Said*. London: Bloomsbury. 208–32.

Said, E., J.A. Buttigieg and P.A. Bové (1993 [2004]) 'Culture and Imperialism.' In G. Viswanathan (Ed.) *Power, Politics and Culture: Interviews with Edward W. Said*. London: Bloomsbury. 183–207.

Said, E., B. Robbins, M.L. Pratt, J. Arac and R. Radhakrishnan (1994) 'Edward Said's *Cultural and Imperialism*: A Symposium.' *Social Text*, 401–37.

Salusinszky, I. (1987) *Criticism in Society*. New York; London: Methuen.

Sarkar, S. (1996 [2000]) 'Decline of the Subaltern.' In V. Chaturvedi (Ed.) *Mapping Subaltern Studies and the Postcolonial*. London; New York: Verso. 300–23.

Schaub, U.L. (1989) 'Foucault's Oriental Subtext.' *Publications of the Modern Languages Association of America*, 104(3), 306–16.

Scott, D. (1995) 'Colonial Governmentality.' *Social Text*, 43, 191–220.

Sidaway, J.D. (2000) 'Postcolonial Geographies: An Exploratory Essay.' *Progress in Human Geography*, 24(4), 591–612.

Sidaway, J.D., T. Bunnel and B.S.A. Yeoh (2003) 'Geography and Postcolonialism.' *Singapore Journal of Tropical Geography*, 24(3), 269–72.

Slemon, S. (1989) *After Europe: Critical Theory and Post-colonial Writing*. Sydney; Coventry: Dangaroo Press.

Smith, N. (1994) 'Geography, Empire and Social-Theory.' *Progress in Human Geography*, 18(4), 491–500.

Spivak, G.C. (1985) 'Subaltern Studies: Deconstructing Historiography.' In R. Guha (Ed.) *Subaltern Studies IV*. Delhi: Oxford University Press. 330–63.

Spivak, G.C. (1988 [2000]) 'Can the Subaltern Speak?' In D. Brydon (Ed.) *Postcolonialism: Critical Concepts in Literary and Cultural Studies*. London: Routledge. 1427–77.

Spivak, G.C. (2000) 'The New Subaltern: A Silent Interview.' In V. Chaturvedi (Ed.) *Mapping Subaltern Studies and the Postcolonial*. London; New York: Verso. 324–40.

Spurr, D. (1993) *The Rhetoric of Empire: Colonial Discourse in Journalism, Travel Writing, and Imperial Administration*. Durham, NC: Duke University Press.

Stoler, A.L. (1995) *Race and the Education of Desire: Foucault's History of Sexuality and the Colonial Order of Things*. Durham, NC; London: Duke University Press.

Stoler, A.L. (2002) *Carnal Knowledge and Imperial Power: Race and the Intimate in Colonial Rule*. Berkeley; Los Angeles; London: University of California Press.

Suleri, S. (1992) *The Rhetoric of British India*. Chicago: University of Chicago Press.

Terdiman, R. (1985) *Discourse/Counter-Discourse: The Theory and Practice of Symbolic Resistance in Nineteenth-century France.* Ithaca; London: Cornell University Press.

Trigo, B., Ed. (2002) *Foucault and Latin America: Appropriations and Deployments of Discourse Analysis.* New York; London: Routledge.

Vaughan, M. (1991) *Curing their Ills: Colonial Power and African Illness.* Cambridge: Polity Press.

Watts, M. (2003) 'Development and Governmentality.' *Singapore Journal of Tropical Geography,* 24(1), 6–34.

Young, R.C. (2001) *Postcolonialism: An Historical Introduction.* Oxford: Blackwell.

Chapter 25

Foucault, Sexuality, Geography

Philip Howell

Introduction

The purpose of this chapter is three-fold. I want firstly to reconsider Foucault's *History of Sexuality* and related writings on the subject, many of them newly published (Foucault 1980, 1985, 1986; of the recently published texts, the most important for the question of sexuality is Foucault 2003b). I want to emphasize his theorization of sexuality as a discourse and to address some of its implications for geography. I consider this particularly important not merely because it is a text largely neglected by geographers, but also because it is generally susceptible to misreadings bordering on caricature. A quarter of a century on, the *History*'s ambiguities and complexities are worth revisiting, most particularly in the light of the ongoing publication of Foucault's seminars at the *Collège de France*. This is the most unapologetically exegetic section of this chapter, but it is followed by a more empirical discussion of some geographies of sexuality that have been directly inspired by Foucault's wider discussion of modernity. Here, and secondly, the focus is on the kind of 'work' that sexuality accomplishes within modern culture. We shall see that whilst many geographers have found Foucault's work extraordinarily stimulating, it is all too easy to embed an analysis of sexuality within his discussion of the development of a disciplinary society. By contrast, I will argue, the role of space in sexual normalization and in the 'tolerance' of sexual 'freedoms', have been far less analyzed. Again, though, the ambiguities in Foucault's work – some of them productive, some frustrating – are very much to the fore, and I have not tried to gloss over the difficulties in this notion of sexuality as a *dispositif*. Thirdly, I conclude with a discussion of the importance of Foucault's work for research into the geographies of sexual subjectivity, including its major importance for queer theories and the related political critique of heterosexism and homophobia. Geographers for whom sexual subjectivity is a concern have found in – or perhaps it is more accurate to say *after* – Foucault a series of resources, theories, formulations and suggestions that have contributed to major developments in our understanding of sexuality and space. In this section, though the distance from Foucault's writing is the greatest, and his influence the most indirect, I have nevertheless found the most unambiguously to celebrate.

Throughout, I have tried to clarify what I understand Foucault to have actually said and meant, and to trace in this a significance for geography that is as much potential as it is actual. Much of Foucault's work on sexuality is, sadly, barely acknowledged by geographers, although it is to be hoped that the publication of the lecture series will stimulate a new wave of interest. I cannot however claim that this is a comprehensive survey of either past or potential work; labourers in the small vineyard that the geography of sexuality represents will easily recognize debts, glosses and omissions (if hopefully not too many errors). I have only stressed a small number of topics here, and the phrase 'for instance' crops up all too frequently. Nevertheless, my hope is that this chapter is representative of some of the work in geography following Foucault's lead, and some of the roads that might be taken. I have tried in fact to trace what might be thought of as 'Foucault effects' – lines of influence rather than simply derivation, avenues of enquiry for which Foucault is the condition of possibility rather than the origin. This chapter is inevitably partial, in every sense of the word, but I have striven for a balance between criticism and generosity, and with a recognition that the best of this work in geography has been done *with* and *after* Foucault rather than simply *for* or *against* him.[1]

Geographies in the *History of Sexuality*

The first volume of Foucault's *History of Sexuality* is by most accounts his most-read work, at least as far as the English-speaking world is concerned. Powerful, provocative, though perhaps above all, short, the *Introduction* is often used to encapsulate the key elements in Foucault's philosophy, particularly for new students (Mills 2003, 130). It is treated as an introduction not just to Foucault's last, protracted, and ultimately unfinished project, but also to his entire philosophical oeuvre. Given this popularity for students of Foucault, it is somewhat surprising that so few geographers have considered it in its specifics – rather than, say, as a relatively accessible summary of Foucault's analysis of power. Yet even for those who have commented on Foucault's theorization of sexuality, attention to it, and to the actual texts of the *History of Sexuality* has been notably uneven. Quite apart from the fact that the later volumes in the series, concerned as they are with the ancient and early Christian world, are largely unread by generalists, even the hundred or so pages of the *Introduction* are too often scanned, in conjunction with and confirmation of the various commentaries. Readers will quickly take on board Foucault's critique of the 'repressive hypothesis', his argument that 'sexuality' is a nineteenth-century discourse to which we are still in thrall, and his suspicion at any suggestion that sex can be 'liberated' from power/ knowledge. But there are still many elements in this very short book – frustratingly both dense and repetitive – that have been neglected, or distorted. The *Introduction* is actually replete with problems for student and scholar alike. It occupies a more

---

1   I would like to add, without burdening them with my errors, that I have particularly benefited from discussing Foucault, over many years at Cambridge, with Jim Duncan, Gerry Kearns, Stephen Legg and Andy Tucker.

than usually transitional place in the development of Foucault's ideas, and it betrays ambiguities that become ever more obvious when the volume is brought into a comparison with the work that remained unpublished in Foucault's lifetime. It is hard now not to see it as more than usually provisional, particularly in the complex context of the *History of Sexuality* series. In this regard, a careful account of the relation of the published work to Foucault's original and subsequent plans for the series can be found in Elden (2005, 23–41). The reasons for the abandonment of the original plans for the *History* have attracted much speculation, which we might call idle if it were not so elaborately and unremittingly hostile; by contrast Elden offers a persuasive if not definitive solution for the reorientation of Foucault's researches into sexuality.

That said, let me try to summarize the *Introduction* in the following way, taking my cue from Foucault's threefold interest in types of understanding (the formation of domains of knowledge), forms of normality (the formation of rationalities for the control, manipulation and general government of individuals), and modes of relation to oneself and others (the formation of subjectivities) (Foucault 1985, Preface, reprinted in Rabinow 1991, 333–9, and 1997, 199–205). In the first place, in terms of discourse and domains of knowledge, Foucault wants to insist that sexuality is *produced*. It is not a universal biological fact, something that stands outside individuals and societies to a greater or lesser extent directing or determining them. It is, instead, an historical product – a historically and culturally specific discourse through which a new and insidious form of power, the 'truth' of sex, assumes a locus in the body and its pleasures. 'Desire,' as Thomas Laqueur puts it, 'is discursively created in order to be the locus of control' (Laqueur 2003, 271). Secondly, and in terms of these 'rational and concerted coercions' of modernity, Foucault claims that the great process of transforming sex into discourse was a product of the early nineteenth century (in Europe), originating with the hegemonic mission of the bourgeoisie; in crude terms, the middle class pioneered the discursive '*deployment*' of sexuality, first upon themselves, before exporting and generalizing it to cover the entire social body (Foucault, 2003; note that there are severe problems with the translation of *dispositif* as 'deployment' [see Halperin 1995, 188–9, note 6, and Elden 2001, 110–11: 'deployment' – variously alternatives are 'apparatus', 'construct', 'grid of intelligibility', 'device', 'network', 'formation' – chimes of course rather too readily with a thesis of social control]). Sexuality thus becomes fundamental to the biopolitical order of the modern polity – 'biopolitics' meaning here the governmental preoccupation with social welfare and security, the large-scale management of life and death in the interests of the state. Sex – that is as 'sexuality' – becomes a proper concern for government; and it is in fact redundant or even oxymoronic to talk about the 'regulation' of sexuality. Thirdly, in terms of ethics, politics and the history of the present, Foucault suggests that we have now to abandon the self-congratulatory view of the nineteenth century as a period of sexual repression, and see it instead as characterized by 'a regulated and polymorphous incitement to discourse' (Foucault 1980, 34). In the modern age the loquacious discourse of sexuality proliferated sexual identities, figures and types, with 'perverse' sexualities being as central and fundamental as 'normal' and legitimate ones. From the nineteenth century through to our own day, we have witnessed, he says, both 'the proliferation of specific pleasures and the multiplication of disparate sexualities' (Foucault 1980, 49). This is a phenomenon that cannot be understood through a Freudian model of sexual liberation from repression; indeed, since the latter is complicit in the incitement of sexual discourse, it is a hindrance to our understanding of the significance of sexuality in the modern era. '[P]ower in our societies functions primarily not by repressing spontaneous sexual drives but by producing multiple sexualities, and that through the classification, distribution, and moral rating of those sexualities the individuals practicing them can be approved, treated, marginalized, sequestered, disciplined, or normalized' (Bersani, quoted in Halperin 1995, 20). This third axis therefore directs us to critically examine the question of sexual *identities* (and what can be done to resist them).

The *production* of sexuality, the notion of sexuality as a 'deployment' or *dispositif*, the proliferation and interrelatedness of sexual *identities*: this is a very bald summary, but it is worth emphasizing these key elements, not least because several general commentaries are plainly misleading even in their summary of Foucault's argument. Sara Mills' recent introduction to Foucault, for example, bizarrely confuses Foucault's critique with the comforting narrative of Victorian sexual repression that he sets out to challenge (Mills 2003, 84–5). In her defence it could be noted that Foucault's presentation is oblique; and, in addition, as we shall see, he does not entirely discount the notion of 'repression'. There is no equivocation on the Freudian discourse of repression: but on the silencing of sexuality, 'It is quite false if you speak of language in general, but it is quite true when you distinguish carefully between types of discursive formation or practice' (Foucault 2003, 70). However, this ventriloquism is a gross mistake, and it reminds us that it is all too easy to misrepresent Foucault's views on sexuality. To take one further instance, consider the rather sterile argument over whether we should consider sexuality to be 'socially constructed' (as introduced in Stein 1990). Foucault's insistence on sexuality as a *discourse* is usually taken to be an endorsement of the radical constructionist position – which is accurate enough as far as this goes, though what is meant by construction or production might be debated further.[2] But Foucault is as often considered to have suggested as a result that sexuality did not exist before the nineteenth century. Taken to extremes – not uncommon – the suggestion is

---

2 On constructionism, in general, see the insightful discussion in Hacking 1999. However, as a counterweight to Hacking's as it were 'weak' construction of constructionism, see the comments of Butler 1993, particularly in relation to the sex/gender distinction. For Butler, construction should be thought of not in terms of idealism or nominalism, voluntarism or determinism, but rather as a constitutive constraint, a process that involves discursive reiteration and the materialization of regulatory norms. Both Hacking and Butler are indebted to Foucauldian ideas, and though they are not easily reconciled, perhaps each does not in the end entirely contradict the other.

absurd, both historically and theoretically.[3] We would be better off perhaps to echo Foucault's own rejection of blanket, universalizing statements about sexuality. When asked outright whether homosexuality was determined by nature and biology *or* by social conditioning, Foucault responded by declaring that the issue was beyond his expertise and declined to offer an opinion on the question (Halperin 1995, 4; Foucault 1997a). This was not simply evasion, but rather a strategic refusal to be intellectually blackmailed (Foucault 1997b). Foucault refused to be drawn into making unwise and impolitic statements about the relative balance of nature versus nurture, and instead preferred to offer in his histories of sexual discourse warnings about the dangers of the science of sexuality. Foucault's response has never seemed more sensible; we need not deny the reality of the biological and the somatic (their specification is in any case a discursive performance, as Judith Butler has argued) to accept the power of the discourse of sexuality and to trace its history. Judith Butler sidesteps or displaces the essentialist/constructionist debate by arguing that the setting of limits to discourse is itself a discursive and materialized *performance*; this is not to say that there is nothing but discourse, nor to deny the reality/materiality of the non-discursive world, but simply to note that 'the extra-discursive is delimited … by the very discourse from which it seeks to free itself' (Butler 1993, 11). To put it another way, with reference to the material and extra-discursive body: 'To claim that discourse is formative is not to claim that it originates, causes, or exhaustively composes that which it concedes; rather, it is to claim that there is no reference to a pure body which is not at the same time a further formation of that body' (10).

If we can sidestep some of the wilder assertions in such a way, what does the *Introduction* to the *History of Sexuality* offer to geographers? Foucault's discussion and foregrounding of sexuality as a *discourse* remains in many ways forbiddingly abstract. But Foucault does not abandon the attention to space and spatiality that has led Stuart Elden to characterize his work as a 'spatial history' (Elden 2001). For one thing, the *Introduction* follows on very closely from his analysis of disciplinary power, so replete with architectural and spatial referents; the *Introduction*, at least, is not in any sense a repudiation of his earlier genealogical emphasis on bodies and spaces. In linking the deployment of sexuality to the rationale of modern regimes, the emphasis on the disciplinary surveillance of the social field is in many ways straightforwardly recapitulated. Foucault adds that the deployment of sexuality has its very reason for being in the processes of 'proliferating, innovating, annexing, creating, and penetrating bodies in an increasingly detailed way, and in controlling populations in an increasingly comprehensive way' (Foucault 1980, 107). This represents the intensification of the body as a target for power, a further and far subtler colonization of the body by discipline. The state apparatus of disciplinary institutions – schools, workhouses, prisons, as well as the more obvious magdalens and lock hospitals – indeed become saturated with the discourse of sexuality. It is impossible

---

3 For confirmation of Foucault's notion of sexuality as a modern institution, see Halperin, Winkler and Zeitlin (1990). Hull (1996) concurs. For a different view, however, see for the ancient world, Skinner (2005), and for the medieval, Karras (2005).

to imagine their operation, and their role in modernity, without understanding the proliferation of sexual discourse. This is a theme that I return to in the second section of this chapter, but it is enough for now to acknowledge the very evident link with *disciplinary spaces* of modernity.

It is also important to take literally Foucault's invocation of *sites of sexuality*. Foucault argues very forcefully for instance that the family is the most active of these 'sites', with one of the most significant consequences of the deployment of sexuality being 'the affective intensification of the family space' (Foucault 1980, 109). On the one hand, Foucault is clearly keen to confront the notion that the space of sexuality is confined to the marriage bed; when he writes that '[a] single locus of sexuality was acknowledged in social space as well as at the heart of every household, but it was a utilitarian and fertile one: the parents' bedroom' it is only to set up the straw man for his critique of the repressive hypothesis (1980, 3). However, Foucault does argue that the family 'anchors' sexuality; it is where the deployment of an earlier system of *alliances* (concerned with marriage, kinship, consanguinity, legitimacy, property, the reproduction of an elite, and so on) became intertwined with the new deployment of *sexuality*. As John Ransom puts it, 'Whereas the old deployment of alliance (basically, kinship) was tied to the broader social system by facilitating the ordered circulation of wealth and the reproduction of an elite, the new deployment of sexuality helped to intensify awareness of the body and its rhythms of production and consumption' (Ransom 1997, 69). Sexual bodies were now productive (or unproductive), normal (or abnormal), proper (or perverse). The family becomes a crucial social and spatial formation that effectively conceals the significance of sexuality by claiming to be its source (Foucault 1980, 111). It appears to banish sexuality from the public and social sphere by locating it in the family and the private sphere. I have taken some liberties with this last suggestion, as Foucault rarely refers to the private and the public as such, but it is a legitimate one, given the proliferation of spatial imagery surrounding the concept of the family. Foucault asks (in a nod to the carceral archipelago): 'Was the nineteenth-century family really a monogamic and conjugal cell? Perhaps to a certain extent. But it was also a network of pleasures and powers linked together at multiple points and according to transformable relationships'; it is 'a complicated network, saturated with multiple, fragmentary, and mobile sexualities' (Foucault 1980, 46). And far from simply underwriting a normalized sexuality, the centrality of the family also provokes an intense awareness of the other sexualities (and spatialities) to which it is intimately connected. One can think here, if one wanted to concentrate on Victorian sexual narratives, of the straying husband whose dalliances with prostitutes threatens his wife and unborn child with disease and death; or, if one wanted a more contemporary equivalent, that of the closeted husband who journeys from the safety of the suburbs to explore his sexual desires in leather bars and S/M clubs (Califia 2000). In each case the family is not isolated from literal or metaphorical 'contagion', and its place in the modern discourse of sexuality is characteristically paradoxical. We should also consider, therefore, alongside the approved space of the family, the 'perverse spaces' that are inseparable from the proliferation of other sexual identities: 'those *devices of sexual saturation* so characteristic of the space and the social rituals of the nineteenth century', is how Foucault puts it (1980, 45, italics in original). One key space of perversity for example – a 'much narrower space' than the family, as Foucault describes it elsewhere, is the frame of reference for the figure of the masturbating child: 'It is the bedroom, the bed, the body; it is the parents, immediate supervisors, brothers and sisters; it is the doctor: it is a kind of microcell around the individual and his body' (Foucault 2003, 59). If Foucault had completed the series on sexuality according to his original plan, there would no doubt have been an analysis of the role that the crusade against masturbation played in installing sexuality in the most intimate domestic spaces: not only requiring children to confess their sexual trangressions but at the same time attaching the taint of perversity to those parents who too thoroughly supervise their offspring. Foucault's history of sexuality (both published and prospective) directs us in this way to investigate the various sites, scales and spaces in and through which the techniques of sexuality are propagated.

Thirdly, there is a concern, embedded in Foucault's discussion of biopolitics, with what we might call the *geopolitics of sexuality*. Biopolitical rationality necessarily links the body with the body politic, so that sexual conduct becomes a proper domain of government, and of 'governmentality'.[4] Sex and sexual subjectivity become biopolitical issues because 'It was essential that the state know what was happening with its citizens' sex, and the use they made of it, but also that each individual be capable of controlling the use he [sic] made of it' (Foucault 1980, 26). Here, Stephen Legg's recent contribution to the retheorization of population geography, which makes much use of unpublished or recently translated material from Foucault's lecture series, makes it crystal clear why this should be a concern for geographers. It is obvious from Foucault's discussion of the Malthusian couple as an object, target and anchor of sexual knowledge why demographers should be concerned with the *History of Sexuality* (Foucault 1980, 105; Clifford 2001, 110–11). But the *History* speaks to population geography in a different register. Drawing attention to 'the different scales and spaces in which populations are conceived and governed', Legg demonstrates that sexuality is implicitly or explicitly inscribed into the modern biopolitical state; the problematization of 'population' and the technologies of new 'governmental spaces' are indebted to the emergence of sexuality (Legg 2005, 137–56, 144). Simply put, the modern state and its delineation of its field of operations cannot be divorced from sexuality, from a concern with reproduction, disease and deviancy; in short, biopolitics *is* geopolitics. Equally important, however, is the fact that the biopolitical state can be separated neither from the history of *sexuality* nor the construction of *race*. In her pioneering analysis of Foucault's discussion of race in the *Introduction* to the *History of Sexuality*, Ann Laura Stoler has demonstrated the significance of sexuality for students of both colonialism and imperialism. Stoler is no mere exegete, but she rightly points out how much of the *Introduction*'s focus

---

4   For a good commentary on governmentality see Dean 1999; within geography see Hannah (2000). For examples of work on sexual themes, see Brown (2000, 88–115, note that this chapter was written with Paul Boyle), and Howell (2004, 229–48).

on race has simply been ignored by commentators and critics (Stoler 1995; 2002). Stoler particularly emphasizes the 1976 lecture series at the *Collège de France* over and above the *Introduction* to the *History of Sexuality* where racialized sexuality is recognized as only one domain amongst many. She shows by contrast how in Foucault's account biopower was not only bourgeois in origin, but also imperial. The bourgeois model of the self – productive, respectable, normal – was secured through a comparison both with the sexualities of colonial others and with those of 'internal enemies' at home, so that '[t]o be truly European was to cultivate a bourgeois self in which familial and national obligations were the priority and sex was held in check – not by silencing the discussion of sex, but by parcelling out demonstrations of excess to different social groups and thereby gradually exorcising its proximal effects' (Stoler 1995, 183–4). In this way, Stoler has convincingly affirmed Foucault's original insight that colonial discourses of sexuality were productive of class and racial power, not mere reflections of them. The *Introduction* thus offers geographers of imperialism an opportunity to focus on the importance of sexuality for the colonial project, and an insistence that the construction of metropolitan sexualities cannot be separated from this imperial history: '[t]here was no unitary bourgeois self already formed, no core to secure, no "truth" lodged in one's sexual identity. That "self", that "core", that "moral essence" ... was one that Europe's external and internal "others" played a major part in making' (Stoler 1995, 194).

There is a great deal, then, even in the relatively few pages of the *Introduction*, that speaks directly to the geographies of sexuality. It is certainly a text that should be read more widely and more carefully than it has been. For all this, though, it remains the case that it is a work with frustrating ellipses, and even the most generous assessment reveals some notable omissions. To focus only on the question of sexuality as a *discourse*, there is a striking lack of geographical specificity in the *Introduction* and the rest of the *History*. Foucault's Eurocentrism may be readily conceded, and Ann Stoler's recovery of a discourse on race and colonialism notwithstanding, we may also simply pass over the unwillingness to concede the discursive contours of regions, nations and communities. Beyond this though, it is a puzzle that Foucault says nothing about the '(hidden) geography' that lies behind the production of discourse – as for instance the concentration of discursive authority and institutions in metropolitan centres (Phillips and Watt 2000, 1). Richard Burton's sexual geographies, to take but one example, as ably discussed by Richard Phillips, illustrate the role that geographical imaginations of sexuality at the colonial margins played in the critique as well as the constitution of 'Victorian' sexuality (Phillips 1999). To take another perhaps more critical example, we might wonder what role, what *place*, there is in the history of sexuality for non-European sexual discourses. In the *Introduction* Foucault blithely contrasts the modern, western *scientia sexualis* with the *ars erotica* of ancient and modern others (including 'China, Japan, India, Rome, the Arabo-Moslem societies'), but the suspicious taint of orientalism here has, for some recent critics, taken on special significance in the context of Foucault's notorious endorsement of the Iranian revolutionary movement, and in the knowledge of the Khomeini regime's subsequent gender and sexual politics.[5]

It is moreover a real limitation that the materiality of discursive agency – including its social and spatial specificity – is barely acknowledged; as if geography and genealogy do not require each other. There is I might add by contrast a great deal of work, hardly or not at all indebted to Foucault, which explores these discursive geographies of sexuality. To take just erotic or pornographic discourse as an example, consider Felicity Nussbaum's discussion of the place of the prostitute in eighteenth-century English pornography: 'Written on the body, on London, and on the world map, sexual geography established an analogy between prostitute and torrid zone, so that one was a geographical displacement (evoking a geographical equivalent in segregating prostitution into confined "stews"), and the other a socioeconomic and moral distinction (evoking a correlative categorization of the geographically displaced Other)' (Nussbaum 1995, 97). In the same period, and equally concerned with gender, Karen Harvey has recently considered how, in English erotica, male and female sexuality were viewed spatially. She identifies 'a widely shared culture of sexualized locations and bodies' whose spatial codes and metaphors allowed the integration of the moral and the physical. In an appealing and provocative formulation, Harvey argues that in this discourse 'sex is a place for men to visit' – that is, sex is rendered spatial, confining it and firmly situating it, separate from the world of masculinity (Harvey 2004, 172, 173). Although this work recapitulates the argument that sexuality is produced in discourse, it owes more to feminist analyses of gender and sexual identity, and there is here a more revealing and rewarding focus on spatial narratives within that discourse than anything to be found directly in Foucault's *History of Sexuality*.

We should not, then, think that the geographical analysis of sexuality, even simply as discourse, begins and ends with Foucault. The *Introduction* contains much that is stimulating, but even in its general discussion misses a great deal that is useful, even essential, for geographers of sexuality. I have, for reasons both of lack of space and expertise, to neglect the other volumes in the series, and I am aware of my hypocrisy in doing so, but I hope that I have said enough to at least suggest the potential as well as the pitfalls of the texts of the *History* for geographers. In the next section, which treats sexuality as a *dispositif* as well as a discourse – that is as a political project, a normative system and apparatus, concerned with the control and regulation of bodies, acts and individuals – we will see that the general discussion of sexuality in Foucault's works is similarly intriguing and ingenious but at the same time equally erratic and misleading.

---

5   On Foucault's orientalism and its significance, see the very critical comments of Afary and Anderson (2005). For a more careful and judicious discussion of the 'otherness' of the non-Western world in sexual discourse, see Bleys (1996). I might note here that there is more careful discussion of non-western sexualities – discourses and practices – in later volumes of the *History*, though these remain fleeting and intriguing rather than substantial.

From the *History* to the Geography of Sexuality

The *History of Sexuality* does not of course exhaust the significance of Foucault's writings on sexuality. Although his work traces the emergence of the sexual as a distinct field, he was at the same time concerned to show how impossible is the segregation of sexual discourse from other forms of power/knowledge; we have seen for instance how in the conceptions of biopower and governmentality sexuality was central to the understanding of the emergence of the 'social', a 'quasi-object' that defined the domain of liberal government (Donzelot 1980). Sexuality and society were indeed never separate or separable; it is hard to see how each discourse could have emerged without the other. The tenor of Foucault's critique is in any event to disassemble sexuality as a 'quasi-object' in itself, and to trace how the discourse of sexuality was implicated in the wider regime that modernity represents. Bearing this in mind, it becomes necessary to place the notion of sexuality as a *dispositif* within the context of Foucault's discussions of discipline and disciplinary power, all the more so since geographers have by and large taken more from the discussion of discipline than from any other part of Foucault's corpus of work (Driver 1994).

Geographers have of course been quick to identify in Foucault's disciplinary genealogies a distinctive place for space. The discussion of panopticism, with its emphasis on the spatial distribution of bodies in the service of surveillance and the project of individual reformation, remains arguably the most well known element of Foucault's work. Geographers have rightly been stimulated by Foucault's dictum that 'discipline is above all an analysis of space' (Elden 2001, 139). Historical geographers, in particular, have localized, detailed, developed, and extended Foucault's insights into the emergence and spread of a new, disciplinary power in the modern era. I am thinking here, *inter alia*, of Felix Driver's careful mapping of poor law reform in nineteenth-century England and Chris Philo's thorough discussion of madness and moral management (Driver 1993; Philo 2004).[6] If we ally such work to that of historians and historical sociologists, our understanding of disciplinary power in Britain, at least, has been substantially developed over the last quarter century.

Having said this, the thematic of sexuality *within* the discussion of discipline has not been particularly well developed. This is as much a criticism of Foucault as of his interpreters. It is rather striking, re-reading Foucault's *Discipline and Punish*, the extent to which not just gender but also sexuality is neglected. It does not take too much imagination however to register the extent to which 'discipline-blockades' like the military camp, the school, the monastery, the hospital and the prison were pervaded with the disciplining of sexuality. Foucault admitted as much in the *Introduction* to the *History of Sexuality*, when he wrote that 'On the whole, one can have the impression that sex was hardly spoken of at all in these institutions. But one only has to glance over the architectural layout, the rules of discipline, and their whole internal organization: the question of sex was a constant preoccupation'

---

6   In my own work on the regulation of prostitution I have also drawn critically upon the theme of discipline (Howell 2000).

(Foucault 1980, 27). He was speaking here of eighteenth-century secondary schools, as it happens, but the point is a general one. These disciplinary institutions were saturated with the new discursive understanding of sexuality, as were their theories and practices of space. Elsewhere, in fact, Foucault speaks more persuasively about this link between sexuality and disciplinary technologies. In the 1974–75 *Collège de France* lectures, for instance, Foucault states that 'if instead of the army, workshops, and primary schools et cetera, we consider these techniques of penance and practices in the seminaries and colleges that derive from them, then we see an investment of the body at the level of desire and decency rather than an investment of the useful body at the level of aptitudes. Facing the political anatomy of the body there is a moral physiology of the flesh' (Foucault 2003, 193). Consider for instance the installation of pastoral care and the confessional into disciplinary institutions (226–7), and the simultaneous silencing and incitement of sexuality in the spatial partitioning and control of bodies in colleges, seminaries and schools (232–3).

There is then, in the lectures at least, a much closer engagement between discipline and sexuality, particularly in the discussions of 'normalization' and 'regulation'; but it is still the case that the disciplinary society Foucault begins to trace is curiously asexual. Consider the English workhouse. From the Foucauldian account of disciplinary power, the new model workhouse is a straightforwardly panoptical institution, based on the unprecedented Malthusian principle of the separation of husband and wife. This was a device to prevent procreation, but it was also a statement about the workhouse's deliberately alienating environment, an essential element of the deterrent workhouse test of need. But to reduce sexuality to a question of design is wholly misleading. As Seth Koven has recently pointed out in his marvellous book *Slumming*, the workhouse occupied a central place in the imagination of sexual deviance in Victorian England (Koven 2004). In James Greenwood's sensational account, 'A Night in the Workhouse', published in *The Pall Mall Gazette* in 1866 and influential for nearly a century or more, the casual ward of the Lambeth workhouse in London was transformed into something little short of a male brothel. The workhouse, supposed then as now to be a harshly policed institution where sexual intimacy was effectively banished, became in the sensation journalism of the time virtually the opposite – a space of sexual excess and perversity. The workhouse was not merely (allegedly) a site of abominated sexual practices, however; it was in addition a space in which heterodox sexual desires could be represented and explored. Koven's wider discussion of the spatial and epistemological practice of 'slumming' is a signal contribution to accounts of *flânerie* and urban rambling. These were 'spaces in which social investigators, clergymen, reformers, philanthropists, social workers, and writers could explore and represent heterodox sexual desires and practices' (Koven 2004, 27). Indebted in part to Foucault's linkage of discourse to the construction of sexualized identities, it also challenges the seemingly monolithic account of the 'deployment' of sexuality in the service of bourgeois hegemony, the surveillance state, and disciplinary society. Of course this might be read as another demonstration of the lack of silence about sex, and perhaps also simply as another incitement to power, in this case to control the sexual behaviour of male casuals, but it was not easy for the machinery of local and central government to address such an issue, and the implications were never very reassuring. It is rather easier to accept Koven's conclusion that 'The sodomitical subtext of "A Night" threw into disarray the social scientific categories underpinning sanitary and poor-law reform' (Koven 2004, 57). Koven also notes that Jeremy Bentham's essay of 1785 argued that sex between men should not be considered a crime.

The problem of prostitution/sex work is just as pertinent an issue. Although Foucault devotes relatively little attention to the prostitute, she certainly figures in his roster of discursively exemplary sexual figures, and it is easy enough to relate the management of prostitutes and prostitution to the discussion of disciplinary society.[7] Miles Ogborn's discussion of the London Magdalen Hospital, for instance, suggests that women, and in particular female prostitutes, might just be the first modern subjects (Ogborn 1998, 39–74). The Magdalen, the eighteenth-century metropolitan reformer Jonas Hanway's home for penitent prostitutes, was a pioneer in the use of space to produce an autonomous, self-reflexive, individualized – and thus *modern* – subjectivity. It anticipated the disciplinary technologies attendant upon the birth of the prison, and it clearly laid out the belief that regulated and disciplined behaviour could produce moral subjects. This account is suggestive rather than conclusive, but these conclusions have been largely approved by Kevin Siena's recent history of the relationship between sex, disease and social welfare in early modern London. Siena's narrative is rightly cautious, and demonstrates that London's venereal disease institutions put medical treatment above moral reformation – salivation above salvation – but by the end of the eighteenth century there was an evident eagerness to bring the two together, and to use segregation and discipline to effect projects of reform (Siena 2004). Sexuality was therefore fundamental to discipline and to the spaces of modernity; Ogborn is absolutely right to claim the need to construct 'an understanding of modernity's disciplinary armoury and its characteristic subjectivities that gives full weight to gender and sexuality' (Ogborn 1998, 73). In the project of penitentiary reform of prostitutes we have not only a clearly gendered but also a sexualized form of disciplinary power and authority.

We might take the question of prostitution further, however, because it sheds light on the limitations as well as the advantages of a concern for disciplinary power. In the first place, Miles Ogborn and others have also related the nineteenth-century regulation of prostitution to the development of a disciplinary society. In Britain, the Contagious Diseases Acts (1864–1889) represented the closest equivalent to the regulationist regimes in Europe that attempted to combat venereal disease by subjecting female sex workers to disciplinary surveillance, regular medical inspection, and incarceration in 'lock' hospitals if found to be in a contagious state. Ogborn has written here that the project of regulation represented a fundamentally disciplinary

---

7   Stuart Elden points out (personal communication) that the female prostitute would surely have figured large in the projected volume in the *History of Sexuality* devoted to the hysterical woman. I say 'she', but there is brief discussion of male prostitution, or rather anxieties about male prostitutes in the ancient polis (Foucault 1985, 217–18).

apparatus (Ogborn 1993). Moreover, the foremost historian of British regulationism, Judith Walkowitz, has acknowledged the Acts in terms of a Foucauldian 'technology of power' (Walkowitz 1980, 4–5). If regulationism is modern, it is characteristically a disciplinary modernity. Its essentials – inscription, inspection, and incarceration – are symptomatic of the disciplinary society. The police registration of prostitutes clearly recapitulated the extension of surveillance technologies, whilst their intimate medical inspection embodied the penetrative power of the medical gaze. The incarceration of prostituted women similarly invokes the theme of disciplinary enclosure and institutionalization, and these can be elaborated even beyond the key regulationist institutions of the lock hospital and the prison; for the red-light district, the brothel, and the Magdalen also appropriate this distinctive concern for enclosure, confinement and surveillance. We are directed to an acknowledgement that disciplinary technologies were appropriated for the policing of sex and the regulation if not always reformation of deviant sexual subjects.

However, on closer inspection, there are some oddities and ambiguities about such a reading of prostitution regulation. It is puzzling, for instance, that Walkowitz's pioneering and influential history of the Contagious Diseases Acts referred to the then recently published first volume of the *History of Sexuality*, but not at all to *Discipline and Punish*. This is surprising because in that book Foucault explicitly discusses prostitution and its regulation. Clearly indebted to the work of Alain Corbin on regulationism in France – particularly Paris – Foucault considers the function of prostitution in its relation to wider society (Corbin 1990). It is principally, he says, as a form of useful delinquency that nineteenth-century prostitution must be understood:

> Delinquency, controlled illegality, is an agent for the illegality of the dominant groups. The setting up of prostitution networks in the nineteenth century is characteristic in this respect: police checks and checks on the prostitutes' health, their regular stay in prison, the large-scale organization of the maisons closes, or brothels, the strict hierarchy that was maintained in the prostitution milieu, its control by delinquent-informers, all this made it possible to canalize and to recover by a whole series of intermediaries the enormous profits from a sexual pleasure that an ever-more insistent everyday moralization condemned to semi-clandestinity and naturally made expensive; in setting up a price for pleasure, in creating a profit from repressed sexuality and in collecting this profit, the delinquent milieu was in complicity with a self-interested Puritanism: an illicit fiscal agent operating over illegal practices. (Foucault 1979, 279–80)

Sidestepping for the moment the references here to sexual repression, this is rather suggestive of the ways in which prostitution might be, under regulationist authority, controlled, managed, and administered. In this reading, regulationist practices constitute 'an instrument for administering and exploiting illegalities' (1979, 280). Colonized by the dominant illegality of class and class power, prostitution could be made into a useful delinquency only through the development of a sophisticated technology of administrative and police surveillance. As Foucault puts it, 'The organization of an isolated illegality, enclosed in delinquency, would not have been possible without the development of police supervision' (1979, 285).

Historians and geographers have been authorized then to read regulationism as an element – not a marginal one – of the disciplinary society whose development Foucault traces in *Discipline and Punish*. As Corbin notes in his pioneering work on Parisian regulation, a thesis developed seemingly in tandem with Foucault's study of discipline, 'The desire for panopticism ... finds expression in a quasi-obsessional way in regulationism'; the 'enclosure' of commercial sex work under the impress of regulationist policy constitutes 'a tireless effort to *discipline* the prostitute, the ideal being the creation of a category of "enclosed" prostitutes' (Corbin 1990, 9, emphasis in original). He singles out four enclosed spaces – the lock hospital, the Magdalen, the prison and the brothel – that serve as the disciplinary institutions from which disciplinary power could radiate outwards into the wider society. Regulationism, to echo Foucault's analysis of the prison, 'isolates, outlines, brings out a form of illegality that seems to sum up symbolically all the others, but which makes it possible to leave in the shade those that one wishes to – or must – tolerate' (Foucault 1979, 277). Given that in the discourse of sexuality prostitution was 'tolerated', that brothels were 'tolerances', prostitution policy in much of nineteenth-century Europe appears to be a species of the disciplinary genus.

It is worth contrasting this emphasis on discipline, however, to the comments that Foucault makes in the first volume of the *History of Sexuality*:

> Such was the hypocrisy of our bourgeois societies with its halting logic. It was forced to make a few concessions, however. If it was truly necessary to make room for illegitimate sexualities, it was reasoned, let them take their infernal mischief elsewhere: to a place where they could be reintegrated, if not in the circuits of production, at least in those of profit. The brothel and the mental hospital would be those places of tolerance: the prostitute, the client, together with the psychiatrist and the hysteric – those 'other Victorians', as Steven Marcus would say – seem to have surreptitiously transferred the pleasures that are unspoken into the order of things that are counted. Words and gestures, quietly authorized, could be exchanged there at the going rate. Only in those places would untrammelled sex have a right to (safely insularized) forms of reality, and only to clandestine, circumscribed, and coded types of discourse. Everywhere else, modern Puritanism imposed its triple edict of taboo, nonexistence, and silence. (Foucault 1980, 4)

I hope that I may be forgiven another long quotation, and especially one that seems simply to recapitulate Foucault's description of regulationism in *Discipline and Punish*. The point here is that this last statement is *not* Foucault speaking. It repeats the loaded words of *puritanism* and *repression* from *Discipline and Punish*, and adds *hypocrisy* for good measure; but these are concepts that are not authorized by the *History of Sexuality*. This is in fact a statement of the kind of discourse – about the Victorian *repression* of sexuality – that the *History of Sexuality* is famously formulated to oppose. In this view – again it is a straw man of sorts – the underworld of prostitution exists only to cater for the 'other Victorians', and thus it acts as a kind of index of bourgeois hypocrisy, the repression of sexual instincts, and their transference to special, hidden places of sexual transgression. None of these views is endorsed by Foucault's mature analysis. It is as if Foucault's focus on the disciplinary function of prostitution has become transformed, somewhere along the line, into something more complex and considered. It is not hard to see why Foucault might come close to repudiating the focus on discipline, or on discipline alone: for one thing, emphasizing the disciplining of prostitution might remind one too much of a sexually repressive regime; on the other hand, the licensing and toleration of prostitution might suggest only hypocrisy. And, withal, the emphasis on enclosure, isolation and secrecy does not fit very well with the remarkable volubility of the Victorian discourses of sexuality.

None of this is meant to take Foucault to task for inconsistency. It is well known that Foucault changed his mind or happily thought better of earlier formulations. But we may reasonably ask what *was* Foucault's view of regulated prostitution in the formation of modern society. It seems to me that whilst the later Foucault would surely repudiate the clumsy emphasis on puritanism, repression and bourgeois hypocrisy, the significance of 'places of tolerance' would not be lost. If they are certainly not the necessary concessions to irrepressible sexual instinct, regulated brothels and tolerated zones of prostitution do at least exist as heterotopia or counter-sites within modern societies, as 'other places' that testify to the spatial differentiation of modernity. But it is better to say, perhaps, that regulationism is not a classic 'discipline' at all. Sexual regulation, as it is expressed in the regulation of sex work, is better seen as a branch of biopolitical rationality, and of a related 'governmentality' that supplements if not succeeds disciplinary power. Unlike discipline, which is focused on the training of the individual body, biopolitical techniques are aimed at bodies as they relate to the health of entire populations; governmentality, moreover, at least in its 'liberal' form, conceded certain 'freedoms', spaces and domains alien to state control, as a necessary adjunct to the process of government. It is in this sense that regulation of prostitution – and perhaps of sexuality in general – might be understood. The publication of the 1975–1976 lecture course certainly suggests that this is how Foucault conceived it. In these lectures, Foucault stated that there are two 'series' – the first, that of 'body-organism-discipline-institutions', and the second, that of 'population-biological processes-regulatory mechanisms-State' (Foucault 2003, 250). The second axis or series was not about controlling the individual to the fullest, disciplinary extent, but about assuring the biopolitical security of the State; this was 'a matter of taking control of life and the biological processes of man-as-species and of ensuring that they are not disciplined but regularized' (2003, 253). 'Regularization' – a word virtually synonymous in Foucault with regulation, and linked to 'normalization' – is here quite distinct from discipline. The normalization of sexuality is not to be confused with the general technique of discipline, and in many ways cuts across it. Regulationism in particular, and regulation in general, is certainly spatial, but it is not identical with the kinds of panoptical and carceral technologies that geographers have well examined. The distinctive spaces of regulated prostitution should thus be seen in terms of a calculated disposition rather than simply a carceral discipline, that is as 'a matter for discipline, but also a matter for regularization' (2003, 252).

I have spent some time on this one example because it is necessary to disentangle the Foucauldian notion of sexuality as a *dispositif* from the largely discredited thesis about ever increasing disciplinary power and the spread of social control mechanisms throughout the social body. Regulated brothels were places in which a certain sexuality was 'tolerated' rather than simply disciplined; it engenders, literally, certain spaces of sexual 'freedom'. 'Men are permitted to make love much more often and under less restrictive conditions. Houses of prostitution exist to satisfy their sexual needs' (Foucault in Rabinow 1997, 146). It is not about the disciplining of sex, but about the production of sexual subjects and the combating of certain dangers and irregularities (disease, perversity, disorder, and so on). A Foucauldian perspective on the regulation of sexual practices and sexual identities should not then be confined to a focus on disciplinary technologies. Deployment (faulty translation as it is) is not a synonym for discipline. If we open up the question of sites for the deployment of sexuality, and its *disposition* – by which I mean 'the spatial and strategic arrangement of things and humans and the ordered possibilities of their movement within a particular territory' – then the conception of biopolitics and governmentality offer I think a more productive Foucauldian resource for geographers, though one whose use is still in its infancy (Dean 1999, 105, quoted in Joyce 2003, 3). It moves Foucault's discussion of power away from a concern with the coercive techniques of the disciplinary institutions to the more insidious, dangerous and 'commonplace' geographies of 'normalization'. Biopolitics, in 'liberal' states at least, speaks to spaces of 'freedom' colonized by power in the modern era – that is to say, to those spaces in which the exercise of personal freedom is authorized and indeed *required*. Sexual 'freedoms', of course, are fundamental here, and it is to this question that the final part of this chapter is directed.

Geography and Sexuality *after* Foucault

Looking at Foucault's work on sexuality, one generation on, we must acknowledge its limitations for the study of both the history and the geography of sexuality. In terms of the *discourse of sexuality*, even Foucault's most famous statement – his rejection of a Victorian 'repression' of sexuality – seems, on closer inspection, a much more guarded and specific statement, even to the point of being platitudinous; and, in any event, several historians have recently challenged his account both empirically and theoretically. For instance, Michael Mason is both concessionary and critical towards Foucault's work in his discussion of the rise of an anti-sensual mentality in Victorian Britain (Mason 1994). Hera Cook on the other hand straightforwardly rejects Foucault's account, noting that he himself concedes a repression of sexual discussion within the social constraints of a 'restrictive economy', and more importantly arguing that in an age before adequate contraceptive techniques the fear of conception meant that women disciplined and repressed their own sexuality (Cook 2004). This does not represent a return of 'repression', exactly, but it does signal a more considered understanding of Victorian sexuality, one that neither takes the latter as an index of our own liberation nor removes the gulf between them and us. Within geography, whilst it is productive to emphasize the role that geographical representations play in mapping sexual identity, there may well be alternatives to discursive models of sexual subjectivity; Phil Hubbard has pointed to 'psychosocial' models, for instance, and has suggested that the two might after all be reconciled. For many, the somatic and the psychic have not been and can not be folded into the discourse of sexuality; Foucauldianism is not the only game in town (Hubbard 2002). This is subject to the criticism, however, that the psychic and the somatic, in being discursively defined as non-discursive, remain cultural and social constructions. Likewise, in terms of the sexuality as a *dispositif*, we have seen that it is too easy a temptation to talk of a disciplining of sexuality, a social and spatial regulation that leaves little room for real histories and geographies.

However, there is one area where a much more unequivocally positive appreciation of Foucault's significance for geographers of sexuality is called for, and that is his discussion of *sexual identity*; or, rather, *sexual subjectivity*. For Foucault's approach, elaborated in the later volumes of the *History of Sexuality*, but also in a variety of statements elsewhere on contemporary gay politics and culture, make it clear that whilst sexual identities were indeed produced and 'deployed', there was nevertheless considerable room for manoeuvre, a space in which individuals and communities could explore alternative kinds of sexual subjectivity (Foucault 1997a–d). It is not simply that identities might be resisted and reclaimed, through a 'reverse discourse', though this is important. It is moreover that sexual radicals could experiment with desire and pleasure, culture and community, in short a whole field of ethical relationships with oneself and others that the term 'sexuality' does not even begin adequately to represent. It is for this reason that Foucault dedicated such effort to an exploration of antiquity, for he found there, in an age as it were 'before sexuality', ethical models of conduct that challenge our own culture's social and political norms (Larmour, Miller and Platter 1998). Whether or not we should endorse any or all of these models – Foucault's discussion of *askesis* is the most prominent – the point remains that Foucault redirects our concerns away from the issue of sexual *identity* towards a discussion of the variety of practices by which we recognize ourselves as sexual *subjects*. This sounds very abstract, but it has clear implications for contemporary sexual politics. Foucault's work has suggested to many for instance that a politics based on essentialized sexual identities – 'gay', say – is limiting, exclusive, and politically vulnerable (Watney 2000, 50–62; Halperin 1995). Foucault makes it clear that he is fully appreciative of the gains made by marginalized sexual minorities in the contemporary era, and concedes that an emphasis on a (natural) sexual identity has been politically useful in the past. But he has forcefully made the case that resistance should entail a refusal of the discourse of 'sexuality' that we have inherited from (at least) the Victorians. It is for this reason that, after initial suspicion and denigration, Foucault has become of such central importance to contemporary sex radicals. In gay politics, and in queer theory or theories in particular, Foucault's work has been strikingly influential. Geographers of sexuality have also made significant contributions to this politics and this theory (Bell and Binnie 2000; Binnie 2004; Binnie and Valentine 1999; Brown 2000).

Consider, firstly, the concept – or rather the discourse – of the 'closet'. Although I am not aware that Foucault spoke directly about the term, his discussion of the role that 'silence' and 'silencing' plays in the discourse of sexuality is the direct inspiration for Sedgwick's highly influential *Epistemology of the Closet* (Sedgwick 1990). Sedgwick extends Foucault's epistemological analysis of sexuality to the cultural effects of the regime of ignorance that is implied in the proliferation of sexuality as a discourse. The closet is the inevitable result: 'by the end of the nineteenth century, when it had become fully current – as obvious to Queen Victoria as to Freud – that knowledge meant sexual knowledge, and secrets sexual secrets, there had in fact developed one particular sexuality that was distinctively constituted *as* secrecy: the perfect object for the by now insatiably exacerbated epistemological/sexual anxiety of the turn-of-the-century subject' (1990, 73). Part of what Sedgwick points to is the way in which the 'closet' functions discursively in relation to power; and her analysis necessarily throws into question the emphasis in gay politics on coming out of the closet, of refusing to be silenced, of being truthful about one's sexuality. Whilst this discourse certainly has been empowering, it is problematic if it envisages the world 'outside' as being somehow outside of power. And thus the notion of gay 'liberation' is also fundamentally flawed. As David Halperin puts it, *'Coming out is an act of freedom ... not in the sense of liberation but in the sense of resistance'* (Halperin 1995, 30, italics in original). In Sedgwick's remarkable essay in political epistemology, the closet is a place of contradictions whose impossibility helps rather than hinders, its working in the interests of a homophobic and heteronormative society. Following from this, however, we can note that the closet is more than simply a spatial metaphor. In this respect, Sedgwick is not particularly helpful; her absolute unwillingness to respond to the spatial formations of the 'epistemologically-cloven culture' she describes means that injunctions to attend to 'the contingencies and geographies of the highly permeable closet' remain rhetorical (Sedgwick 1990, 12, 165). A geographical analysis of the closet also recognizes the limitations of a politics based on the spatial narrative of 'coming out'. George Chauncey's recovery of the 'gay male world' of early twentieth-century New York, for instance, is a critique of the spatial metaphor of the closet, countering the myths of gay men's isolation, invisibility, and internalization (of society's heterosexist norms) by pointing to the ways in which gay men appropriated public space and reterritorialized the city long before 'Stonewall' (Chauncey 1994, for similarly impressive recent accounts of geographies of male homosexuality in London, see Cocks 2003; Houlbrook 2005, 40–134). This 'gay world', in contrast to the established narrative of the pre-Stonewall 'closet', was diverse, creative and remarkably public, even spectacular; it was nothing less than an 'open secret' (Miller 1988, 192–220). The point here is not to affirm a 'freedom' that simply did not exist, nor to deny the real achievements of the 1970s, but to explore the contours of a gay (male) culture outside of the discursive politics of either sexology or gay liberation; this was a world in which the gendered identity of the 'fairy' prevailed, where effeminacy was a cultural survival strategy, and one which structured gay men's sexual and social encounters. Again, the debt to Foucault is very striking – Chauncey can describe, persuasively and in great detail, a very recent historical geography before and in the early transition to 'homosexuality'/'heterosexuality'. It is a world that the discourse of the 'closet' paradoxically (or maybe not) serves to silence and make invisible.

Beyond this, though, there is a closet or rather series of closets that are enacted and performed in *material* space. Michael Brown's studies in his remarkable book *Closet Space* show how contemporary society maps and remaps a heteronormative and homophobic spatiality; the materiality of the closet, as it operates at a variety of scales, mediates and indeed is fundamental to the experience of oppression (Brown 2000). In such ways, geographers have thus begun to reveal the metaphorical and material nature of the closet and to reflect on the strategies of liberation and resistance necessary to combat it. The questions that the institution of the closet raises concerning visibility and invisibility, isolation and concentration, privacy and publicity are of course also very current when thinking about gay spatialities in general. We have clearly come a long way from the pioneering explorations of gay men's urban geographies, with their affirmation of the need for gay men (at least) to congregate in gay neighbourhoods in order to survive, publicize and politicize. Recent work, influenced to a greater or lesser degree by Foucault, has been more sceptical about the role of 'gay ghettoes' like San Francisco's Castro, particularly with the question of lesbian spatialities in mind (Mitchell 2000, 184–93; for a recent unapologetic prioritization of gay identity politics, however, see Armstrong 2002). But we might as geographers go much further than this. Rather than see space as fundamental to the production of sexual identities like 'gay' – or even 'queer' at least in the sense of 'Queer *Nation*' – we might view space in terms of its contribution to the kinds of self-fashioning that Foucault considers in the later volumes of *History of Sexuality*, in which the practice of (sexual) subjectivity is foregrounded. For instance, Foucault spoke intriguingly about the bathhouses of San Francisco and New York as 'laboratories of sexual experimentation' in which communities of pleasure might be enacted; the bathhouses allowed, he suggests, for the possibility of desubjectivization and desubjugation, for the affirmation of non-identity (Foucault 1997a, 151; Halperin 1995, 94). There has been much written, from a pretty openly homophobic standpoint, about the bathhouses, but there has been little genuine consideration of their role in the making and unmaking of gay subjectivities (Altman 1982, 79–80; Chauncey 1994, 207–25; Corber 1997, 142; Foucault 1997a, 146–7). Were the bathhouses really emancipatory and utopian spaces, creative of an 'empty space of new relational possibilities' (Foucault 1997c, 160)? It is a moot point, but the role of places and spaces such as bathhouses, bathrooms and bars (and bedrooms too of course) in the creation of new forms of subjectivity surely deserves greater consideration by geographers.

Again, the question of queer is central. Jon Binnie has recently referred to Michel Foucault as 'the daddy of queer theory', and whilst this seems a peculiarly inappropriate remark, it is clear that he has been vitally if complexly influential in the development of queer theory and queer politics (Binnie 2004, 70). This has perhaps been largely through the work of other theorists and commentators. Eve Sedgwick's analyses are clearly important here, even if her own work has largely been directed to textual criticism. Perhaps of more importance – certainly for geography – has been Judith Butler's critique of the sex/gender distinction, and her development of a theory of the performance of both gender and sex (Butler 1990; 1993). Butler's appropriation and extrapolation of Foucauldian insights remain highly influential, taking what is a constructionist approach to sexuality to a logical extension. By denaturalizing not just 'sexuality' but sex, Butler can demonstrate that the binary order of heterosexuality is conventional, arbitrary, and confused; heterosex is just as much a *performance* as, famously, is drag. Such a strong constructionist position makes Butler's affinities with Foucault perfectly clear (Bristow 1997, 215, 218). Despite the fact that Butler herself does not readily endorse the geographical emphasis on the spatial context of sex/gender performances, and despite the fact that geographical work influenced by queer theory tends not to reference Foucault, this emphasis on performance is one of the main conduits through which Foucault's work on sexuality has flowed into the discipline of geography (Bell, Binnie, Cream and Valentine 1994; Binnie 1997). By drawing attention to 'the spatial specificity of the performance of gender identities' geographers have demonstrated how all space is sexualized. It is nevertheless notable how little direct reference to Foucault is made in geographical work on performativity, nor indeed on queer. The recent *Environment and Planning D: Society and Space* theme issue (21(4), 2003) on 'Sexuality and space: queering geographies of globalization' contains not a single reference by any contributor to Foucault's *History of Sexuality*. It would be better to say that all space carries traces of heteronormative spatiality, and/or that all space is actually *queer*, in the sense that the norms of heterosexual society are unstable and incoherent. This means, for one thing, that the identification of spaces as either *gay* or *straight* is fundamentally mistaken. Speaking of places such as Greenwich Village or the Castro, for instance, Jean-Ulrich Désert has written:

> The general perception and belief, in part from the mistaken notion that most other places are really straight, is that these are gay/queer zones. This implication, seductive as it is, invites the occupant or observer into a complicit act of faith. Queer space is in large part the function of wishful thinking or desires that become solidified: a seduction of the reading of space where queerness, at a few brief points and for some fleeting moments, dominates the (heterocentric) norm, the dominant social narrative of the landscape. The observer's complicity is key in allowing a public site to be co-opted in part or completely. So compelling is this seduction that a general consensus or collective belief emerges among queers and nonqueers alike. (Désert 1997, 21)

The implications for a queer politics are diverse, and contested, but queer theory is extremely powerful in contesting not just *gay* identity, but all gendered and sexual identities – *all* identities, in fact. It may be that heterosexuality as a *material* practice is underplayed by queer theorists but by revealing that heterosexuality, as the absence of abnormality, depends on homosexuality, and by demonstrating that a logic of exclusion lies at the heart of all questions of identity queer theory has revealed a whole new landscape of power, previously barely glimpsed, let alone understood (on the general neglect of heterosexuality, and its problematization within geography, see Hubbard 2000).

Conclusions

Geography's first, very belated, engagement with sexuality seems to have been the publication of Richard Symanski's *The Immoral Landscape* in 1981, a study of female prostitution indebted to sociobiology, fully approving of capitalist individualism, and thoroughly masculinist in substance and tone (Symanski 1981). That this book appeared more or less coincidentally with the English edition of the *Introduction* to the *History of Sexuality* is something of an embarrassment, but the work that has appeared in the last twenty-five years in geography represents some kind of sustained apologia. For geographers of sexuality, Foucault's writings do not amount to anything like a programme – together with their suggestions, cautions, and brilliant insights, there are ambiguities, false trails and frustrating ellipses – but they do represent the most powerful inspiration to examine the spatial history of sexuality. As David Halperin notes (2002, 42–68), Foucault's work is valuable precisely because it does not have a theory of sexuality. I have tried in this chapter to suggest how Foucault's focus on the discourse of sexuality is both revealing and limiting, and in addition how the concern for sexuality as a *dispositif* ought to be read as something more than sexuality's cooption within the development of a disciplinary society. It is clear enough that considerable problems remain with regard to the place of sexuality within Foucault's archaeologies and genealogies. It is equally clear that we should not ask Foucault to assume responsibility for work that remains to be accomplished. The geography of sexuality remains a very young field, whose introduction into the academy is still contested in many places, but there are plenty of indications of the rewards of that engagement and encounter. Amongst the most important has been the work of queer geographers in revealing the spatial codes by which heterosexism is inscribed into all of our lives, and I would like to reiterate that Foucault's most unequivocally positive influence has been in this analysis of sexual identity and subjectivity. Now a paternity test might not, after all, reveal Foucault to be the 'father' of queer theory; and feminism's contribution – (I cannot resist asking: as its *mother*?) – has in any case been routinely downplayed and neglected. But it would be difficult to deny that Foucault anticipated as well as inspired a queer epistemology and a queer politics. It is Foucault who most forcefully problematizes the modes by which individuals recognize themselves as sexual subjects, and it is Foucault who has himself become a kind of rallying cry for those who wish to oppose this kind of oppression through ascription. David Halperin rightly draws attention to a 'Foucault effect' by which his life and theories have become part of the fabric of social and political resistance (Halperin 1995, 13–14; and literally so in terms of the AIDS Names Project quilt, Halperin 1995, 123–5). It is hard to see how any contemporary critique of sexual normalization can ignore Foucault's life, his work, his example. Geographers have been less effusive, and hagiography has not formed much of its engagement with Foucault, but geography too has been marked by his noble and unwavering politicization of the sexual.

References

Afary, Janet and Kevin B. Anderson (2005) *Foucault and the Iranian Revolution: Gender and the Seductions of Islamism*. Chicago: University of Chicago Press.

Altman, Dennis (1982) *The Homosexualization of America: The Americanization of the Homosexual*. New York: St. Martin's Press.

Armstrong, Elizabeth A.(2002) *Forging Gay Identities: Organizing Sexuality in San Francisco, 1950–1994*. Chicago: University of Chicago Press.

Bell, David and Jon Binnie (2000) *The Sexual Citizen: Queer Politics and Beyond*. Cambridge: Polity Press.

Bell, David, Jon Binnie, Julia Cream and Gill Valentine (1994) 'All Hyped Up and No Place to Go.' *Gender, Place and Culture*, 1(1), 31–47.

Binnie, Jon (1997) 'Coming out of Geography: Towards a Queer Epistemology?' *Environment and Planning D: Society and Space*, 15, 223–37.

Binnie, Jon (2004) *The Globalization of Sexuality*. London: Sage.

Binnie, Jon and Gill Valentine (1999) 'Geographies of Sexuality – A Review of Progress.' *Progress in Human Geography*, 23(2), 175–87.

Bleys, Rudi (1996) *The Geography of Perversion: Male-to-Male Sexual Behaviour Outside the West and the Ethnographic Imagination*. London: Cassell.

Bristow, Joseph (1997) *Sexuality*. London: Routledge.

Brown, Michael (2000) *Closet Space: Geographies of Metaphor from the Body to the Globe*. London: Routledge.

Butler, Judith (1990) *Gender Trouble: Feminism and the Subversion of Identity*. London: Routledge.

Butler, Judith (1993) *Bodies That Matter: On the Discursive Limits of 'Sex'*. London: Routledge.

Califia, Pat (2000) *Public Sex: The Culture of Radical Sex*. San Francisco: Cleis Press.

Chauncey, George (1994) *Gay New York: The Making of the Gay Male World, 1890– 1940*. London: Flamingo.

Clifford, Michael (2001) *Political Genealogy After Foucault: Savage Identities*. London: Routledge.

Cocks, H.G. (2003) *Nameless Offences: Homosexual Desire in the Nineteenth Century*. London: I.B. Tauris.

Cook, Hera (2004) *The Long Sexual Revolution: English Women, Sex, and Contraception 1800–1975*. Oxford: Oxford University Press.

Corber, Robert J. (1997) *Homosexuality in Cold War America: Resistance and the Crisis of Masculinity*. Durham: Duke University Press.

Corbin, Alain (1990) *Women For Hire: Prostitution and Sexuality in France after 1850*. Cambridge: Harvard University Press.

Dean, Mitchell (1999) *Governmentality: Power and Rule in Modern Society.* London: Sage.

Désert, Jean-Ulrick (1997) 'Queer Space.' In Gordon Brent Ingram, Anne-Marie Bouthillette and Yolanda Retter (Eds) *Queers in Space: Communities/Public Places/Sites of Resistance.* Seattle: Bay Press. 17–26.

Donzelot, Jacques (1980) *The Policing of Families.* London: Hutchinson.

Driver, Felix (1993) *Power and Pauperism: The Workhouse System 1834–1884.* Cambridge: Cambridge University Press.

Driver, Felix (1994) 'Bodies in Space: Foucault's Account of Disciplinary Power.' In Colin Jones and Roy Porter (Eds) *Reassessing Foucault: Power, Medicine and the Body.* London: Routledge. 113–31.

Elden, Stuart (2001) *Mapping the Present: Heidegger, Foucault and the Project of a Spatial History.* London: Continuum.

Elden, Stuart (2005) 'The Problem of Confession: The Productive Failure of Foucault's *History of Sexuality.*' *Journal for Cultural Research*, 9, 23–41.

Foucault, Michel (1979) *Discipline and Punish: The Birth of the Prison.* New York: Vintage.

Foucault, Michel (1980) *The History of Sexuality, Volume I: An Introduction.* New York, Vintage.

Foucault, Michel (1985) *The History of Sexuality, Volume II: The Uses of Pleasure.* London, Penguin.

Foucault, Michel (1986) *The History of Sexuality, Volume III: The Care of the Self.* London: Penguin.

Foucault, Michel (1997a) 'Sexual Choice, Sexual Act.' In P. Rabinow (Ed.) *Ethics, Subjectivity and Truth.* New York: The New Press. 141–56.

Foucault, Michel (1997b) 'What is Enlightenment?' In P. Rabinow (Ed.) *Ethics, Subjectivity and Truth.* New York: The New Press. 303–19.

Foucault, Michel (1997c) 'The Social Triumph of the Sexual Will.' In P. Rabinow (Ed.) *Ethics, Subjectivity and Truth.* New York: The New Press. 157–62.

Foucault, Michel (1997d) 'Sex, Power, and the Politics of Identity.' In P. Rabinow (Ed.) *Ethics, Subjectivity and Truth.* New York: The New Press. 63–173.

Foucault, Michel (2003a) *"Society Must Be Defended": Lectures at the Collège de France, 1975–1976.* New York: Picador.

Foucault, Michel (2003b) *Abnormal: Lectures at the Collège de France, 1974–1975.* London: Verso.

Hacking, Ian (1999) *The Social Construction of What?* Cambridge: Harvard University Press.

Halperin, David M., John J. Winkler and Froma I. Zeitlin, Eds (1990) *Before Sexuality: The Construction of the Erotic Experience in the Ancient Greek World.* Princeton: Princeton University Press.

Halperin, David M. (1995) *Saint Foucault: Towards a Gay Hagiography.* Oxford: Oxford University Press.

Halperin, David M. (2002) 'Forgetting Foucault: Acts, Identities and the History of Sexuality.' In Kim M. Phillips and Barry Reay (Eds) *Sexualities in History: A Reader.* London: Routledge. 42–68.

Hannah, Matthew G. (2000) *Governmentality and the Mastery of Territory in Nineteenth-Century America.* Cambridge: Cambridge University Press.

Harvey, Karen (2004) *Reading Sex in the Eighteenth Century: Bodies and Gender in English Erotic Culture.* Cambridge: Cambridge University Press.

Houlbrook, Matt (2005) *Queer London: Perils and Pleasures in the Sexual Metropolis, 1918–1957.* Chicago: University of Chicago Press.

Howell, Philip (2000) 'A Private Contagious Diseases Act: Prostitution and Public Space in Victorian Cambridge.' *Journal of Historical Geography*, 26, 376–402.

Howell, Philip (2004) 'Race, Space and the Regulation of Prostitution In Colonial Hong Kong.' *Urban History*, 31, 229–48.

Hubbard, Philip (2000) 'Desire/disgust: Moral Geographies of Heterosexuality.' *Progress in Human Geography*, 24, 191–217.

Hubbard, Phil (2002) 'Sexing the Self: Geographies of Engagement and Encounter.' *Social & Cultural Geography*, 3, 4, 359–81.

Hull, Isabel V. (1996) *Sexuality, State and Civil Society in Germany, 1700–1815.* Ithaca: Cornell University Press.

Joyce, Patrick (2003) *The Rule of Freedom: Liberalism and the Modern City.* London: Verso.

Karras, Ruth Mazo (2005) *Sexuality in Medieval Europe: Doing Unto Others.* New York: Routledge.

Koven, Seth (2004) *Slumming: Sexual and Social Politics in Victorian London.* Princeton: Princeton University Press.

Laqueur, Thomas W. (2003) *Solitary Sex: A Cultural History of Masturbation.* New York: Zone Books.

Larmour, David H.J., Paul Allen Miller and Charles Platter, Eds (1998) *Rethinking Sexuality: Foucault and Classical Antiquity.* Princeton: Princeton University Press.

Legg, Stephen (2005) 'Foucault's Population Geographies: Classifications, Biopolitics and Governmental Spaces.' *Population, Space and Place*, 11, 137–56.

Mason, Michael (1994a) *The Making of Victorian Sexuality.* Oxford: Oxford University Press.

Mason, Michael (1994b) *The Making of Victorian Sexual Attitudes.* Oxford: Oxford University Press.

Miller, D.A. (1988) *The Novel and the Police.* Berkeley: University of California Press.

Mills, Sara (2003) *Michel Foucault.* London: Routledge.

Mitchell, Don (2000) *Cultural Geography: A Critical Introduction.* Oxford: Blackwell.

Nussbaum, Felicity A. (1995) *Torrid Zones: Maternity, Sexuality, and Empire in Eighteenth–Century English Narratives.* Baltimore: Johns Hopkins University Press.

Ogborn, Miles (1993) 'Law and Discipline in Nineteenth Century English State Formation: The Contagious Diseases Acts of 1864, 1866 and 1869.' *Journal of Historical Sociology*, 6, 28–55.

Ogborn, Miles (1998) *Spaces of Modernity: London's Geographies 1680–1780*. New York: Guilford Press.

Phillips, Richard (1999) 'Writing Travel and Mapping Sexuality: Richard Burton's Sotadic Zone.' In James S. Duncan and Derek Gregory (Eds) *Writes of Passage: Reading Travel Writing*. London: Routledge. 70–91.

Phillips, Richard and Diane Watt (2000) 'Introduction.' In Richard Phillips, Diane Watt and David Shuttleton (Eds) *De-Centring Sexualities: Politics and Representation Beyond the Metropolis*. London: Routledge. 1–17.

Philo, Chris (2000) 'The Birth of the Clinic: An Unknown Work of Medical Geography.' *Area*, 32, 1.

Philo, Chris (2004) *A Geographical History of Institutional Provision for the Insane from Medieval Times to the 1860s in England and Wales: The Space Reserved for Insanity*. Lampeter: Edwin Mellen Press.

Rabinow, Paul, Ed. (1991) *The Foucault Reader: An Introduction to Foucault's Thought*. London: Penguin.

Rabinow, Paul, Ed. (1997) *Essential Works of Foucault, 1954–1984, Volume I: Ethics, Subjectivity and Truth*. New York: The New Press.

Ransom, John S. (1997) *Foucault's Discipline: The Politics of Subjectivity*. Durham: Duke University Press.

Sedgwick, Eve Kosofsky (1990) *Epistemology of the Closet*. Berkeley: University of California Press.

Siena, Kevin P. (2004) *Venereal Disease Hospitals and the Urban Poor: London's "Foul Wards", 1600–1800*. Rochester: University of Rochester Press.

Skinner, Marilyn B. (2005) *Sexuality in Greek and Roman Culture*. Oxford: Blackwell.

Smart, Barry (2002) *Michel Foucault*, revised edition. London: Routledge.

Stein, Edward, Ed. (1990) *Forms of Desire: Sexual Orientation and the Social Construction Controversy*. New York: Garland.

Stoler, Ann Laura (1995) *Race and the Education of Desire: Foucault's History of Sexuality and the Colonial Order of Things*. Durham: Duke University Press.

Stoler, Ann Laura (2002) *Carnal Knowledge and Imperial Power: Race and the Intimate in Colonial Rule*. Berkeley: University of California Press.

Symanski, Richard (1981) *The Immoral Landscape: Female Prostitution in Western Societies*. Toronto: Butterworths.

Walkowitz, Judith R. (1980) *Prostitution and Victorian Society: Women, Class, and the State*. Cambridge: Cambridge University Press.

Watney, Simon (2000) *Imagine Hope: AIDS and Gay Identity*. London: Routledge.

Chapter 26

The Problem with *Empire*

Mathew Coleman and John A. Agnew

Introduction

There has been a resurgence of interest in the logic and substance of geopolitical practice at the global scale. This renewed interest can, in part, be traced to the September 11 2001 terrorist attacks and to the Bush administration's subsequent invasion of Afghanistan as well as its unilateral and pre-emptive war against Iraq (Arrighi 2005). Indeed, much of this literature focuses explicitly on the US and on its geopolitical interests. For example, it is said: that the international stage is an increasingly dense network of imperial interactions centred on Washington DC – a 'world state' in which US geopolitical and geoeconomic power is more concentrated and far-reaching than it was in 1945 (Barkawi and Laffey 1999; 2002; Shaw 2000); that the 'grand strategy' of the current Bush administration is to exploit the post-9/11 geopolitical climate of fear to shore up support for the Anglo-American model of free trade and for US capitalism more specifically (Callinicos 2002); that the punitive financial innovations engineered by the IMF, the World Bank, Wall Street bankers, and other international financial institutions headquartered in the US allows the US Treasury to restructure trade relations with LDCs around exploitative market access politics (Gowan 1999); that US neoimperial might in the world economy is the product of a successful post-Bretton Woods drive to break down national controls on finance and to redirect global savings to the US (Panitch and Gindin 2003); that, although imperial, US foreign policy is caught between economic, political, ideological, and military projects which play out incoherently on the ground and which portend a crisis of leadership (Mann 2003); and, among other things, that the ongoing US war in Iraq is a unilateral attempt to secure access to the Middle East oil spigot in order to give energy-dependent US-based firms a competitive advantage over overseas rivals (Klare 2001, Harvey 2003; Jhaveri 2004; more generally on the resource basis of US geopolitical practice, see Cohen 2003).

Our goal here is not to adjudicate between these positions but rather to suggest that, on the whole, and despite some important differences (see discussion in Agnew, 2003), there is widespread agreement in this literature that global scale geopolitical practice centres on the US, and moreover, that the US is a (neo)imperial power with identifiable territorial and/or strategic ambitions. It is in the context of these ongoing debates, then, that Michael Hardt and Antonio Negri's *Empire* (2000) is noteworthy. For *Empire* marks a decisive conceptual break in the (loosely speaking) political economy literature on the spatiality of (neo)imperial power between those who would continue to see states and state interests – and specifically, US foreign policy – as geopolitically relevant (as above) and those, following Hardt and Negri's lead, who would see the problem of government in a quite different light as absent any place-specific identities or territorial strategies. *Empire*'s apparent legacy, then, is that it has shifted the terms of debate about what might constitute the spatiality of geopolitical power at the global scale. For example, whereas the scholars grouped together above would presumably disagree about whether contemporary world geopolitics is about an inter-national geography of states or a more complex scalar geography of power, as well as about whether or not US geopolitical practice should be considered coherent or representative of certain interests, Hardt and Negri, on the one hand, refuse the relevance of any political geography (*i.e.*, states, regions, cities, *etc.*) to the exercise of power, with the exception of the undifferentiated and engulfing space of the global; and on the other hand, downplay the geopolitical importance of the exercise of power via boundaries, borders, and regulations on the movement of peoples, products, and monies. The upshot is a model of global government which neither insists on the territoriality of power nor on the existence of a set of located interests (coherent or otherwise) that might be cobbled together and referred to generally under the rubric of US foreign policy. Hardt and Negri see the global scale of geopolitical practice as an instance of imperialism without an emperor, and moreover without an empire, if by the latter term we mean some geography over which imperial influence is exerted.

We should state upfront that we find *Empire* to be a much-needed intervention in the literature on the spatiality of (neo)imperial power. While legions of scholars have broad-sided *Empire*, usually for the authors' lack of attention to the class problematic, it is our feeling that the book has prompted a good deal of useful debate about the what and how of contemporary government which at the very minimum has pointed to the difficulty of explaining contemporary world politics using state-centric mappings of power. As Walker (2002a, 339) puts it, despite its omissions and shortcomings, *Empire* nonetheless insists that 'international relations is not a synonym for world politics' (see also Barkawi and Laffey 2002). We also appreciate Hardt and Negri's attempt to articulate an imperialism in which power is not neatly centred in one location and extended coherently outwards toward the periphery. At the same time we find that Hardt and Negri's inclination to develop a general model of power 'without regard to the specific modalities of the exercise of different kinds of power in different kinds of contexts' (Jessop 2003, 54) to be extraordinarily problematic and deserving of interrogation.

A central claim we will make here, then, is that *Empire* offers a suspect account of the spatiality of power that erases the geographical particularities of geopolitical practice. We will describe Hardt and Negri's discussion of power as a spatialized calendar of successive modes of government in which the transition from modernity to postmodernity is absolute. Our basic critique is that Hardt and Negri's spatialized calendar of power leaves us with an unproductively polarized account of how power might operate spatially. In *Empire*'s terms, power either functions according to a (now defunct) sovereign model of power and its strict boundaries between self and other, inside and outside; or, it operates according to an antithetical postmodern logic which does away with boundaries between self and other, inside and outside. This unqualified either/or mapping of political power – as either centred on states or de-centred in networks – prohibits a much more complex appreciation of the re-territorialization and/ or re-scaling of state power in late modernity, a topic which has been taken up recently by political geographers (Brenner 1998; Brenner et al. 2003; Mansfield 2005).[1]

But given that this is a book about the French political theorist and philosopher Michel Foucault and his importance to the discipline of geography, a closely related goal in this chapter is to point out how Hardt and Negri's spatial calendar of power draws on a selective – and we think, deeply misleading – interpretation of Foucault's work on subjectivity and government. The base line for us is that although Hardt and Negri's analysis pays lip-service to Foucault, it in fact does a great disservice to his nuanced understanding of when and where we might expect to find alterations to the Hobbesian model of state power. Indeed, exemplary of perhaps the bulk of writing about Foucault when it comes to the question of power, Hardt and Negri employ Foucault's insights about sovereign-juridical, disciplinary, and biopolitical modes of government to discuss how one overcomes and replaces the other in a temporal succession of modes of government. Arguably, this periodization of power owes more to thinkers such as Carl Schmitt for whom the 20th century is marked by a decisive transition from sovereign (state) to legal (global) government than it does to Foucault, for whom such epochal transitions would be fundamentally ungenealogical. In this sense, we will suggest that Foucault is best interpreted not as an historian of great epochs but as a philosopher; and following from this, that his philosophical interrogation of power and subjectivity proceeds more spatially than temporally, or in other words proceeds on the basis that relations of power are made manifest more clearly in space rather than sequentially in time. Our project here, then, is to review Hardt and Negri's assumptions about contemporary Empire and juxtapose these with Foucault's genealogy of power, in order to offer what we find to be a much more complex account of power and its spatialities. At the broadest level, our goal is to look again in more detail at the theoretical claims and equivalences skirted over in *Empire* (also see Wainwright 2004) with an eye to how we might re-conceptualize the geography of contemporary (neo)imperial government.

Foucault's Geosociology of Political Power

Taken as a whole, Foucault's studies can be considered as a more or less sustained attempt to treat in general the question of the constitution of subjectivity. As Foucault

---

1   Although in *Multitude* (2004) Hardt and Negri place emphasis on the corruption and mutation of forms of state governance rather than on their immediate bankruptcy, the upshot is ultimately the displacement of state territorial forms of rule and the emergence of a newly networked model of global power as charted in *Empire*. In this chapter we choose to focus on what seems to be much the most widely circulating and influential of the two books – *Empire*.

himself explains, his work seeks to 'create a history of the different modes by which, in our culture, human beings are made subjects' (Foucault 1982, 208). This focus on how subjects come to know themselves and others led him to a career-long consideration of what has been called the 'analytics of government', or the relatively stable 'regimes of practices' through which subjects are both realized and self-realized (Dean 1999, 20–27). Portions of this work on government are frustratingly underarticulated in spatial terms, and at times Foucault appears equivocal about the importance of thinking geographically about power and subjectivity (Foucault 1980a). Indeed, for some geographers, by virtue of his emphasis on the topologically dispersed quality of power relations, Foucault's analyses obfuscate the grounded scalar spatiality of power and operate generally via an 'evacuation of the spatial' (Allen 2003, 191). We tend to agree, however, with Elden's (2001) argument that Foucault's various attempts to come to terms with the diversity of subjectivity-constituting techniques of power can be read as so many mappings of power, particularly as they probe explicitly into the complex and sometimes convoluted spatiality of the practices that produce modern subjects. From this perspective we might note that for Foucault *'les questions d'espace'* guided a career-long interrogation into modes of government and the production of modern subjectivity (see in particular Foucault 2000a; 2004, 13).[2]

For us, Foucault's crucial political geographic argument in his exploration of government and subjectivity is that we rethink relations of power beyond the state, or more accurately, beyond the state as an already and always centralized apparatus of interests and strategies which is at best tangentially concerned with individuals and their everyday lives. For example, a recurring target in Foucault's work is the model of government articulated in Hobbes' *Leviathan*. Foucault finds this model wanting because it amalgamates subjects into a contractual mass and then essentially forgets about them, or at least assumes their consent, calculability, and/or obedience as the collective, unified body of the state. For Foucault (1980b, 90), if politics is a sort of Clausewitzian war inscribed 'in social institutions, in economic inequalities, in language, in the bodies themselves of each and everyone of us,' the question of political order and stability necessarily exceeds a static territorial relation between sovereign subjects (conceived as a singular, coherent body) and their sovereign. As Foucault argues (1982, 334):

> I don't think that we should consider the modern state as an entity that was developed above individuals, ignoring what they are and their very existence, but on the contrary, as a very sophisticated structure in which individuals can be integrated.

Those forgotten in canonized theories of the state become, for Foucault (1980b, 98), subjects through which – rather than over which – power is exercised:

---

2   Foucault (2000a): 'I think it somewhat arbitrary to dissociate the effective practice of freedom by people, the practice of social relations, and the spatial distributions in which they find themselves. If they are separated, they become impossible to understand. Each can only be understood through the other.'

We should try to grasp subjection in its material instance as a constitution of subjects. This would be the exact opposite of Hobbes' project in the *Leviathan* ... Think of the scheme of *Leviathan*: insofar as he is a fabricated man, Leviathan is no other than the amalgamation of a certain number of separate individualities who find themselves reunited by the complex of elements that go to compose the State; but at the heart of the State, or rather, at its head, there exists something which constitutes it as such, and this is sovereignty, which Hobbes says is precisely the spirit of Leviathan. Well, rather than worry about the problem of the central spirit, I believe that we must attempt the study of the myriad of bodies which are constituted as peripheral subjects as a result of the effects of power.

From this comes Foucault's (1980b, 102) widely cited claim that 'we must eschew the model of the Leviathan in the study of power'. However, Foucault's call to 'cut off the King's head' (121) is not a dismissal of the state or an argument for its dissolution, as is frequently claimed (Curtis 1995; Kerr 1999; Bartelson 2001). Rather, Foucault's appeal is to theorize subjectivity in terms beyond either the coercive downwards exercise of force or the consensual upwards transfer of rights from autonomous unified individuals that have dominated both liberal and Marxist accounts of power. For Foucault, the problem of modern subjectivity requires understanding power differently as an all-pervasive field of structuration which percolates below and outside the sovereign's sanctioned reach but which nonetheless intersects with sovereign-juridical power. Foucault's principal project, we might say, is to reintroduce the problem of subjectivity to that of government, or in other words to restate the relation between the sovereign and his/her subjects such that the binary sovereignty/obedience mapping of power can be complicated with alternate explanations of the way that power works and subjects are formed (Hindess 1996; Allen 1997; Sharp et al. 2000; Herod and Wright 2002).

For us, then, Foucault's work should be valuable for political geographers because by interrogating the conditions of possibility of subjectivity it avoids conflating general questions about the spatiality of power with much more (historically and geographically) specific questions regarding the territorial powers of the state. Otherwise said, in his exploration of the way that modern individuals come to self-recognize as subjects, Foucault looks askance at the modern geopolitical imagination, which insists rather simplistically that power is about coercion exercised monopolistically and coherently by practitioners of statecraft between (and over) undifferentiated blocks of subjects fixed in absolute spaces referred to as states. What he offers instead is what we might call a 'geosociology of political power' or an understanding of the complex sociological contexts of overlapping and discontinuous spatialities of power in the plural (Agnew 2005).

It is in this general spirit of problematizing the spatial operation of sovereign power via the study of subjectivity that Foucault discusses two principal techniques of subject-constituting power above and beyond the sovereign's exceptional power to take life, or the sovereign's 'power of the sword' (Foucault 1979). Schematically, we can say that for Foucault subjects are produced by two additional, generally overlooked, technologies: 1) by micro-practices that divide, isolate, and objectivize; and 2) in the midst of less determinate participatory configurations of self-examination. The former are disciplinary technologies of power centred directly on individual bodies, and the latter are biopolitical technologies of power exercised indirectly through and by means of populations of bodies. For the most part, political geographers interested in Foucault's work on subjectivity have zeroed in on the first (Driver 1985; 1993; Soja 1989; Herbert 1996; Hannah 1997; Pallot 2005), and as such have examined in detail Foucault's panoptic techniques of spatial segregation in *Discipline and Punish* (1978). The tendency, then, has been to interpret Foucault's work on power from the standpoint of subjects as malleable bodies socially produced under duress and surveillance.

The disciplinary mode of subjectivity-constituting government involves a very specific spatiality, which we will review here only briefly. On the one hand, disciplinary power requires an absolute spatial configuration which Foucault calls the 'figure of the camp [*la figure du camp*]' (Foucault 2004, 18). In the form of 17th and 18th century schools, penitentiaries, hospitals, barracks, etc., the figure of the camp functions via the strict placement of bodies in closed, partitioned, empty, and transparent spaces (*quadrillages*) of examination and correction (Foucault 1978, 170–94). And on the other hand, as Dreyfus and Rabinow (1982, 153) argue, these absolute spaces of correction occur at a micro-scale, and accord to place-specific norms and punishments: 'Scale is crucial [to disciplinary power]; the greatest, most precise, productive, and comprehensive system of control of human beings will be built on the smallest and most precise of bases.' Accordingly, the spatiality of disciplinary power is best described as an intensive, institutional time-space geometry of local and concretized power relations whose uniformity or relationality over space cannot be assumed (Philo 1992).

The question of the comparative variability of localized disciplinary exercises of power is of key importance in Foucault's work. For example, although the panopticon is about the centralization of power it is at once about the localization of power, and shows how the constitution of productive 'docile bodies' does not require the uniform exercise of juridical power by the sovereign across an even, isotropic surface. As Foucault (1978, 170) argues:

> Instead of bending all its subjects into a single, uniform mass, [disciplinarity] separates, analyzes, differentiates, carries its procedures of decomposition to the point of necessary and sufficient single units ... It is not a triumphant power, which because of its own excess can pride itself on its omnipotence; it is a modest, suspicious power, which functions as a calculated, but permanent, economy. These are humble modalities, minor procedures, as compared with the majestic rituals of sovereignty or the great apparatuses of the state.

As such, Foucault describes disciplinarity as a radical departure from the generalizing spatial logic of sovereign-juridical power:

> The perpetual penalty that traverses all points and supervises every instant in the disciplinary institutions ... is opposed, therefore, term by term, to a judicial penalty [of the sovereign] whose essential function is to refer, not to a set of observable phenomena, but to a corpus of laws and texts that must be remembered; that operates not by differentiating

individuals, but by specifying acts according to a number of general categories; not by hierarchizing, but quite simply by bringing into play the binary of the permitted and the forbidden. (1978, 183)

That the operation of disciplinary power accords to locally specific norms and punishments does not mean, however, that the figure of the camp exists somehow outside the jurisdiction of the more abstract and macro-scale territorial powers of the sovereign. Indeed, Foucault is careful to describe disciplinary power as the *relocation* and *transposition* of sovereign power to substate centres of incarceration (Foucault 2003). In effect, Foucault elaborates on the problem of discipline to demonstrate how state power can be (formally and informally) subcontracted to remote, local, and specialized authorities. So, although Foucault has been rightly criticized for expelling law from his analysis of power (Hunt 1992; Hunt and Wickham 1994), his basic project in *Discipline and Punish* is to move away from purely legal theories of sovereign power to contemplate how power might be exercised through localized relations of measurement, normalization, treatment, and rehabilitation typically thought irrelevant to, or at least strategically removed from, the exercise of sovereign authority.

Judged by some as a 'bleak political horizon on which the subject will always be an effect of power relations, and on which there is no possibility of escape from domination of one sort or another' (Patton 1998, 64; see also Gordon 1991, 4–8), Foucault's emphasis on discipline was significantly retooled in *The History of Sexuality* (1979). Here, we find an emphasis on a more general or global technology of power which explores how individuals might resist or otherwise navigate (although never escape) relations of domination. Foucault describes this second 'nondisciplinary' technology of government – which on the whole has been side-stepped by political geographers (although see Philo 1992; Ó Tuathail 1996; Herod et al. 1998; Sharp et al. 2000) – as a biopolitical model of power which operates in a spider-like fashion through populations rather than over individual bodies or in a strictly territorial fashion.

The diffused spatiality of this second technology of power is altogether dissimilar to the centralized operation of power under disciplinarity. Gordon (1991, 20; emphases added) describes the difference as one between the 'police conception of order as a *visible grid* of communication' and the 'necessarily opaque, dense, autonomous character of the *processes of the population*'. Whereas the former concerns a fixity of power relations in specific localities or places, the latter is about more extensive 'spaces of dispersion: spaces where things proliferate in a jumbled-up manner on the same level as one another' (Philo 1992, 139). It is appropriate, then, that against the disciplinary figure of the camp Foucault describes biopolitical power as active in an 'aleatory space [*espace aléatoire*]' or 'field of intervention [*un champ d'intervention*]' (Foucault 2004, 22, 23). The goal of biopolitical government is not the direct management of the individual through the concentrated scrutiny and correction of his/her actions in a carceral space. Rather, what Foucault has in mind is a sort of *laissez-faire* regulation in which vast circulations of things and people are managed in the aggregate according to cost-benefit statistical analyses, loosely delineated bands of acceptable conduct, and measurements of risk – what Foucault (2003, 7–8) sums up as 'mechanisms of security [*dispositifs de sécurité*]'. For example, Foucault suggests that the production of sexual subjectivity in the late eighteenth century depended not on corrective means of punishment in institutions in specific locations but on the de-centred, network-like circulation of specialist biopolitical knowledges about the dangers of female sexuality, child masturbation, and procreative behaviour, as well as of the conjoined qualities of pleasure and perversion (Foucault 1978). With these knowledges, the practice of government shifted from the *conduct of others* under surveillance by an authority to the *conduct of the self* – an exhaustive 'government of the living' in which individuals, as members of a larger accumulation of beings said to possess certain tendencies and characteristics, self-scrutinize and then confess piecemeal their apparent deviances to an authority (Foucault 1997a). This is a technology of power built on the 'slow surfacing of confidential statements' rather than on a penal system of hard and fast rules which define exactly what is permitted and prohibited (Foucault 1979, 63).

As with his discussion of discipline, Foucault's elaboration of biopower as a mechanism of security is inherently spatial in that it unsettles the simplistic state-centric model of power at the heart of the *Leviathan*. Foucault's (1997a; 1997b; 2000b; 2003) provocative reinterpretation of 17th and 18th century *raison d'état* and the police science of *Polizeiwissenschaft* is illustrative. *Raison d'état*'s shepherd-like power over the flock *omnes et singulatim* – over all and each – was, for Foucault, operative not simply through the top-down royal prerogative of the King, but also through myriad bottom-up technologies of self-scrutiny and confession on the part of individuals compelled to obey the sovereign. Thus, on the one hand, obedience under *Polizeiwissenschaft* stemmed in part from the sovereign-juridical threat of sanction outlined in legal code, from the sovereign's exceptional power of the sword, as well as from the sovereign's attempt to generate a thorough accounting of the peoples and places under his/her territorial control. However, on the other hand, this last mercantilist goal of total geopolitical and geoeconomic knowledge – about a population's size, strengths, weaknesses, resources, wealth, and health – was as much the development of a 'government of individuals by their own verity' (Foucault 2000b, 312) as it was in any easy sense an expression of the all-powerfulness of sovereign knowledge. In short, Foucault suggests that *Polizeiwissenschaft* – due to the sheer size of state populations about which knowledge was needed, and the impossibility of accumulating with any enduring certainty knowledge about their shifting properties – was a much more *participatory* phenomenon than is acknowledged in standard theories of state sovereignty. The upshot is that Foucault effectively does away with the dialectic identities of the governor and the governed, and replaces them with a fluid, diagrammatic conceptualization of power in which subjects are caught up in constitutive, collusive webs of non-localizable and co-extensive relations (Deleuze 1988, 23–44).

What we can extrapolate from this brief survey of discipline and biopower (for more, see Huxley and Philo, chapters 20 and 27 in this volume) is that Foucault sought – via an examination of prisons, hospitals, madness, and sexuality, throughout a period taken as exemplary of sovereign territorial power – to explore the subjectivity-structuring power relations missed by the oftentimes simplistic sovereign-juridical model of power employed by historians of the state. The point here is that if through his discussion of disciplinarity Foucault relocates sovereign power to a local, carceral site at some remove from the direct authority of the sovereign, then through his discussion of biopolitical power he discusses yet another *dislodgement* of the sovereign-juridical model of power, for example by looking to an array of 'mechanisms [of security] through which it becomes possible to link calculations at one place with action at another, not through the direct imposition of conduct by force, but through a delicate affiliation of a loose assemblage of agents and agencies into a functioning network' (Miller and Rose 1990, 9–10; Rose and Miller 1992). In this sense, Foucault's insistence on the 'micro-diversity' of power rather than on the 'macro-necessity' of state power can be compared favourably to the discussions initiated by thinkers such as Gramsci and Poulantzas on hegemonic networks of social reproduction and strategic relations (see in particular Jessop 1987; 1990; 2005). Indeed, Gramsci's discussion of hegemony in *The Prison Notebooks* (1971, 206–76) – which suggests that the labour of government is performed as much through the self-government of individuals as through official moments of statecraft – can be considered an important pre-cursor to Foucault's contention that scholars of government look again, and with care, at the individuals and populations typically displaced in Hobbesian accounts of sovereign power.[3] At any rate, what Foucault's double displacement of the sovereignty/obedience mapping of state power offers political geographers is a much more complicated story about the spatiality of power than we might get from, say, Cold War-era political sociology, political geography, and political science scholarship whose treatises on the conditions of possibility of political community were for the most part based on a narrow sovereign-juridical interpretation of power (Walker 1993; Agnew 1994; Edkins 1999).

The Multitude and the Thermidor

Hardt and Negri hitch their thesis about the shape and substance of the changing properties of global imperial government to Foucault's displacement of sovereign state power. This is counter-intuitive insofar as Foucault has very little of substance to

---

3   Foucault's work on subjectivity and government is not *sui generis*. Consider, for example, how Gramsci's (1971, 268) discussion of 'legislation' in *The Prison Notebooks* turns to the problem of self-government, in ways remarkably similar to *The History of Sexuality*: 'The assertion that the State can be identified with individuals (the individuals of a social group), as an element of active culture (i.e. as a movement to create a new civilization, a new type of man and of citizen), must serve to determine the will to construct within the husk of political society a complex and well-articulated civil society, in which the individual can govern himself without his self-government thereby entering into conflict with political society – but rather becoming its normal continuation, its organic complement.'

say about imperialism. Indeed, as Ann Laura Stoler (1995, 14) points out, Foucault's genealogy of 'bourgeois identity [is] not only deeply rooted in a self-referential western culture but [is] bounded by Europe's geographic parameters'. Nonetheless, Foucault's problematization of sovereign-juridical government is the inspiration for *Empire* insofar as Hardt and Negri, like Foucault, seek to unravel the Hobbesian model of power found in much writing about world politics. Let us briefly, then, review the argument in *Empire*.

For Claude Lefort, in his *Democracy and Political Theory* (1988), modernity's democratic revolutions shifted questions of law, power, and knowledge from a monarchical seat above the social world to a public site defined by uninterrupted social contest and material labour. Lefort describes this new location of authority as an uncertain 'empty place' born from local sociological contingencies, and he argues that anxiety with its indeterminacies – particularly among rulers – gave rise to totalitarian forms of government which sought to restore order in a violent 'fantasy of the People-as-One' (1988, 20). A similar dynamic lies at the root of the many arguments presented in *Empire*. For Hardt and Negri, modernity is about a radical refounding of questions of knowledge and authority in the immanent, material practices of the human multitude – that seething mass of creative, restless, and dynamic populations which in the modern period make up the living stuff of states. Indeed, for the authors of *Empire*, modernity is constituted by the rejection of transcendental knowledge and growing awareness of the partiality or situatedness of truth claims in various technical, political, social and historical contexts. And as with Lefort, Hardt and Negri (2000, 75) argue that this growing appreciation for the localization of knowledge in the everyday was met with repeated counter-revolutionary attempts on the part of the sovereign to bury immanent knowledges, re-establish 'ideologies of command and authority, and thus deploy a new transcendent power by playing on the anxiety and fear of the masses, their desire to reduce the uncertainty of life and increase security'. However, there is a key difference between the analysis presented by Lefort and that in *Empire*. Whereas Lefort sees a singular, totalitarian response to modernity's democratic revolutions, Hardt and Negri see a more multi-faceted encounter that, although born in modernity, continues to be relevant today in a period identified as the postmodern. From modernity, then, according to Hardt and Negri, we have inherited an ongoing struggle of (literally) history-giving proportions between the creative forces of immanence and the restorative forces of transcendence, between the generativity of the multitude (plural) and the Thermidor (singular) who seeks to channel, divide, and/or extinguish the revolutionary agitations of the former.

If Hardt and Negri's historical narrative returns us again and again to a recurring and spiralling struggle between the forces of immanence and transcendence, regardless of time and space, it does at the same moment present us with a rather interesting calendar of power. In other words, *Empire* presents a provocative account of different spatialities of power and resistance, or revolution and counter-revolution, predominant in various epochs of constitutive encounter between the Thermidor and the multitude. Hardt and Negri's specifically spatial account of the differences marking transcendent imperialism (modernity) from immanent Empire (postmodernity) is what is of primary interest to us here. We will look first at the spatiality of power and resistance in what Hardt and Negri describe as the modern disciplinary society, and then at the altogether different spatiality of power and resistance characteristic of what they call the postmodern society of control.

In modernity, according to Hardt and Negri, the generative powers of the multitude were besieged by a many-sided, combative, and elusive constellation of conservative knowledges and practices which the authors refer to as the 'sovereignty machine'. The 'sovereignty machine', which brought together the extensive power of capital and the intensive police power of the state, sought the replacement of princely command or abstract sovereign-juridical power with the state's territorial power of administration and 'arrangements of discipline', which exerted 'a continuous, extensive, and tireless effort to make the state always more intimate to social reality, and thus produce and order social labour' (2000, 89). On the one hand, the state apparatus imposed a penal order on newly territorialized populations and thereby prevented the multitude 'from organizing itself spontaneously and expressing its creativity autonomously' (2000, 83). This found a complement in nationalist and colonial articulations of identity and difference which gave citizens – rather than the multitude – what we might identify as a noncosmopolitan sense of belonging which complemented the state's institutional and administrative territoriality. And on the other hand, the expansion of capitalist relations of production across the globe brought peoples and places into a common economic field which – albeit geographically uneven – uniformly appropriated the multitude's material labour. Merging for a brief spell in the contradictory institutions and passageways of civil society, the countervailing territorial logic of the state and the networked logic of capital worked together to 'accomplish the miracle of the subsumption of singularities in the totality, of the will of all into the general will' (2000, 87–88). In other words, the organization of capital under the disciplinary umbrella of the state cut violently into and across the generative, networked material energies of the multitude.

The scope of the contemporary Thermidorian counter-revolution is substantially more bewildering and at least in terms of spatial organization, owes very little if anything at all to the territorializing disciplinary logic sketched out above. Following the work of prominent Marxist theorists on the boundary-dissolving cultural politics of late modernity, Hardt and Negri suggest that the 'sovereignty machine' has recently been replaced by a maze-like postmodern paradigm of 'imperial sovereignty' signposted by the politics of difference and structurally undergirded by the growth of radically de-centred regimes of accumulation and modes of regulation. This new condition of imperial sovereignty – in which the sovereign's counter-revolutionaries have abandoned the fort and have circled round to join the marching masses from the rear in a confusing mass of political networks, as in Seattle – is, for Hardt and Negri, in substantial part the product of the globalization of the American constitutional experiment, which governs according to checks and balances rather than by executive fiat.

In this sense, Hardt and Negri's notion of newly emergent imperial sovereignty can be likened to Carl Schmitt's Weimar Republic-era discussion of Wilsonian internationalism. Schmitt warned after the end of WWI and repeatedly thereafter of the arrival of a new form of American imperialism dependent not on the simple military might of the Allied powers but on the erection of global legal and commercial networks, which operate by deterritorializing the existential commitments and institutional functions once monopolized by states (Schmitt 1976 [1932]; 1985 [1933]; 1987; 1996 [1938]; see also Ulmen 1987). For Schmitt, the 20th century diffusion of the American republican experiment – in the name of global peace and human rights – was, then, really about the geopolitical production of a de-centred, supranational 'empty space' (Schmitt 1996 [1938], 49) of depoliticized, spectacular consumption and cultural difference governed through the privatizing and pluralizing tendencies of democratic government and constitutional law. The point here is that Hardt and Negri present more or less the same story, arguing that the 'contemporary idea of Empire is born through the global expansion of the internal US constitutional project' (2000, 182). The difference, however, is that whereas Schmitt saw the US as an undisclosed force behind the global liberal project, Hardt and Negri see no one orchestrating agent, and certainly not the US.[4] As they explain in an addendum to *Empire* (Hardt et al. 2002b, 210–11):

> The US government is not the centre of Empire, and its president is not the Emperor. The primary principle of Empire … is that its power has no actual and localizable centre. Imperial power is distributed in networks and through articulated mechanisms of control … The centre of Empire, if it still makes sense to speak of that, resides in no place but in the virtuality of its power. The long 20th century, then, is not really an American century, but an imperial century.

Indeed, from Hardt and Negri's perspective, Empire cannot be about US power because contemporary imperial government is the *antithesis* of state power; because in postmodernity the counter-revolution has abandoned 'the tired transcendentalism of modern [state] sovereignty, presented either in Hobbesian or Rousseauian form' (Hardt and Negri 2000, 161; Hardt et al. 2002a, 179–80). Now arranged in democratic, open, and consensus-based networks in which horizontal processes of self-regulation – rather than vertical disciplinary tactics – thwart the material creativity of everyday lives, the Thermidor has forsaken the quest for transcendence via the state. Instead, 'it' now pursues a differently imperial strategy – a parasitical one that simultaneously reproduces and taps the unruly powers of the multitude while at once providing a minimum of functional balances, limits, and

---

4   The key point of departure between *Empire* and Schmitt is the latter's celebration of the state and sovereign exceptional authority. Although critical of Wilsonian liberal internationalism, Hardt and Negri do not embrace Schmitt's 'Hobbesian existentialism' (Strong 1996). Indeed, they move in the opposite direction and embrace deterritorialized non-state networks, in the manner suggested by Deleuze and Guattari (1987), and moreover, celebrate these networks as a potentially revolutionary geography (see in particular the exchanges in Negri and Zolo 2003 and Veroli and Mudede 2002). For the centrality of Schmitt to *Empire*, see Balakrishnan (2000) and Hardt and Dumm (2000).

equilibria necessary to keep the anarchic potentialities of the multitude in check. This is an immanentist mode of network power which – in spite of our reference to Schmitt – at least nominally owes its formulation to Foucault's discussion of biopolitical power, or what we described above as a shepherd-like network of forces active in an 'aleatory space' of flows and circulations. And it is this biopolitical notion of imperialism which sets the analysis in *Empire* apart from other current theories about the territorial imperial strategies adopted by the US in the global political economy. For the biopolitical logic of Empire is inclusionary and democratic rather than exclusionary and authoritarian, and its spatiality is an unbroken 'field of interventions' rather than either a world of sovereign states or of border-drawing disciplinary tactics.

The Multiplication – Not Polarized Periodization – of Government

The most remarkable argument in *Empire* is not that power might operate via disciplinary techniques or biopolitically, but that these modes of government have a specific temporality. For Hardt and Negri, we have passed neatly from a modern – and now defunct – mode of counter-revolutionary resistance to the multitude (the 'sovereignty machine') to a postmodern mode of Thermidorian government (the 'unitary machine') in which it is 'no longer possible to identify a sign, a subject, a value, or a practice that is "outside"' (2000, 385). In other words, we have moved from an era of sovereign and disciplinary power to an era of biopolitical power in which there are little if any traces of the former mode of government. Indeed, although disavowing that their work is teleological in the sense of a necessary and pre-ordained series of (modern and then postmodern) regimes of government, Hardt and Negri are nonetheless quite clear that their broad-brushed goal is to periodize these two polar opposite forms of sovereignty as temporally *distinct* and *successive* phenomena:

> [O]ur hypothesis is that since Empire is itself a universalizing phenomenon, it can be conceived adequately only in a global perspective. The object of study itself demands this large-scale framework ... [W]e understand the nature of Empire though a periodizing argument, as the successor to the modern, imperialist form of capitalist power. Such periodizing arguments always require a significant historical sweep and, specifically, our notion requires that we theorize in some detail the modern period that we claim has come to an end. Our book seeks to give new names to a series of phenomena that can no longer be conceived adequately using our old categories.

From this universalizing perspective, then, the brand new moment of Empire is marked by everything that the former was not: the collapse of borders, the withering away of states and national economies, the smoothing of global space, the eclipse of centres of power, the effacement of local particularities, the collapse of relations of identity/difference, and the obliteration of the 'outside' more generally; it is the inverted – Ohmaean, neoliberal – image of the modern world's spatiality of power defined perhaps by what has been called the 'death of geopolitics' (Blouet 2001, 159). Territoriality is now not simply on the wane but is altogether irrelevant to the political: the contemporary world is a 'smooth space of uncoded and deterritorialized flows' such that the 'place of politics has been de-actualized' (Hardt and Negri 2000, 333, 188).

The upshot is a deeply problematical temporalization of the spatiality of power which, paradoxically, reifies rather than problematizes the Hobbesian model of sovereign government. By this we mean that sovereignty and state territoriality remain a (hidden and unproblematized) benchmark in Hardt and Negri's work against which changes in the spatiality of (neo)imperial power are plotted.[5] The result is an oddly polarized spatial calendar of power in which either the state orders social relations powerfully and effectively (then) or it does not (now); either power is territorialized in the form of the state (then) or it is thoroughly deterritorialized in webs which have nothing to do with state government (now); and, finally, power is either transcendental (in the form of the state's counter-revolutionary powers) or immanent (in the form of networks without centres). The central underlying problem with *Empire*'s immanentist reframing of global power, then, is that in taking the sovereign state as the paragon of modern government transcended in postmodernity, Hardt and Negri leave conventional accounts of the spatially coherent operation of state power uninterrogated and in turn can pose nothing except the negative spatial image of Hobbes' binary sovereignty/obedience mapping of power (i.e., the network) to explain the contemporary spatiality of global government.

Our point, then, is twofold: first, that Hardt and Negri present a state-centric account of how power works, which requires that power be either territorial or networked, and nothing in between; and second, that they then go on to annex these two distinct spatialities of power to two distinct temporalities – modernity and postmodernity. The result is what we might call a polarized periodization of power. In this respect, Hardt and Negri do represent something of a departure from much scholarship about world politics making reference to Foucault. For example, Foucault's influence in international relations has generally been to justify a continuing focus on the territoriality of juridical and disciplinary power at the expense of attending to the attrition of sovereign power and to the rise of resolutely non-sovereign (as well as non-disciplinary) modes of global biopolitical government (Hutchings 1997). The problem is, however, that Foucault's insights about subjectivity, power, and the spatiality of government do not lend themselves easily to Hardt and Negri's periodical project, if at all. As we argue below, with reference to Foucault, we prefer to see modes of government differently as *spatialities of power* which might re-

---

5   We are reminded of Walker's (2002b) criticism of post-sovereign accounts of power which assume the state as an unproblematic centre of power which has been avoided, evaded, and/or transcended in late modernity. From this perspective, state sovereignty remains the 'assumed foundation' of scholarship which disavows the contemporary relevance of the sovereign-juridical spatiality of power.

circulate through, and co-occur in, different geopolitical periods and with varying intensities (Galli 2001; Agnew 2005).

Hardt and Negri's interpretation of Foucault is not unique. Foucault's work on discipline and biopower is typically understood by his interpreters in terms of discrete phases of scholarship on subjectivity and power, comprising 'early' studies on the specificity of power relations and the subjugation of individuals under oppressive techniques of government, and 'later' studies focused more on nebulous networks underpinning relations of domination (Moss 1998). For example, Dreyfus and Rabinow (1982) remark that Foucault's oeuvre is often divided up into a preliminary 'genealogy of the modern individual as object' and a subsequent 'genealogy of the modern individual as subject'. This sort of distinction has led many scholars to construe Foucault's 'early' account of oppression and 'later' account of domination not as complementary approaches to an overarching interrogation about the how of government and subjectivity but rather as different parts of a disconnected body of thinking which can be emphasized at will by an author. For example, Philo (1992) makes the very useful point that Foucault's explorations have been used by political geographers in a sort of schizophrenic way to discuss either the strict spatial ordering of panoptic power relations (as in Soja 1989) or the fragmented and disordered nature of power (closer to Philo's stance), as if the two are disconnected approaches to the problem of subjectivity and government. We are adding to this insight that it is often the case that Foucault's examination of sovereign-juridical, disciplinary, and biopolitical governmentalities are understood not only as disconnected phenomena but also as periodic, or as together comprising a history of successive moments and modes of government: disciplinary government replaces sovereign-juridical government; biopolitics replaces disciplinarity. Our position here, following in the spirit of Dreyfus and Rabinow (1982; see also Burchell et al. 1991; Rabinow 1994; Dillon 1995; Moss 1998) is that this break between the 'early' and 'later' Foucault detracts from a larger argument Foucault makes about how various mechanisms of power *build on one another* and *interpenetrate* to produce multiple new modes of subjectivity-constituting government in which different relationships of power are stressed.

Nowhere does Foucault definitively periodize sovereign-juridical, disciplinary, and biopolitical modes of government. Although a less generous interpretation would be that this discloses a general pattern of exaggerated and imprecise argumentation in Foucault's work,[6] the point we wish to draw attention to is that Foucault's *genealogical* investigation of subjectivity, power, and government – which is weary of fixed identities and temporalities, and instead is attuned to history's 'moments of intensity, its lapses, its extended periods of feverish agitation, its fainting spells' (Foucault 1977, 145) – would by definition refuse such a tidy historical periodization

---

6   Two prominent Foucauldian scholars note that Foucault's empirical work – plagued sometimes by an 'interpretive exaggeration' – is characterized by 'areas of unclearness and sketchiness which can be read either as confusion, or more sympathetically, as problems he has opened up for further exploration' (Dreyfus and Rabinow 1982, 126).

of power and subjectivity. We can go as far as to say that Foucault's entire account of subjectivity is developed in order to refute such a chapter-like, beginning-and-end form of 'global history' in which the social world is divided temporally by distinct ontologies of power (Young 1990, 69–87).

A recently published collection of his 1976 lectures at the Collège de France entitled *Society Must be Defended* (2003) illustrates this point nicely. Here, Foucault denies the possibility of periodizing government by positing the superimposition and mutual constitution of disciplinary, biopolitical, and sovereign-juridical diagrams of power. Foucault argues (2003, 249), for example, that due to an inability to govern increasingly complex economic and political relations, the sovereign-juridical model of government underwent a double *adjustment* rather than an *erosion*; namely, an *incorporation* of, on the one hand, a disciplinary 'anatamo-politics of the human body', and on the other hand, a regulatory 'bio-politics of the population' (2003, 249–50):

> It is as though power, which used to have sovereignty as its modality or organizing schema, found itself unable to govern the economic and political body of a society that was undergoing both a demographic explosion and industrialization. So much so that things were escaping the old mechanism of the power of sovereignty, both at the top and the bottom, both at the level of detail and at the mass level. A first adjustment was to take care of the details. Discipline had meant adjusting power mechanisms to the individual body by using surveillance and training ... And then at the end of the eighteenth century, you have a second adjustment; the mechanisms are adjusted to phenomena of population, to the biological or biosociological processes characteristic of human masses.

Thus, the local surveillance of individual bodies as well as the biological regulation of the population are introduced *into* the very fabric of sovereign-juridical government as, respectively, first- and second-cut attempts to rejuvenate the sovereign-juridical model of government. As a result, the concept of sovereignty understood by Foucault – which is shot through with disciplinary and biopolitical practices, rather than replaced by them – is an uneasy and potentially conflictual combination of the sovereign-juridical 'right of the sword', disciplinary-carceral surveillance, and a biopolitical-capillary power 'to make live and let die'.[7] This is a clarification of what Foucault elsewhere refers to in passing as the need to reconfigure the '*monstre froid*' of the state as 'a triangle, sovereignty – discipline – government, which has as its primary target the population' (1991, 102). The point is that the spatiality of state power is shaped by changes of accent rather than by wholesale reconfigurations that can be neatly marked out on a calendar. The result is a complex process by

---

7   Foucault sees this combination at the heart of Nazi geopolitics. For Foucault, Nazism's corporeal identification of threat drew on biopolitical articulations of disturbance to the health and vitality of the population. But at the same moment such biological identifications were an 'indispensable [warring] precondition that [allowed] someone to be killed', expelled, rejected, or exposed to certain potentially lethal risks – either via the (disciplinary) concentration camp or inter-state warfare (Foucault 2003, 256).

which modes of subjectivity-constituting government are *multiplied, reactivated,* and *transformed* (Foucault 2004, 3–29; Fontana and Bertani 2003). Indeed, as Foucault argues in his 1977–1978 lectures, *Sécurité, Territoire, Population* (2004, 10, our translation), sovereign-juridical, disciplinary, and biopolitical techniques of government do not define a fragmented temporal series but a constellation of overlapping and complex relationships:

> There is not the age of the legal, the age of disciplinarity, the age of security. You do not have mechanisms of security which take the place of disciplinary mechanisms, which themselves would have taken the place of juridico-legal mechanisms. In fact, you have a series of complexes [of power] in which what changes, of course, are the techniques themselves which become more refined, or in any case become more complicated, but especially what changes, is the dominant mechanism or more exactly the system of correlation between juridico-legal mechanisms, disciplinary mechanisms and mechanisms of security. Said otherwise, you have a history of techniques themselves.

In short, what Hardt and Negri give us – despite their claim to a Foucauldian genealogy of government, which we can now consider suspect – is an overarching, universalizing, and cyclical struggle between forces of light (the multitude) and darkness (the Thermidor) which in various epochs – modern and postmodern – actualize in distinct and successive arts of government and spatialities of power. What Foucault gives us – as a philosopher of subjectivity interested primarily in the spatiality of power – is poles apart: a consideration of how disciplinary and biopolitical arts of government combine and recombine with sovereign-juridical powers to produce a topology of power that doesn't quite fit either the traditional *realpolitik* map of inter-national politics or the 'end of geography' thesis (for an discussion of the interstitial spatiality of power in Foucault, which anticipated *Empire*, see Dillon 1995; see also Dillon and Reid 2000; 2001). So, if Hardt and Negri owe an intellectual debt to Foucault's work on subjectivity, power, and government, then they do so by means of 'teleotranscendental' logic (Connolly 1991, 181) that confounds the intellectual spirit that animates Foucault's rethinking of the state as a monolithic apparatus of power. In fact, we can argue that Hardt and Negri – in describing the colonization of all space and time by an enveloping struggle between the multitude and the Thermidor, in which state power is either present or absent – embark on the sort of historical attunement that Foucault's genealogy of subjectivity sets out to unsettle, disrupt, and open to further scrutiny (on genealogy and attunement, see Connolly 1985). What we see in *Empire*, from this perspective, is an inversion of Foucault that might be described as a paradoxical use of the ideas of a key and very public poststructural sceptic of 'global history' to reinscribe a certain overarching identity and purpose to History.

Conclusion

The philosophical problem for *Empire*, therefore, is that at the same time it provides an analysis of world politics based on an opposition between immanence and transcendence in which the former is explicitly favoured, it implicitly reinstates a transcendentalist view of history – the 'view from nowhere' with all of its fallibilities. In this regard it departs fundamentally from the spatially nuanced understanding of the workings of power forwarded by Michel Foucault in our interpretation of his works. It is Foucault's presumed take on power that is invoked by Hardt and Negri as one of their most important theoretical inspirations; but in their hands the either/or logic of a singular mechanism of power prevalent in a particular historical epoch transcends the and/or logic of multiple mechanisms of power (if in different relative balance between them over time) that we emphasize from the writings of Foucault. The unfortunate larding of Foucault's biopolitics of power with Schmitt's 'total replacement' logic of history produces a problematic compounding that leaves *Empire* without the compelling account of world politics that a more thorough affiliation with Foucault's writings would have found immanent in his approach. If the writers we referred to in our introduction tend to see little but ontological continuity from the past in the geopolitics of contemporary world politics, Hardt and Negri see nothing but the totally new. The particular geographical problem for *Empire* is that its either/or periodization of the spatiality of power cannot do justice to the complexities of the spatial workings of power to which Foucault's work draws explicit attention.

References

Agnew, J.A. (1994) 'The Territorial Trap: The Geographical Assumptions of International Relations Theory.' *Review of International Political Economy*, 1 (1), 53–80.

Agnew, J.A. (2003) 'American Hegemony Into American Empire? Lessons from the Invasion of Iraq.' *Antipode*, 35(5), 871–75.

Agnew, J.A. (2005) *Hegemony: The New Shape of Global Power*. Philadelphia: Temple University Press.

Allen, J. (1997) 'Economies of Power and Space.' In J. Wills and R. Lee (Eds) *Geographies of Economies*. London: Arnold. 59–70.

Allen, J. (2003) *Lost Geographies of Power*. Malden MA: Blackwell.

Arrighi, G. (2005) 'Hegemony Unravelling –I.' *New Left Review*, 32(March/April), 23–80.

Balakrishnan, G. (2000) 'Virgilian Visions.' *New Left Review*, 5(September/October), 142–8.

Barkawi, T. and M. Laffey (1999) 'The Imperial Peace: Democracy, Force and Globalization.' *European Journal of International Relations*, 5(4), 403–34.

Barkawi, T. and M. Laffey (2002) 'Retrieving the Imperial: *Empire* and International Relations.' *Millennium Journal of International Studies*, 31(1), 109–27.

Bartelson, J. (2001) *The Critique of the State*. Cambridge: Cambridge University Press.

Blouet, B. (2001). *Geopolitics and Globalization in the Twentieth Century*. London: Reaktion.

Brenner, N. (1998) 'Between Fixity and Motion: Accumulation, Territorial Organization, and the Historical Geography of Spatial Scales.' *Environment and Planning D: Society and Space*, 16(4), 459–81.

Brenner, N., B. Jessop, M. Jones and G. MacLeod (2003) 'State Space in Question.' In N. Brenner, B. Jessop, M. Jones and G. MacLeod (Eds) *State Space*. Oxford: Blackwell. 1–26.

Burchell, G., C. Gordon and P. Miller, Eds (1991) *The Foucault Effect: Studies in Governmentality*. Chicago: University of Chicago Press.

Callinicos, A. (2002) 'The Grand Strategy of the American Empire.' *International Socialism*, 97(Winter), 3–38.

Cohen, S.B. (2003) 'Geopolitical Realities and United States Foreign Policy.' *Political Geography*, (1), 1–33.

Connolly, W.E. (1985) 'Taylor, Foucault, and Otherness.' *Political Theory*, 13(3), 365–76.

Connolly, W.E. (1991) *Identity/Difference: Democratic Negotiations of Political Paradox*. Minneapolis: University of Minnesota Press.

Curtis, B. (1995). 'Taking the State Back Out: Rose and Miller on Political Power.' *The British Journal of Sociology*, 46(4), 575–89.

Dean, M. (1999) *Governmentality: Power and Rule in Modern Society*. London: SAGE.

Deleuze, G. (1988) *Foucault*. Minneapolis: University of Minnesota Press.

Deleuze, G. and F. Guattari (1987) *A Thousand Plateaus: Capitalism and Schizophrenia*. Minneapolis: University of Minnesota Press.

Dillon, M. (1995) 'Sovereignty and Governmentality: From the Problematics of the "New World Order" to the Ethical Problematic of the World Order.' *Alternatives*, 20(3), 323–68.

Dillon, M. and J. Reid (2000) 'Global Governance, Liberal Peace, and Complex Emergency.' *Alternatives*, 25(2), 117–43.

Dillon, M. and J. Reid (2001) 'Global Liberal Governance: Biopolitics, Security and War.' *Millennium Journal of International Studies*, 30(1), 41–66.

Dreyfus, H.L. and P. Rabinow (1982) *Michel Foucault: Beyond Structuralism and Hermeneutics (2nd Edition)*. Chicago: University of Chicago Press.

Driver, F. (1985) 'Power, Space and the Body: A Critical Assessment of *Discipline and Punish*.' *Environment and Planning D: Society and Space*, 3(4), 425–46.

Driver, F. (1993) 'Bodies in Space: Foucault's Account of Disciplinary Power.' In C. Jones and R. Porter (Eds) *Reassessing Foucault: Power, Medicine, and the Body*. London: Routledge. 113—31.

Edkins, J. (1999) *Poststructuralism and International Relations*. Boulder: Lynne Rienner.

Elden, S. (2001) *Mapping the Present: Heidegger, Foucault and the Project of a Spatial History*. London: Continuum.

Fontana, A. and M. Bertani (2003) 'Situating the Lectures.' In A. I. Davidson (Ed.) *Society Must be Defended*. New York: Picador, 273–94.

Foucault, M. (1977) 'Nietzsche, Genealogy, Power.' *Language, Counter-Memory, Practice: Selected Essays and Interviews by Michel Foucault*. Edited by D. F. Bouchard. Ithaca: Cornell University Press. 139–64.

Foucault, M. (1978) *Discipline and Punish: The Birth of the Prison*. New York: Vintage Books.

Foucault, M. (1979) *The History of Sexuality: An Introduction*. New York: Vintage Books.

Foucault, M. (1980a) 'Questions on Geography.' *Power/Knowledge: Selected Interviews and Other Writings 1972–1977*. Edited by C. Gordon. New York: Pantheon. 63–77.

Foucault, M. (1980b) 'Two Lectures.' *Power/Knowledge: Selected Interviews and Other Writings 1972–1977*. Edited by C. Gordon. New York: Pantheon Books. 78–108.

Foucault, M. (1982) 'The Subject and Power.' *Michel Foucault: Beyond Structuralism and Hermeneutics*. Edited by H.L. Dreyfus and P. Rabinow. Chicago: University of Chicago Press. 208–26.

Foucault, M. (1991). 'Governmentality.' In G. Burchell, C. Gordon and P. Miller (Eds) *The Foucault Effect: Studies in Governmentality*. Chicago: University of Chicago Press. 87–104.

Foucault, M. (1997a) 'On the Government of the Living.' *Ethics, Subjectivity, and Truth – The Essential Works of Michel Foucault 1954–1984*. Edited by P. Rabinow. New York: The New Press. 81–85.

Foucault, M. (1997b) 'The Birth of Biopolitics.' *Ethics, Subjectivity, and Truth – The Essential Works of Michel Foucault 1954–1984*. Edited by P. Rabinow. New York: The New Press. 73–9.

Foucault, M. (2000a) 'Space, Knowledge and Power.' *Power – The Essential Works of Michel Foucault 1954–1984*. Edited by J.D. Faubion and P. Rabinow. New York: The New Press. 349–64.

Foucault, M. (2000b) 'Omnes et Singulatim: Toward a Critique of Political Reason.' *Power – The Essential Works of Michel Foucault 1954–1984*. Edited by J.D. Faubion and P. Rabinow. New York: The New Press. 298–325.

Foucault, M. (2003) *Society Must be Defended — Lectures at the Collège de France 1975–1976*. New York: Picador.

Foucault, M. (2004) *Sécurité, Territoire, Population: Cours au Collège de France 1977–1978*. Paris: Hautes Études/Gallimard Seuil.

Galli, C. (2001) *Spazi Politici: L'Eta Moderna e L'Eta Globale*. Bologna: Il Mulino.

Gordon, C. (1991) 'Governmental Rationality.' In G. Burchell, C. Gordon and P. Miller (Eds) *The Foucault Effect: Studies in Governmentality*. Chicago: University of Chicago Press. 1–52.

Gowan, P. (1999) *The Global Gamble – Washington's Faustian Bid for World Dominance*. London: Verso.

Gramsci, A. (1971) *(Selections from) The Prison Notebooks*. New York: International Publishers.

Hannah, M.G. (1997) 'Space and the Structuring of Disciplinary Power: An Interpretive Review.' *Geografiska Annaler*, 79B(3), 171–80.

Hardt, M. and T. Dumm (2000) 'Sovereignty, Multitudes, Absolute Democracy: A Discussion between Michael Hardt and Thomas Dumm about Hardt and Negri's *Empire* (Harvard University Press, 2000).' *Theory and Event*, 4(3), 1–40.

Hardt, M. and A. Negri (2000) *Empire*. Cambridge: Harvard University Press.

Hardt, M. and A. Negri (2004) *Multitude – War and Democracy in the Age of Empire*. New York: Penguin Press.

Hardt, M., A. Negri, N. Brown and I. Szeman (2002a) 'The Global Coliseum: On *Empire*.' *Cultural Studies*, 16(2), 177–92.

Hardt, M., A. Negri, N. Brown and I. Szeman (2002b) 'Subterranean "Passages of Thought": *Empire*'s Inserts.' *Cultural Studies*, 16(2), 193–212.

Harvey, D. (2003) *The New Imperialism*. Oxford: Oxford University Press.

Herbert, S. (1996) 'The Geopolitics of the Police: Foucault, Disciplinary Power, and the Tactics of the Los Angeles Police Department.' *Political Geography*, 15(1), 47–59.

Herod, A. et al., Eds (1998) *Unruly World? Globalization, Governance and Geography*. London: Routledge.

Herod, A. and M.W. Wright, Eds (2002) *Geographies of Power: Placing Scale*. Malden, MA: Blackwell.

Hindess, B. (1996) *Discourses of Power: From Hobbes to Foucault*. Oxford: Blackwell.

Hunt, A. (1992) 'Foucault's Expulsion of the Law: Toward a Retrieval.' *Law and Social Inquiry*, 17(1), 1–39.

Hunt, A. and G. Wickham (1994) *Foucault and Law: Towards a Sociology of Law as Governance*. London: Pluto Press.

Hutchings, K. (1997) 'Foucault and International Relations Theory.' In M. Lloyd and A. Thacker (Eds) *The Impact of Michel Foucault on the Social Sciences and Humanities*. New York: St. Martins. 102–27.

Jessop, B. (1987) 'Poulantzas and Foucault on Power and Strategy.' *Ideas and Production*, 6(1), 59–84.

Jessop, B. (1990) *State Theory: Putting the Capitalist State in its Place*. University Park, PA: Pennsylvania State University Press.

Jessop, B. (2003) 'Informational Capitalism and Empire: The Post–Marxist Celebration of US Hegemony in a New World Order.' *Studies in Political Economy – A Socialist Review*, 71/72(Autumn/Winter), 39–58.

Jessop, B. (2005) 'Gramsci as a Spatial Theorist.' *Critical Review of International Social and Political Philosophy*, 8/4(December), 421–37.

Jhaveri, N.J. (2004) 'Petroimperialism: US Oil Interests and the Iraq War.' *Antipode*, 36(1), 2–11.

Kerr, D. (1999) 'Beheading the King and Enthroning the Market: A Critique of Foucauldian Governmentality.' *Science and Society*, 63(2), 173–202.

Klare, M.T. (2001) *Resource Wars: The New Global Landscape of Conflict*. New York: Metropolitan Books.

Lefort, C. (1988) *Democracy and Political Theory*. Oxford: Polity Press.

Mann, M. (2003) *Incoherent Empire*. London: Verso.

Mansfield, B. (2005) 'Beyond Rescaling: Reintegrating the "National" as a Dimension of Scalar Relations.' *Progress in Human Geography*, 29(4), 458–73.

Miller, P. and N. Rose (1990) 'Governing Economic Life.' *Economy and Society*, 19(1), 1–31.

Moss, J. Ed. (1998) 'Introduction: The Later Foucault.' *The Later Foucault: Politics and Philosophy*. London: SAGE. 1–17.

Negri, A. and D. Zolo (2003) 'Empire and the Multitude: A Dialogue on the New World Order of Globalization.' *Radical Philosophy*, 120(July/August), 23–37.

Ó Tuathail, G. (1996) *Critical Geopolitics*. Minneapolis: University of Minnesota Press.

Pallot, J. (2005) 'Russia's Penal Peripheries: Space, Place, and Penalty in Soviet and Post-Soviet Russia.' *Transactions of the Institute of British Geographers*, 30(1), 98–112.

Panitch, L. and S. Gindin (2003) 'American Imperialism and Eurocapitalism: The Making of Neoliberal Globalization.' *Studies in Political Economy – A Socialist Review*, 71/72(Autumn/Winter), 7–38.

Patton, P. (1998). 'Foucault's Subject of Power.' In J. Moss (Ed.) *The Later Foucault*. London: SAGE. 64–77.

Philo, C. (1992). 'Foucault's Geography.' *Environment and Planning D: Society and Space*, 10(2), 137–61.

Rabinow, P. (1994) 'Introduction.' In J.D. Faubion (Ed.) *Power*. New York: The New Press. xi–xli.

Rose, N. and P. Miller (1992) 'Political Power Beyond the State: Problematics of Government.' *British Journal of Sociology*, 43(2), 173–205.

Schmitt, C. (1976 [1932]) *The Concept of the Political*. New Brunswick: Rutgers University Press.

Schmitt, C. (1985 [1933]) *Political Theology*. Cambridge: MIT Press.

Schmitt, C. (1987). 'The Legal World Revolution.' *Telos* 72 (Summer), 73–89.

Schmitt, C. (1996 [1938]) *The Leviathan in the State Theory of Thomas Hobbes*. Westport: Greenwood Press.

Sharp, J.P., , P. Routledge, C. Philo and R. Paddison (2000) 'Entanglements of Power.' In J.P. Sharp, P. Routledge, C. Philo and R. Paddison (Eds) *Entanglements of Power: Geographies of Domination/Resistance*. London: Routledge. 1–42.

Shaw, M. (2000) *Theory of the Global State*. Cambridge: Cambridge University Press.

Soja, E. (1989). *Postmodern Geographies*. London: Verso.

Stoler, L.A. (1995) *Race and the Education of Desire: Foucault's History of Sexuality and the Colonial Order of Things*. Durham: Duke University Press.

Strong, T.B. (1996). 'Dimensions of the New Debate Around Carl Schmitt.' In G. Schwab (Ed.) *The Concept of the Political*. Chicago: University of Chicago Press. ix–xxvii.

Ulmen, G.L. (1987). 'American Imperialism and International Law: Carl Schmitt on the US in World Affairs.' *Telos*, 72(Summer), 43–71.

Veroli, N. and C. Mudede (2002) 'Challenging Empire: An Interview with Michael Hardt.' *Radical Society*, 29(4), 59–73.

Wainwright, J. (2004) 'American Empire: A Review Essay.' *Environment and Planning D: Society and Space*, 22(5), 465–8.

Walker, R.B.J. (1993) *Inside/Outside: International Relations as Political Theory*. Cambridge: Cambridge University Press.

Walker, R.B.J. (2002a) 'On the Immanence/Imminence of Empire.' *Millennium Journal of International Studies*, 31(2), 337–45.

Walker, R.B.J. (2002b) 'After the Future: Enclosures, Connections, Politics.' In R. Falk, L.E.J. Ruiz and R.B.J. Walker (Eds) *Reframing the International: Law, Culture, Politics*. London: Routledge. 3–25.

Young, R. (1990) *White Mythologies: Writing History and the West*. London: Routledge.

Chapter 27

'Bellicose History' and 'Local Discursivities': An Archaeological Reading of Michel Foucault's *Society Must be Defended*

Chris Philo

> You might ask: Why all these details, why locate these different tactics within the field of history? (Foucault 2003a, 207)

The first English-language volume of Michel Foucault's lecture courses at the Collège de France appeared as *Society Must be Defended* (Foucault 2003a), covering lectures given between 7th January and 17th March, 1976.[1] As Surokiecki (2005, no pagination) notes, the book covers 'a topic' – in short, the relationship between history, war, politics and power – that is 'not [one] Foucault wrote on at length in any of his previously published work, so the lectures include a lot of new, compelling material'. An exception for an Anglophone audience is the first two lectures, which were translated as the 'Two Lectures' chapter in the *Power/Knowledge* collection (Foucault 1980a; in Gordon 1980). The course was given at an intriguing moment in Foucault's thinking, sandwiched between the publications of *Surveiller et punir* (Foucault 1975; translated as *Discipline and Punish*, 1977) and the first volume of *Histoire de la sexualité* (Foucault 1976; translated as *The History of Sexuality, Vol.1*, 1979a).[2] Its contents reflect a significant shift in his understanding of power from a disciplinary version to one concerned with the various levels of biopower, individual and collective, operating alongside – note, *not* instead of – the swarming disciplinary mechanisms. What also occurs is a return to Foucault's older fascination with

---

1   The French edition, *'Il faut défendre la société'*, appeared in 1997. The second English-language volume of these lectures (8th January to 19th March, 1975) is *Abnormal* (Foucault 2003b) (*Les Abnormaux* 1999); the third (6th January to 24th March, 1982) is *The Hermeneutics of the Subject* (Foucault 2005) (*L'herméneutique du sujet* 2001); and the fourth (7th November, 1973 to 6th February, 1974) is *Psychiatric Power* (Foucault 2006) (*Le pouvoir psychiatrique* 2003).

2   Elden (2002; 2005; this volume) argues that the book carries material that Foucault probably envisaged being explored further in the projected sixth volume of *The History of Sexuality* on the theme of *Population et races* ('population and races').

questions of discourse and knowledge, in which this book might be seen as a hinge between the conventionally demarcated 'archaeological' and 'genealogical' phases of Foucault's *oeuvre*. Indeed, this book arguably embodies continuity between these two stances on intellectual inquiry, rather than discontinuity, and Stone (2004, 79) suggests that 'the[se] lecture courses offer the archaeological analysis that is implicit (or sometimes completely missing) from the published works'. A further implication is that Foucault sketches here the ground of what he terms 'political historicism', a politicized approach to history that necessarily tracks between the archaeological and the genealogical, and one usefully complementing what have since been cast as his 'critical and effective histories' (Dean 1994).

My intention below is to offer a reading of these elements within the book, providing a text-based exegesis with the focus squarely upon the book itself (see also Elden 2002). As such, the chapter will be a fragmentary intervention in the genealogy of Foucault's own thinking, but at the same time will take seriously what Foucault himself claims about the utility of 'genealogical fragments' (2003a, 11). The book certainly enhances what Foucault brings to the spatialized analysis of power (Elden 2001; 2003), and speaks to recent attempts at enlarging what might be taken from Foucault *beyond* the portrayal of a too-simplistic panoptic power (wherein the figure of Bentham's Panopticon has perhaps become an impediment to progressing inquiry: see also Driver 1985; 1993a; 1993b; Elden 2001, esp. 133–50; 2003; Hannah 1997a; 1997b; Philo 1989; Robinson 1999). More particularly, the book may also contribute to the project of showing the centrality of population, and by extension population geography, within Foucault's later writings on 'biopower', 'biopolitics' and 'bioregulation' (Elden 2006; 2007; Legg 2005; Philo 2001; 2005).[3] Additionally, because the book touches upon the eclipse of sovereign power, seemingly replaced historically by an admixing of disciplinarity and biopower, it has also figured in debates provoked by Agamben (1998) about the possible continuation of sovereign

---

3   The final (eleventh) lecture, covering ground similar to that in *The History of Sexuality* (Foucault 1979a, Part V), identifies a shift, gradually materializing in the Early Modern era, from 'the power of sovereignty' – the power of the sovereign to *take* life – to a 'power over life' – the power of the state and related institutions, those operating in the field of 'governmentality' (Foucault 1979a; 1979b), to *make* life. Running alongside the development of 'disciplinary power', directed at individual bodies through the manipulation of institutional-spatial forms, Foucault detects a gathering 'biopower', concerned more with collectivities (masses of population) and how to govern their demographics, overall health and hence productivity. As he says: 'we have two series: the body-organism-discipline-institutions series, and the population-biological processes-regulatory mechanisms-State [series]' (2003a, 250). In the book, we hear about 'the emergence of something that is no longer an anatamo-politics of the human body, but what [Foucault] would call a "biopolitics" of the human race' (2003a, 243), and various aspects of this new object, 'population', are discussed: 'It is these processes – the birth rate, the mortality rate, longevity and so on – together with a whole series of related economic and political problems ... which, in the second half of the eighteenth century, become biopolitics' first objects of knowledge and the targets it seeks to control' (2003a, 243).

power in more contemporary forms of biopolitical socio-spatial exclusion (Ojakangas 2005 [and responses]; also Cadman 2006).

In the face of a tendency of too many geographers to treat Foucault solely as 'the geometer of power' (Philo 1992), however, my reading of the book tackles the less-noted connections back to archaeologies of discourse and knowledge. I will argue that these connections are not incidental, but vital supports for the intellectual scaffolding and ethico-political intent of his later work, and will suggest that reconstructing what Lemert and Gillan (1982, 39) term Foucault's 'bellicose history' – itself compellingly elaborated in the book – amounts to a valuable exercise demonstrating the continuing salience of his *oeuvre* to contemporary thought. I will conclude by underlining certain spatial dimensions integral to Foucault's bellicose history, indexed by his constant references to 'the local' and parallel insistence on prioritizing multiplicities. I will also demonstrate that the book provocatively extends Foucault's spatialization of history, complementing the more existential and social-historical variants of his 'spatial history' as elaborated by the likes of Elden (2001, Chapters 4 and 5) and Philo (1986; 1992; 2004), precisely because its ethico-political charge is conveyed through an imagery touched by the bloody spatial juxtapositions of the battlefield.

Society Must be Defended: Power, Politics and War

The book is indeed about 'power', a subject-matter introduced when reflecting on the 'power effects' or the 'power-hierarchy' (2003a, 10) of conventional science, a theme to be revisited later. Foucault asks:

> 'What is power?' Or, rather – given that the question 'What is power?' is obviously a theoretical question that would provide an answer to everything, which is just what I don't want to do – the issue is to determine what are, in their mechanisms, effects, their relations, the various power-apparatuses that operate at various levels of society, in such very different domains and with so many different extensions? (2003a, 13)

This quote encapsulates Foucault's orientation: he is *not* striving for a 'total theory' of power, which he feels is implausible given worldly complexities, but instead wishes to interrogate the many different dimensions of power traversing 'real' societies. This is also why he resists the standard means of analyzing power, smacking either of 'economism' (2003a, 13), reading power as the 'property' of the economically dominant class, or of 'repression' (2003a, 15–18), reading power as 'that which represses nature, instincts, a class or individuals' (2003a, 15; a notion associated with Freud). Alternately, he contrasts a 'contract-oppression schema', one 'articulated around power as a primal right that is surrendered' (2003a, 16–17), with a 'domination-repression schema' wherein the realm of the social is conceived as 'a perpetual relationship of force' (2003a, 17) identical in principle (if not empirics) to the conflicts of war-time. This is not the place to retread Foucault's well-known critique of 'the repressive hypothesis', central to *The History of Sexuality* (Foucault

1979a, Part II), and neither is it necessary to trace the equally well-known relational, circulating and capillary vision of power as something 'productive' – creating things, making things happen – developed in the second lecture and then also in *The History of Sexuality* (Foucault 1979a, esp. Part II, Chapter 2; for a geographical commentary, see Sharp et al. 1999).[4]

Rather, what is most distinctive is how Foucault elaborates a model of power informed by a sense of *war*:

> Power is war, the continuation of war by other means. At this point, we can invert Clausewitz's proposition and say that politics is the continuation of war by other means. ... Politics, in other words, sanctions and reproduces the disequilibrium of forces manifested in war. Inverting the proposition also means something else, namely that within this 'civil peace', these modifications of relations of force – the shifting balance, the reversals – in a political system, all these things must be interpreted as a continuation of war. And they are interpreted as so many episodes, fragmentations and displacements of the war itself. We are always writing the history of the same war, even when we are writing the history of peace and its institutions. (2003a, 15–16)

Foucault extends these claims, exploring what he calls 'bellicose relations' (2003a, 23), and in effect elaborating what Lemert and Gillan (1982, 39) usefully refer to as his 'bellicose history'. The analysis is scattered, including notes on 'all the techniques that are used to fight a war' (2003a, 47),[5] but the key point is that Foucault advances 'as a principle for the interpretation of society and its visible order ... the confusion of violence, passions, hatreds, rages, resentments and bitterness, ... [and] asking the

---

4   The second lecture and Foucault's chapter on 'Method' (in Foucault 1979a) are similar, underscoring that power is not to be conceived as emanating from a 'single centre', nor something to be analyzed in terms of individuals' 'intentions or decisions', nor something to be traced to removed origins, but rather located in 'the places where it implants itself and produces its real effects' (2003a, 28). Most usefully, he writes: 'Do not regard power as a phenomenon of mass and homogeneous domination – the domination of one individual over others, of one group over others, or of one class over others; keep it clearly in mind that, unless we are looking at it from a great height and from a very great distance, power is not something that is divided between those who have it and hold it exclusively, and those who do not have it and are subject to it. Power must, I think, be analyzed as something that circulates, or rather as something that functions only when it is part of a chain. It is never localized here or there, it is never in the hands of some, and it is never appropriated in the way that wealth or a commodity can be appropriated. Power functions. Power is exercised through networks, and individuals do not simply circulate in those networks; they are in a position to both submit to and exercise this power' (2003a, 29).

5   One passage reflects upon the whole context and content of war: 'War in the sense of the distribution of weapons, the nature of the weapons, fighting techniques, the recruitment and payment of soldiers, the taxes earmarked for the army; war as an internal institution, and not the raw event of a battle. ... War is a general economy of weapons, an economy of armed people and disarmed people within a given State, and with all the institutional and economic series that derive from that' (2003a, 159–60).

elliptical god of battles to explain the long days of order, labour, peace and justice' (2003a, 54). Putting it another way:

> [This bellicose history] is interested in defining and discovering, beneath the forms of justice that have been instituted, the order that has been imposed, the forgotten past of real struggles, actual victories, and defeats which may have been disguised but which remain profoundly inscribed. It is interested in rediscovering the blood that has dried in the codes ... (2003a, 56)

Revealingly, he also sees a connective tissue between bellicose history and his conceptualizing of power:

> So what is the principle that explains history? First, a series of brute facts, which might already be described as physico-biological facts: physical strength, force, energy ... A series of accidents, or at least contingencies: defeats, victories, the failure or success of rebellions, the failure or success of conspiracies or alliances; and finally, a bundle of psychological and moral elements (courage, fear, scorn, hatred, forgetfulness, *et cetera*). Intertwining bodies, passions and accidents, according to this discourse, that is what constitutes the permanent web of histories and societies. And something fragile and superficial will be built on top of this web of bodies, accidents and passions,[6] this seething mass which is sometimes murky and sometimes bloody; a growing rationality. The rationality of calculations, strategies and ruses; the rationality of technical procedures that are used to perpetuate the victory, to silence, or so it would seem, the war ... (2003a, 54–5)

This account, resonating with claims in *Histoire de la folie* (Foucault 1961, translated as *Madness and Civilization*, 1967),[7] implies that the rational calculus of modern institutions, including the machinations of disciplinary power, should not be regarded as the expression of enlightened 'truth' – as the working of some anonymous 'law' moored in a rightfully constituted ruling force – but rather as an outgrowth of murk and blood: 'of wild dreams, cunning and the wicked', the latter 'hav[ing] won a temporary victory' (2003a, 55). Thus we find why Foucault looks to war, to the base

---

6 Foucault hints at a gradual 'cleaning up' of history, with it becoming less openly bellicose through time – with less violent scrapping between fairly random groups of people encountering one another through invasions, pillages, raiding parties, etc. – and the increasing role of the (monarchical) state as the only body owning the legitimate means of waging war: 'gradually, the entire social body was cleansed of the bellicose relations that had permeated it through and through during the Middle Ages' (2003a, 48).

7 This text can be read in part as the story of how Reason, the forces, imperatives and orders of rationality, progressively wins dominion over Unreason, the chaotic and embodied passions of humanity untamed by mental, behavioural and social 'norms' of conduct (see Philo 1999; 2004, esp. Chapter 2). In *Society Must be Defended*, though, Foucault arguably allows more continuity between Reason and Unreason, if we continue to deploy these terms, such that the supposedly rational (cool, calm, truth-based) strategies of those in charge of modern states, for instance, are portrayed as actually shaped by the low cunning, wickedness and spiteful passions of 'unreasonable' people unleashed in war-time.

passions but also the grubby tactics of both the battlefield and those who persuade others to fight, as a guide for further theorizing of power.

The Insurrection of Subjugated Knowledges

Society Must be Defended is not just about power, however, since it also addresses questions of discourse and knowledge. Foucault reflects on why he does what he does *as an historian*, acknowledging the traps of a 'fevered laziness' (2003a, 4):

> It's a character trait of people who love libraries, documents, references, dusty manuscripts, texts that have never been read, books which, no sooner printed, were closed and then slept on the shelves and were only taken down centuries later. All this quite suits the busy inertia of those who profess useless knowledge. (2003a, 4–5)

Even so, Foucault remains persuaded of some worth in recovering the discourses that might stand, as it were, tangential to the major currents of historical change; and much of the book is written in praise of what such dusty alternatives, such 'counterknowledge[s]' (2003a, 130), can bring to the table of critical scholarship. In crafting this justification, he cites the present-day emergence of an 'immense and proliferating criticisability of things, institutions, practices and discourses', achieved in large measure by 'the astonishing efficacy of discontinuous, particular and local critiques' (2003a, 6). He means 'local' in various ways to be considered later, but all implicate the production of critiques – counterknowledges – emerging from particular people, settings, sites, points and maybe networks, their effect being to challenge the coherence of 'totalitarian theories' and the assumed 'theoretical unity of their discourses' (2003a, 6). Returning to the political imperatives of *Madness and Civilization* (Foucault 1967), an example is given of discourses railing against the orthodoxies of a medical-psychiatric establishment, and reference is made to 'the strange efficacy, when it came to jamming the workings of the psychiatric institution, of the discourse, the discourses – and they really were very localized – of antipsychiatry' (2003a, 5). The hint here about a geography of antipsychiatry, given its localization in given places and not others (Jones 1996; 2000), is worth remark, and is wholly consistent with what he asserts about the thorough-goingly 'local character of the critique' integral to 'what might be called the insurrection of subjugated knowledges' (2003a, 6–7).

Foucault explores two senses of what he means by subjugated knowledges:

> When I say 'subjugated knowledges', I mean two things. On the one hand, I am referring to historical contents that have been buried or masked in functional coherences or formal systematizations. To put it in concrete terms if you like, it was certainly not a semiology of life in the asylum or a sociology of delinquence that made an effective critique of the asylum or the prison possible; it really was the appearance of historical contents. Quite simply because historical contents alone allow us to see the dividing lines in the confrontations and struggles that functional arrangements or systematic organizations are designed to mask. (2003a, 7)

The scholarly skill of the historian, exposing the everyday combats of the past, is what thereby liberates a subjugated knowledge, isolating an ethico-political purpose for the (critical) historian and anticipating the larger canvas on which the later chapters paint a picture of historical writing, war, politics and power. But what of the second sense of this term? Foucault continues:

> When I say 'subjugated knowledges', I am also referring to a whole series of knowledges that have been disqualified as nonconceptual knowledges, as insufficiently elaborated knowledges: naïve knowledges, hierarchically inferior knowledges, knowledges that are below the required level of erudition or scientificity. And it is thanks to the reappearance of these knowledges from below, of these unqualified or even disqualified knowledges, ... the knowledge of the psychiatrized, the patient, the nurse, the doctor, that is parallel to, marginal to, medical knowledge, the knowledge of the delinquent, what I would call, if you like, what people know (and this is by no means the same thing as common knowledge or common-sense but, on the contrary, a particular knowledge, a knowledge that is local, regional or differential, incapable of unanimity and which derives its power solely from the fact that it is different from all the knowledges that surround it), it is in the reappearance of what people know at a local level, of these disqualified knowledges, that made the critique possible. (2003a, 7–8)

According to Foucault, a critical stance on the world is cultivated in the soils of 'these singular, local knowledges, the noncommonsensical knowledges that people have, and which have in a way been left to lie fallow, or even kept in the margins' (2003a, 8). They are the stuff out of which the first kind of subjugated knowledges can emerge, comprising the raw material to be fashioned in a scholarly manner that is 'historical, meticulous, precise' (2003a: 8); and so the two forms of this knowledge – its local inception and eruption, on the one hand, and its scholarly recovery and representation, on the other – are positioned as intimately paired in the orbit of broader, social-critical ambitions. As Foucault explains, 'it is the coupling together of the buried scholarly knowledge and knowledges that were disqualified by the hierarchy of erudition and sciences that actually give the discursive critique of the last fifteen years its essential strength' (2003a, 8).

Foucault reflects upon the coupling mentioned here, bringing into the same frame his twin meta-projects of archaeology and genealogy:

> Both the specialized domain of scholarship and the disqualified knowledge ... have contained the memory of combats, the very memory that had until then been confined to the margins. And so we have the outline of what might be called a genealogy, or of multiple genealogical investigations. We have both a meticulous rediscovery of struggles and the raw memory of fights. ... If you like, we can give the name 'genealogy' to this coupling together of scholarly erudition and local memories, which allows us to constitute a historical knowledge of struggles and to make use of that knowledge in contemporary tactics. That can, then, serve as a provisional definition of the genealogies that I have been trying to trace with you over the last few years. (2003a, 8)

Moreover, he goes on:

... geneaology is, then, a sort of attempt to desubjugate historical knowledges, to set them free, or in other words to enable them to oppose and struggle against the coercion of a unitary, formal and scientific theoretical discourse. The project of these disorderly and tattered genealogies is to reactivate local knowledges – Deleuze would no doubt call them 'minor'[8] – against the scientific hierarchicalization of knowledge and its intrinsic power-effects. (2003a, 10)

I will return to the point about struggles against the power-laden hierarchies of science, but for now hear the absolutely explicit link that Foucault then makes:

Archaeology is the method specific to the analysis of local discursivities, and genealogy is the tactic which, once it has described these local discursivities, brings into play the desubjugated knowledges that have been released from them. That just about sums up the overall project. (2003a, 10–11)

This bold statement cannot be ignored as a summation of how, in the mid-1970s at least, Foucault saw the various elements of his *oeuvre* fusing together, with genealogical critique – usually conceived in terms of critiquing power relations – depending upon a prior excavation of diverse subjugated knowledges or local discursivities. There are seeds of such a conception in the early-1970s essay 'Nietzsche, history, genealogy' (Foucault 1986), to be sure, but here Foucault more fully encompasses 'all the fragments of research, all the interconnected and interrupted things I have been repeating so stubbornly for four or five years now' (all of his 'genealogical fragments': 2003a, 11). Revealingly, Foucault concludes the first lecture by musing on the character of power, sketching out some of the more abstract, non-economic and non-repressive conceptualizations of power already mentioned. He also introduces the interest in war and the proposed reversal of Clausewitz's aphorism, but the fact that these materials sit comfortably alongside remarks on subjugated knowledges, local discursivities and the pairing of archaeology with genealogy shows that Foucault *himself* takes all of these as threads tugged from a larger tapestry. In simple outline, this is because he regards the terrains of *both* discourse and everyday social life as striated by what are ultimately the *same* features of struggle, force, domination and repression that mark the battlefields, war cabinets and propagandizing of real war.

'Disqualified Knowledges', Enlightenment and Science

I want now to amplify the relational cast of Foucault's claims about subjugated knowledges, clarifying his view that such knowledges are constituted, sustained and potentially made most effective within the horizon of more prominent, prevailing and one might say 'powerful' knowledges. Foucault's vision here:

---

8   The explicit nod to this Deleuzo-Guattarian notion of 'minor theory' is intriguing, and see the useful explanatory editorial endnote in 2003a, 20–21, Note 5. In the geographical literature, see Katz (1996).

... is a way of playing local, discontinuous, disqualified or nonlegitimized knowledges off against the unitary theoretical instance that claims to be able to filter them, organize them into a hierarchy, organize them in the name of a true body of knowledge, in the name of the rights of a science that is in the hands of the many. (2003a, 9)

The ethico-political imperative of such a vision is evident here, but so too is a broader critical stance on how certain knowledges legitimated under the banner of 'science' (Science even) become orthodoxies guiding the major currents of thought-and-action. Genealogies retrieving, reconvening and re-presenting knowledges that do *not* get so legitimated are, 'primarily, an insurrection against the centralized power-effects ... bound up with the institutionalization and workings of any scientific discourse organized in a society such as ours' (2003, 9). It is in this sense that Foucault describes genealogies as 'antisciences', not so much because they are opposed to the intellectual *content* of the sciences, in whatever guise, but because they resist the tendency – often concealed beneath the institutional structures and broader governmental claims accompanying and made on behalf of Science – for everything that does not conveniently 'fit' with mainstream agendas to be sidelined. In short, genealogies remain sceptical about 'the aspiration to power that is inherent in the claim of being a science' (2003a, 10); they must ask 'What theoretical-political vanguard are you trying to put on the throne [by naming something as a "science", as part of Science] in order to detach it from all the massive, circulating and discontinuous forms that knowledge can take?' (2003a, 10).

More concretely, Foucault critiques the Enlightenment and its effective disqualification of many species of knowledge from the table of acceptable wisdom. Lecture eight of the book contains an 'excursus' tackling the 'discourse-power axis' played out in 'the privileged period of the eighteenth century', the aim being to 'outwit the problematic of the Enlightenment' (2003, 178):

... to outwit what was at the time described (and was still described in the nineteenth and twentieth centuries) as the progress of enlightenment, the struggle of knowledge against ignorance, of reason against chimeras, of experience against prejudices, of reason against error, and so on. ... I think that we have to get rid of [this stereotype] when we look at the eighteenth century – we have to see, not this relationship between day and night, knowledge and ignorance, but something very different: an immense and multiple battle, ... not between knowledge and ignorance, but ... between knowledges in the plural – knowledges that are in conflict because of their very morphology, because they are in possession of enemies, and because they have intrinsic power-effects. (2003a, 178–9)

This quote exemplifies Foucault's sense of discourse as itself a bellicose battlefield, as well as highlighting an alertness to 'knowledges in the plural'. He substantiates such claims as follows, gesturing to an uneven geography of the Enlightenment (see also Livingstone and Withers 1999):

It is often said that the eighteenth century was the century that saw the emergence of technical knowledges. What actually happened ... was quite different. First of all, we have the plural, polymorphous, multiple and dispersed existence of different knowledges, which existed with their differences – differences defined by geographical regions, by the size of the workshops or factories, and so on. The differences among them – I am speaking of technological expertise – were defined by local categories, education and the wealth of their possessors. ... At the same time, we saw the development of processes that allowed bigger, more general, or more industrialized knowledges, or knowledges that circulated more easily, to annex, confiscate and take over smaller, more particular, more local and more artisanal knowledges. (2003a, 179)

There was an 'immense economico-political struggle around or over these knowledges' (2003a, 180), with many knowledges, usually the most local, place-specific and craft-bound, losing out in the face of knowledges that, for whatever precise reasons, were more readily standardized, 'universalized' and squared with the emerging 'scientific' claims of the Enlightenment. Foucault wonders too about the role of the state in 'eliminating or disqualifying what might be termed useless and irreducible little knowledges', and in striving to 'normalize' dispersed knowledges so as 'to break down the barriers of secrecy and technological and geographical boundaries' (2003, 180).

Similarly, the Enlightenment sought 'to homogenize, normalize, classify and centralize ... knowledge' (2003, 181),[9] an occurrence in tune with a wider elevation of science – the championing of 'Science in the singular' (2003a, 182) – as a means of excluding or taming other, folk, amateur and misfit 'knowledge that exists in the wild, any knowledge that is born elsewhere'[10] (2003a, 183). Moreover, Foucault identifies 'four operations' – 'selection, normalization, hierarchicalization and centralization' – that 'we [also] see at work in a ... study of what we call disciplinary power' (2003a, 181), thereby detecting disciplinary processes within the realm of discourse that parallel those arising in more material institutional and social spaces (Foucault 1977). Intriguingly, in describing 'the disciplinarization of polymorphous and heterogeneous knowledges' (2003a 182), he proposes that 'statements' were being 'sorted [according to] those that were acceptable ... from those that were unacceptable', imposing 'a control that applies not to the content of statements themselves, to their conformity to a certain truth, but to the regularity of enunciations'[11] (2003a, 184). Putting things like this relates Foucault's claims about the 'disqualifying' of knowledges to passages in *The Archaeology of Knowledge* (Foucault 1972), and so once again the book evidences the links between genealogy and archaeology.

---

9   Foucault is here talking particularly about medical knowledge.

10   By 'elsewhere', Foucault does in part mean other material spaces, those outwith a mainstream 'institutional field – whose limits are in fact relatively fluid but which consists, roughly speaking, of the university and official research bodies' (2003a, 183). The geography of knowledge, and more particularly the historical geography of (what comes to count as) science, is here constituted as central to Foucault's wider argumentation (see also Livingstone 1995).

11   I.e., to the rules governing who, as it were, is 'authorized' to speak with an 'authority' that is widely recognized.

Historical Writing and the Discovery of Power, Struggle and Society

What may surprise readers of the book is its preoccupation with seemingly quite arcane matters to do with the history of France, England and, more broadly, 'Europe'. In fact, six of the eleven lectures provide snippets of a 'big picture' historical geography, speaking of different peoples, 'hairy bands' (2003a, 202), 'nations', even 'races' (see below),[12] moving across the lands of what is now Europe, coming into contact, invading, co-habiting, making alliances, breaking them and going to war with one another or at least entering into war-like relations. In part, the ambition is to demonstrate that the 'politics' of such peoples, struggling for supremacy or, more modestly, jostling for position in these conflicts 'of old', could easily be interpreted as the continuation of war by other means (thus demonstrating the salience of inverting Clausewitz's aphorism). Yet, in a typically Foucauldian manoeuvre, and notwithstanding the mass of often quite indigestible details collected,[13] his main purpose is not to prove in some once-and-for-all fashion that the 'truth' of history *is* that politics is simply war by another name. Rather, it is to trace a process whereby scholars writing about history – the early historians, if you like – began themselves to *conceive* of history in this fashion, and in the process composed a history that took war-like relations, the antagonisms endemic to struggle, conflict, combat and the like, as the model for what needed to be analyzed. As Foucault (2003a, 47) asks, 'How, when and why was it noticed or imagined that what is going on beneath and in power relations is a war? Who, basically, had the idea of inverting Clausewitz's principle?'[14] He declares that this 'is the question I am going to pursue a bit in coming lectures, and perhaps for the rest of the year' (2003, 47).

Although in this context Foucault often uses the curious phrase 'race war',[15] what he is actually charting is the rise of a *social* history – or, to be more precise, a politicized version of social history – taking as its point of departure the fracturing

---

12 'New characters appear: the Franks, the Gauls and the Celts: more general characters such as the peoples of the North and the peoples of the South also begin to appear; rulers and subordinates, the victors and the vanquished begin to appear' (2003a, 75–6).

13 Leading him at one point to the self-reproach quoted at the head of my essay: 'Why all these details?' (2003a, 207).

14 In terms of historical chronology, though, and as Foucault realises, the question is really 'who formulated the principle Clausewitz inverted?', since the principle here – identifying politics as the continuation of war in peace-time – 'was a principle that existing long before Clausewitz' (2003a, 48).

15 Foucault claims that '[t]he war that is going on beneath order and peace, the war that undermines our society and divides it in a binary mode, is basically a race war. At a very early stage, we find the basic elements that make the war possible, and then ensure its continuation, pursuit and development: ethnic differences, differences between languages, different degrees of force, vigour, energy and violence; ... the conquest and subjugation of one race by another. The social body is basically articulated around two races' (2003a, 60). As this quote makes evident, Foucault is using the term 'race' as a shorthand for a range of ethnic differences – overarching a diversity of social, cultural, religious, linguistic and related differences – that arguably should *not* be homogenized in this manner.

of 'society' into groups, classes or social units of some form or another (they are certainly not all 'races') who are commonly in conflict with one another (whether violently or in an ongoing state of attrition likely to trigger moments of physical dispute). His overall argument is that this approach to writing history, one reviving the struggle-based contours of 'biblical history', has gradually emerged from under the yoke of a very different form of history-writing, what might be termed 'Roman history', where elements of conflict were left strangely muted and often glossed over.[16] The aim of such Roman history was to paper over the realities of invasion, subjugation and bloody revolt, in favour of an account stressing fundamental continuities and even 'rights'. A Romanist account said something like this: people X may have invaded region Y, forcibly subduing opposition and creating a new hierarchy of wealth, status and influence, but it can be shown that people X are in fact descendants of people who once lived in region Y, and so are only returning to claim what is rightfully theirs. Such a mode of history legitimated the sovereignty of whoever was ruling at the time of writing, supposedly proving lines of monarchical lineage to 'justify' the rights, apparently sanctioned by deep history, of an invading monarch to take the throne of a subjugated land or an imperial force the government of such a land. This history was of course an imagined history, fostering an imaginary geopolitics of lawful sovereignty, and in particular efforts were made to imagine links between Roman power and that of the kings who 'succeeded' the Romans on the European stage during the Dark Ages. Such a history dispensed with anything that splintered the past, thereby questioning the continuities, rights and justifications of sovereignty, and any alternative 'histories' that might have been told – folk-tales perhaps, telling other stories about invasions and their consequences – were quite literally 'written out of court'. A connection exists to what I said previously about the interest in subjugated knowledges, since Foucault spies the seeds of what he calls 'counterhistory' (e.g. 2003a, 66, 70) in alternative forms of history-writing that began to challenge Romanist accounts by urging a return to the spirit if not the letter of biblical history.

To reiterate, this move – performing 'its counterhistorical function' (2003a, 66) – necessarily entailed a *social* history, emphasizing that the people who ended up living in a particular region would rarely be a unitary body happily living under a time-honoured, historically sanctioned sovereign, but instead a population fractured along at least one basic fault-line: namely, that between conquerors and conquered, or, more subtly, between those who (as the descendants of both invaders and their

---

16 Foucault identifies 'the emergence of something that, basically, is much closer to the mythico-religious history of the Jews than to the politico-legendary history of the Romans. We are much closer to the Bible than to Livy, in a Hebraic-biblical form much more than in the form of the annalist who records, day-by-day, the history and the uninterrupted glory of power. ... [I]t is not surprising that we see, at the end of the Middle Ages, in the sixteenth century, in the period of the Reformation, and at the time of the English Revolution, the appearance of a form of history that is a direct challenge to the history of sovereignty and kings – to Roman history – and that we see a new history that is articulated around the great biblical form of prophecy and promise' (2003a, 71).

collaborators) had locally become the rulers and those who (as the descendants of both the invaded and any other waifs, strays and strangers) had locally become the ruled. As Foucault states, '[t]he two groups form a unity and a single polity only as a result of wars, invasions, victories and defeats, or in other words acts of violence' (2003a, 77). His insistence on a bipolar formulation may now strike us as too simplistic,[17] and arguably stands at odds with his own emphasis on multiplicity elsewhere in the book, but there are resonances with other of his texts where history is depicted as riven by a great divide between the included and the excluded.[18] For the moment, though, we simply need to acknowledge that Foucault 'prais[es] the discourse of race war'[19] (2003a, 65) – this bipolar vision of historical struggle – precisely because its entry into the hallways of history-writing *challenged* the Roman model of history. As he declares, 'the history that appears at this point, or the history of the race struggle, is a counterhistory' (2003a, 70). What occurred in the manoeuvres of these early historians had one further big effect: namely, to insert 'society' into history; or, more accurately, to insert a construct like 'society', acknowledging the presence of groupings within a given region differentiated by access to wealth, status and influence, into the writing of a history bothered about the politics of legitimacy (of who does or does not wield legitimate power in said region). Foucault hence talks about a 'new subject of history' (2003a, 134):

---

17 In their postscript, Fontana and Bertani (2003, 283) remark that a 'binary relationship which is introduced ... by the phenomena of domination, and which the model of war explains, does not really explain the multiplicity of the real struggles that are provoked by disciplinary power or the effects government has on the modes of behaviour produced by biopower'.

18 In fact, much could be said about the provocative historical geography of 'the Same and the Other' (esp. Foucault 1970, xxiv) as a framework and motif that punctuates various of Foucault's texts (for a commentary, see Philo 1986; 2006; also Elden 2001, Chapters 4 and 5). It clearly maps on claims made by him about the spatialized relations between Reason and Unreason, notably in *Madness and Civilization* (Foucault 1967; Philo 1999; 2004, esp. Chapter 2), and also between the Normal and the Abnormal (Foucault 2003b; also Elden 2002; Stone 2004).

19 It must be said with force that in *no* way is Foucault justifying racism. Indeed, at various points (e.g. 2003a, 61–2, 254–61; also Foucault 1979a, Part V; and Elden 2002, esp. 131–3) he starts to explain – and to critique – the emergence of racism within new forms of 'bioregulation', buttressed by new biological theories that pervert Darwin, outlining racism's complicity with both the imperial-colonial project and the most hateful eugenicist programmes of certain late-nineteenth and twentieth century states (with Nazi Germany obviously being the most shocking exemplar). As he observes: 'In the nineteenth century – and this is completely new – war will be seen not only as a way of improving one's own race by eliminating the enemy race (in accordance with the themes of natural selection and the struggle for existence), but also as a way of regenerating one's own race. As more and more of our numbers die, the race to which we belong will become all the purer' (2003a, 257). The whole problematic of dealing with 'abnormals' is also signposted here, referencing many other features of Foucault's *oeuvre* (e.g. Foucault 1967; 2003b) that spiral away from, yet remain connected via notions of race war and bioregulation, to the themes of this book.

It is what a historian of the period calls a 'society'. A society, but in the sense of an association, group or body of individuals governed by statute, a society made up of a certain number of individuals, and which has its own manners, customs and even its own law. (2003a, 134)

It is on this count that 'society' matters, hinting at a politicized sense of something called 'society' worth defending, which takes us towards what has been criticized as the strange title given to this lecture course.[20] True, Foucault's statements about the 'defence of society' surface when critiquing the racist outworking of the above-mentioned race war, wherein those in authority say: 'We have to defend society against all the biological threats posed by the other race, the subrace, the counterrace' (2003a, 61–2). Yet, as implied by his praise for the early historians opening up the race-social conflict paradigm, Foucault also sees something of critical value in the discovery of society, 'this new subject of history', that should be encouraged.

It is difficult to illustrate these claims, given the detail of Foucault's commentary about past history-writing, but it is worth mentioning the contribution of one historian, Boulainviller,[21] who is centralized by Foucault in no less than three of the lectures.[22] This individual was called upon by Louis XIV to abridge the wealth of documentation found in reports requested on the state of France for the benefit of his heir and grandson, the duc de Bourgogne (2003a, 127–8). In 1727 Boulainviller's two volume *Etat de la France* appeared, with a third volume in the following year and related texts, and, despite being commissioned by the monarch, these works of social-historical synthesis effectively disputed a version of history wherein 'the King's knowledge of his kingdom and his subjects "becomes" isomorphic with the State's knowledge of the State' (2003a, 128). Boulainviller wished for a more variegated history, one that could avoid 'saying' exactly what an emerging administrative machinery, into which the King was increasingly embedded, might want it to say. More specifically, his sympathies lay with a bourgeoisie who had lost influence over both monarch and administration, and a goal of his historical writing was to reactivate a bourgeois self-knowledge, a bourgeois imagination even, that could stand against the abuses of executive power under the *ancien regime* (2003a, 154–5). He was hardly an advocate for the most oppressed within society, far from

---

20 'The title of the book is misleading: Foucault does not believe in society as a force for good. Rather, it is the impulse to defend society at all costs that has been the defining force in the evolution of civilization' (Hussey 2005, no pagination). Basically I agree, but it strikes me that Foucault is also finding something more progressive in the discovery of a *notion* of 'society', since it provides a lever for critical analysis of the power-laden relations between the differentiated human fragments comprising a society.

21 There is some confusion over the spelling of this name: in the 2003 translation it is rendered as 'Boulainvilliers', but it is reckoned by authorities that 'Boulainviller' is more accurate (Elden 2002, 131, note 16), the version that will be adopted here.

22 'I want to take Boulainviller simply as an example, because there was in fact a whole nucleus, a whole nebula of noble historicans who began to formulate their theories in the second half of the seventeenth century' (2003a, 144).

it, and Foucault is not claiming Boulainviller as some unexpected champion of the 'Third Estate'. Nonetheless, in the sense detailed above, Foucault *is* crediting Boulainviller with a decisive role in heralding a new form of social history that *could* perform a function critical of the established order.

Corresponding with the wider sweep of the book, the centrality of war to Boulainviller's take on history is what really matters:

> Boulainviller makes the relationship of war part of every social relationship, subdivides it into thousands of different channels, and reveals war to be a sort of permanent state that exists between groups, fronts and tactical units as they in some sense civilise one another, come into conflict with one another, or on the contrary form alliances. There are no more multiple and stable great masses, but there is a multiple war. ... With Boulainviller, ... we have a generalized war that permeates the entire social body and the entire history of the social body; it is obviously not the sort of war in which individuals fight individuals, but one in which groups fight groups. (2003a, 162)

For Boulainviller, and for generations of social historians to follow, 'war [becomes] basically historical discourse's truth-matrix' (2003a, 165), and he thereby 'defined the principle of what might be called the relational character of power' (2003a, 168), showing that '(and this ... is the important point) relations of force and the play of power are the very stuff of history' (2003a, 169). Foucault duly detects here a series of propositions about history, war, politics and power that, at bottom, encapsulate the major themes covered in the book. Furthermore, Boulainviller was 'challenging ... the juridical model of sovereignty which had, until then, been the only way of thinking of the relationship between people and monarch, or between the people and those who govern' (2003a, 168), crafting instead an approach couched 'in historical terms of domination and the play of relations of force' (2003a, 169). Such an approach displayed a measure of commitment to those hoping to resist domination, even if here it was disaffected nobility rather than the dispossessed and marginalized groupings who we might more routinely envisage as the resisters of domination. I will shortly reconsider this politicized imperative in Boulainviller's history, but the salient point is that Foucault's reading of Boulainviller sets up the whole problematic for an alternative version of thinking history, precisely that bellicose history described by Lemert and Gillan (1982).

'Bellicose History' and 'Political Historicism'

Foucault detects in the realm of discourse the *same* play of forces, strategies and tactics, all underpinned by wickedness and cunning, as he finds in the material history of people, places and power; and, indeed, there is a definite sense that the book is as much about the 'war' in discourse – the clash *of* discourses, including different discourses about, or ways of writing, history – as it is anything else. Lemert and Gillan (1982, 37–9) are interesting in this respect:

A bellicose history cannot be read by means of abstractions or systems of thought; nor by meaningful interpretations. History must be read in documents produced by these conflicts. Documents are visible, readable, practices regulated by specific relations at a specific time. … Yes, society is more than discourse. But it is in and by means of discourse that social conflict takes place.

Within the bellicose history of history-writing itself, Foucault identifies – as mentioned – the split between 'the Roman history of sovereignty' and 'the biblical history of servitude and exiles' (2003a, 68), a split then coursing through several centuries of historiography. He stresses that the Roman historians and their Medieval and Early Modern inheritors all tended to operate in the horizon of establishment power, asserting that '[i]n general terms, we can … say that until a very late stage in our society, history was the history of sovereignty, or a history that was deployed in the dimension and function of sovereignty' (2003a, 68). Presumably there were always other versions of history in circulation, residues of the 'biblical history' together with – as also mentioned – folk-histories telling stories of invasion from the viewpoint of the pillaged not the pillagers, but the play of forces would have kept such alternatives largely unheard (except in obscure localities far from the seats of influence). There was always a potential battlefield where differing versions of history *could* have locked into combat, however, and one aspect of Foucault's critical historiography is to identify when this potential to be a battlefield became actualized – in short, to find when mainstream and alternative historical accounts *did* first engage one another in an obvious fashion.

Foucault admits that: 'I do not think that the difference between these two histories is precisely the same as the difference between an official discourse and, let us say, a rustic[23] discourse' (2003a, 78). Nonetheless, he presents the alternative accounts of history as loosely akin to the many, heterogeneous craft knowledges that were, as it were, the 'surplus' of the Enlightenment's disciplinary ambitions (see above), and as thereby standing in an awkward, even oppositional stance to the more officially sanctioned varieties of Early Modern history:

> … history found itself, for different reasons, in the same position as the technical knowledges … . For various reasons, historical knowledge entered a field of struggles and battles at much the same time. … When historical knowledge, which has until then been part of the discourse that the State or power pronounced on itself, was enucleated from that power, and became an instrument in the political struggle that lasted for the whole eighteenth century, the State attempted, in the same way and for the same reason, to take it in hand and disciplinarize it. … [T]here was a perpetual confrontation between the history that had been disciplinarized by the State and that had become the content of

---

23 The manuscript version says 'scholarly and naïve', rather than 'official and rustic' (2003a, 77, footnote). The suggestion that the alternative versions of history might derive from the countryside, what I called a moment ago in the main text 'obscure places', does of course index fascinating possibilities for envisaging an uneven geography of history (of history-writing and -telling) wherein urban-rural contrasts might be significant.

official teaching, and the history that was bound up with the struggles because it was the consciousness of subjects involved in a struggle. (2003a, 185–6)

In the terms defined earlier, this means that Foucault regards the latter form of history as a subjugated knowledge of sorts. Attention has already been paid to what he says about a species of counterhistory (represented by the contribution of Boulainviller) that began to work with notions of race war or, more intelligibly for today's audience, a model of social history fractured by clashing human groupings; and Foucault calls this model a 'counterhistory' precisely because of its adversarial character on the battlefield of historical writing, and also because it held out the promise of a political challenge in its immediate present (i.e. at the time of its writing).[24]

The counterhistory in question here, acknowledging the bellicosity of history, was confronted at different periods by 'the philosophical order' and 'the political order' (2003a, 59), both being hesitant about lending credence to voices from what were previously the sidelines of historical change:

Whatever form it takes, [this discourse of war and history] will be denounced as the discourse of a biased and naïve historian, a bitter politician, a dispossessed aristocracy, or as an uncouth discourse that puts forward inarticulate demands. ... (2003a, 58)

... counterhistory that is born of the story of race struggle will of course speak from the side that is in darkness, from within the shadows. It will be the discourse of those who have no glory, or of those who have lost it and who now find themselves, perhaps for a time – but probably for a long time – in darkness and silence. (2003a, 70)

This kind of history was, and continues to be, 'a disruptive speech' (2003a, 70), telling not 'of the untarnished and uneclipsed glory of the sovereign', less about great 'victories', but more about 'the misfortunes of ancestors, exiles and servitude' (2003a, 71). It was, and is, a history that 'has to disinter something that had been hidden', and, crucially given the broader thrust of the book, a history looking for something 'which has been hidden not only because it has been neglected, but because it has been carefully, deliberately and wickedly misrepresented' (2003a, 72). Such a history strives 'to show that laws deceive, that kings wear masks, that power creates illusions, and that [other] historians tell lies' (2003a, 72), and it is a history acknowledging the depredations of war while claiming rights for the dispossessed and the marginalized. The result is indeed a thoroughly *politicized* sense of history, a parcel of troublesome counterhistories sticking a knife in the belly of establishment histories. Such histories have commonly been opposed, as Foucault reflects:

---

24 '[T]his discourse, which was basically or structurally kept in the margins by that of the philosophers and jurists, began its career – or perhaps its new career in the West – in very specific conditions between the end of the sixteenth and the beginning of the seventeenth centuries and represented a twofold – aristocratic and popular – challenge to royal power' (2003a, 58).

In more general terms, and in the longer term, what had to be eliminated was what I would call 'political historicism', or the type of discourse that we see emerging from the [historiographic] discussions I have been talking about, ... and which consists in saying: Once we begin to talk about power relations, we are not talking about right, and we are not talking about sovereignty; we are talking about domination, and about an infinitely dense and multiple domination that never comes to an end. (2003a, 110–11)

Unsurprisingly, it is again in the example of Boulainviller that Foucault finds the embryo of this 'political historicism',[25] reconstructing this long-dead scholar's 'historico-political field' (2003a, 167) – 'a historico-political continuum' – wherein 'historical narratives and political calculations have exactly the same object' (2003a, 169). Still extrapolating from Boulainviller, but returning to the thematic of war, Foucault then writes:

... we can say that the constitution of a historico-political field is an expression of the fact that we have gone from a history whose function was to establish right by recounting the exploits of heroes or kings, their battles and wars and so on ... to a history that continues the war by deciphering the war and the struggles that are going on within all the institutions of [apparent] right and peace. ... [F]rom the eighteenth century onward, historical knowledge becomes an element of the struggle: it is both a description of struggles and a weapon in the struggle. History gave us the idea that we are at war; and we wage war through history. (2003a, 171–2)

Thus Foucault returns, now armed with a political historicism in which political commitments inform historic inquiry while historical findings are mobilized in political interventions, to a renewed justification for the variety of history – archaeological excavations of subjugated knowledges as the basis for genealogical critiques of reprehensible power-effects – that he has long been endeavouring to practise.

Conclusion: 'Geography Must Indeed Necessarily Lie at the Heart of My Concerns'

The book's sensitivity to questions of geography – to do with space, place, location and environment, however precisely defined – should already be evident from the above, but let me conclude with a few speculative remarks. The first thing to note is Foucault's continual referencing of *the local*, most obviously in his discussion of subjugated or disqualified knowledges, repeatedly accompanied by evocative terms such as 'local knowledges', 'local categories', 'local critiques' and 'local discursivities'. This terminology is used to distinguish the knowledges that he has

---

25 Revealingly, Foucault concludes his fifth lecture by promising that 'next time I would like to both trace the history of this discourse of political historicism and praise it' (2003a, 111). The discussion of Boulainviller is then central to the sixth and seventh lectures, although it is actually in the eighth lecture that claims about 'political historicism' are foregrounded.

in mind from ones that might be described as 'global' or, perhaps more accurately, as *common* in the sense of being in common usage, widespread, prevailing, predominant, known to many, somehow 'filling up' all of a national territory (and maybe beyond). Indeed, Foucault explicitly distances what he calls 'little knowledges' from the 'common knowledge' of the masses, including the taken-for-granted beliefs and assumptions comprising so-called 'common-sense' (2003a, 7); but he also distances these little knowledges from a host of other elite, expert and professional knowledges – ones carrying the sanction of Enlightenment or Science – that may not be common knowledge as such, but which are still widely diffused around many centres of research, learning and calculation. As Foucault clarifies when talking of 'a knowledge that is local, regional or differential', one ambition is simply to underscore that such knowledge 'is different from all the knowledges that surround it' (2003a, 7–8); in which case local becomes a marker of that difference when set against the homogeneity implied by describing something as 'common'. When linking local knowledges to 'disorderly and tattered genealogies' (2003a, 10), moreover, a second ambition is to show how such knowledges, lacking coherence, organization and structure, depart from – and perhaps embody a critical window on or response to – more 'total' or even 'totalizing' knowledges, theories, worldviews and the like. I will return to this claim, but what I must add in passing is that Foucault conceives of alternative knowledges in a materially spatialized manner; so we are *not* merely in the realm of metaphor, as is perhaps the case with the spatial terminologies of some poststructuralist authors (Smith and Katz 1993). He says enough throughout the book to indicate that he really does picture at least some of the knowledges under scrutiny as anchored in quite particular places, with their 'differences defined by geographical regions' (2003a, 179), arising in specific sorts of spaces – certain types of settlement, craft-workshop, school, hospital or asylum – or distributed around definable networks, maybe comprising the people working, living, attending or interned in the spaces just listed. A geographical attentiveness to the knowledges produced in named places and delineated spaces, especially to those that he calls upon us to liberate from their subjugation and disqualification at the hands of knowledges occupying superior positions in the power-hierarchy, is therefore pivotal.[26]

26 A valuable task might be to relate Foucault's claims here to seemingly parallel constructions in certain literatures familiar to human geography. Geertz's notion of 'local knowledge' (1983; see Barnes 2000; Cloke et al. 2004, 319–23) would be the most obvious, not least because Foucault actually uses this term, although a closer inspection suggests that Geertz's focus on local formations of cultural meaning might veer towards a concern for 'common sense' (albeit common sense as locally constituted) that Foucault expressly disavows. The same might be true with respect to the notion of 'indigenous knowledge' (Briggs 2005), although an envisaged – and some might say over-played – binary divide between indigenous knowledges (particular locally-embedded environmental knowledges) and scientific knowledges (particularly Western-derived technological knowledges) does perhaps echo something of Foucault's contrast between dominant and subjugated knowledges. Haraway's (1988; see Barnes 2000; Merrifield 1995; Rose 1997) notion of 'situated knowledges' is

To pick up a point made a moment ago, Foucault's concern for the local is also designed to register the importance of recognizing both differences and partialities in the realm of knowledge. There is a parallel with his formulation of a 'general history' over and against a 'total history', surfacing in the English introduction to *The Archaeology of Knowledge* (Foucault 1972), and with his notes there about thinking history, discourse and knowledge in terms of 'spaces of dispersion' (Philo 1992, 148–50). In *Society Must Be Defended* this concern for the local figures in the conceptualizing of power, which, while not supposed to be simplistically 'localized' in the sense of being 'held' by blocks of the powerful, is always traced through diverse local capillaries where its effects are made and felt. Echoing the formulations in *The Archaeology of Knowledge*, but now stirring in an understanding of power as akin to the fleeting cobble of forces, tactics, passions and events endemic to war, Foucault arrives (in his course summary) at this meta-description of what his scholarly approach resembles:

> We are ... dealing with a discourse [that of bellicose history] that inverts the traditional values of intelligibility. An explanation from below, which does not explain things in terms of what is simplest, most elementary and clearest, but in terms of what is most confused, most obscure, most disorganized and most haphazard. It uses as an interpret[at]ive principle the confusion of violence, passions, hatreds, revenge and the tissue of the minor circumstances that create defeats and victories. (2003a, 269)

This approach is about accenting heterogeneity over homogenization, fragmentation over coherence, multiplicities over singularities: it is about the assault on the theoretical castles of Order, Truth and Reason (with their first letters capitalized). Put like this, Foucault in this book can readily be interpreted, with justification, as expressing a poststructuralist sensibility that is countering a structuralist model of history, language and society (one listing permutations and combinations governed by a set of prior possibilities whose overall order is known).[27] As has now been argued several times, a poststructuralist sensibility is necessarily attuned to space, as what most obviously 'guarantees' the relational play of differences, juxtapositions and contingencies (e.g. Dixon and Jones III 1998a; 1998b; Doel 1999; Massey 2005; Marston et al. 2005; Natter and Jones III 1993; Pratt 2000). Another way of capturing Foucault's stance in this book is to situate it in the orbit of postmodernism (Ley 2000; 2003), as a still more general assault on the ordering pretensions of modernist thought, and such a move was one that I made when aligning 'Foucault's

---

another obvious point of reference, although, in stressing that *all* knowledge cannot but have origins that are situated, partial and thereby local, she also qualifies Foucault's effective pitting of local knowledges (marked by their differences) against more global knowledges (as in a singular Science).

27 The relationship between Foucault and structuralism is often debated, but his vigorous distancing of his own projects from those of structuralism – voiced very strongly in the introduction and conclusion of *The Archaeology of Knowledge* (Foucault 1972, esp. 15, 199–202) – strike me as convincing.

geography' with 'that wider current of thought (or 'attitude') now commonly referred to as postmodernism, in which the certainties of existing (modernist) intellectual projects ... are thrown deeply into question' (Philo 1992, 142; also Cloke et al. 1991, Chapter 6).[28] Symmetrical with claims made about the alertness to space demanded by poststructuralism, so the likes of Gregory (1989) spell out the peculiar affinities between academic geography, with its traditional attachment to 'areal differentiation' (Hartshorne 1939), and postmodernism's insistence on prioritizing differences of all kinds over the alleged *in*difference of modernism. All of these contexts help to make sense of what can be cast as the local obsessions and spatial multiplicities central to these lectures.

I will return to this issue for one last time presently, adding a further slant tied up with the book's bellicose history, but it is important to appreciate the fierce backlash against the emphasis on difference integral to both poststructuralism and postmodernism. In various quarters, albeit not often so fiercely in print (but see Hamnett 2001; 2003), these approaches have been accused of a listless relativism, an 'anything goes' mentality, an ethico-political fecklessness, a lack of serious critical ambition, and an irresponsible celebration of playfulness over a serious engagement with social alternatives to an iniquitous *status quo*. While such a backlash is understandable, it does miss the more radical objectives of postmodernism, 'domesticating' its message so as to render it a prime candidate for critique (Strohmayer and Hannah 1992; also Ley 2003), and it fails to look closely at the precise reasoning that led the likes of Foucault to advance such strong claims on behalf of difference (and the local) in the face of 'total' and even 'totalizing' theories (ones debated in the academy *and* ones mobilized in structuring whole societies, such as fascism and debased versions of socialism: see 2003a, esp. 258–63). It is instructive to consider Foucault's own words on the recovery of subjugated knowledges as the archaeological prelude to genealogical critique, a task which, as already explained, he sees as thoroughly infused with intense ethico-political purpose. Regarding 'multiple genealogical investigations' (2003a, 8), he proposes that their cumulative effect is to leave 'total' theories or discourses – ones as diverse as medical-psychiatric orthodoxy, neo-classical economics or Marxism – 'cut up, ripped up, torn to shreds, turned inside out, displaced, caricatured, dramatized, theatricalized, and so on' (2003a, 6). Stressing the local, difference-seeking, non-homogenizing nature of the critique proposed, he insists:

> So that, if you like, is my ... point: ... the local character of critique; this does not, I think, mean soft eclecticism, opportunism or openness to any old theoretical undertaking, nor does it mean a sort of deliberate asceticism that boils down to losing as much theoretical weight as possible. I think that the essentially local character of the critique in fact indicates something resembling a sort of autonomous and non-centralized theoretical production,

---

28 As Hannah (this volume) remarks, though, this framing of 'Foucault's geography' within the orbit of postmodernism is not one that particularly appeals *now*: it reflected a particular moment when texts by the likes of Gregory (1989), Harvey (1989) and Soja (1989) had suddenly galvanized theoretical geographers into discussing postmodernism.

or in other words a theoretical production that does not need a visa from some common regime to establish its validity. (2003a, 6)

Yet, as many passages make plain, Foucault does not let this 'local character of critique', dealing in multiple possibilities and not requiring the sanction of some singular theoretical court of appeal, remain as a self-satisfied deconstructive mission. Rather, the local critiques, targeted as specific enemies or particular points of weakness in the power-hierarchy,[29] are to be prompted by an ethico-political commitment; something that the central lectures progressively configure as a fully political historicism informed by sustained and detailed encounter with subjugated knowledges, peoples, places, experiences and events.

Finally to close, let me revisit once more that openness to space present throughout the book, whether talking of local discursivities, the local capillaries of power or the local character of critique. It might be objected that Foucault says little here that is as obviously about space as he does elsewhere when writing about the likes of asylums, hospitals, workhouses and prisons,[30] and that as a result it is of less interest to a geographical audience than are many of his other texts. My counterclaim would be that it actually witnesses Foucault developing a still deeper awareness of why spaces matter, precisely because he explores the potential of a bellicose history whose baseline imaginary – the pitted battlefield, a patchwork of mud, rust and blood, an uneven landscape with collapsing tunnels beneath, a site of stressed humanity, strategizing but messing up, often running blindly – is from the very outset spatialized in a variety of ways. It may be a coincidence, but it was in 1976, around the time of these lectures, that Foucault was interviewed by the radical geography journal *Hérodote*, in the course of which he becomes excited about what a geographical perspective might add to his own thinking. What prompts this excitement, though, is his dawning appreciation of the connections between geography and war as a possible framing for his ongoing intellectual endeavours; and it is with his comments on this theme that I will now conclude:

> Now I can see that the problems you put to me about geography are crucial ones for me. Geography acted as the support, the condition of possibility for the passage between a series of factors I tried to relate …

---

29 It is true, even so, that questions arise about whether Foucault envisages different local struggles ever gaining a more generalized coherence – linking hands with struggles over similar issues (e.g. psychiatric reform) in other places or with struggles over other issues (e.g. workers' or women's rights) in the same neighbourhood – and hence about his stance on so-called 'rainbow coalitions' (i.e. what risks attach to such coalitions in imposing singular grids over multiple realities?). In a way, the whole problematic of contemporary 'social movements', including the possibilities but also drawbacks of 'convergence space' (Routledge 2003), is here anticipated.

30 At various moments – deliberately *not* stressed in this chapter, because these were not its focus – such material spaces of institutions and everyday sites do get a mention. Note his reflections on the spatial arrangements of planned towns and housing estates (2003a, 250–51), but also various passing references to schools, asylums and hospitals.

The longer I continue, the more it seems to me that the formation of discourses and the genealogy of knowledge need to be analyzed, not in terms of types of consciousness, modes of perception and forms of ideology, but in terms of tactics and strategies of power. Tactics and strategies deployed through implantations, distributions, demarcations, control of territories and organizations of domains which could well make up a sort of geopolitics where my preoccupations would link up with your methods. One theme I would like to study in the next few years is that of the army as a matrix of organization and knowledge; one would need to study the history of the fortress, the 'campaign', the 'movement', the colony, the territory. Geography must indeed necessarily lie at the heart of my concerns. (Foucault 1980b, 77; this volume, 182)

Acknowledgements

Very substantial thanks to Stuart Elden for his encouragement, advice and patience, and for the quality of his Foucauldian scholarship, and also to Jeremy W. Crampton for his help as well. Thanks as well to the hills, lochans and paths around Arisaig, Mora and Mallaig, where many of the ideas for this chapter were formulated.

References

Agomben, G. (1998) *Homo Sacer: Sovereign Power and Bore Life*. Stanford, CA: Stanford University Press [trans., orig. 1995].

Barnes, T.J. (2000) 'Local Knowledge.' In R.J. Johnston, D. Gregory, G. Pratt and M. Watts (Eds) *The Dictionary of Human Geography (4th ed.)* Oxford: Blackwell. 452–3.

Barnes, T.J. (2000) 'Situated Knowledge.' In R.J. Johnston, D. Gregory, G. Pratt and M. Watts (Eds) *The Dictionary of Human Geography (4th ed.)* Oxford: Blackwell. 742–3.

Briggs, J. (2005) 'The Use of Indigenous Knowledges in Development: Problems and Challenges.' *Progress in Development Studies*, 5, 99–114.

Cadman, L.J. (2006) 'A Genealogy of (Bio)political Contestation during the Reform of the Mental Health Act 1983 in England and Wales.' Unpublished PhD thesis, Department of Geography, University of Sheffield.

Cloke, P., I. Cook, P. Crang, M. Goodwin, J. Painter and C. Philo (2004) *Practising Human Geography: The Construction and Interpretation of Geographical Knowledge*. London: Sage.

Cloke, P., C. Philo and D. Sadler (1991) *Approaching Human Geography: An Introduction to Contemporary Theoretical Debates*. London: Paul Chapman.

Dean, M. (1994) *Critical and Effective Histories: Foucault's Methods and Historical Sociology.* London: Routledge.

Dixon, D.P. and J.-P. Jones III (1998a) 'My Dinner with Derrida, or Spatial Analysis and Poststructuralism Do Lunch.' *Environment and Planning A*, 30, 247–60.

Dixon, D.P. and J.-P. Jones III (1998b) 'For a *supercalifragilisticexpialidocious* Scientific Geography.' *Annals of the Association of American Geographers*, 86, 767–79.

Doel, M.A. (1999) *Poststructuralist Geographies: The Diabolical Art of Spatial Science.* Edinburgh: Edinburgh University Press.

Driver, F. (1985) 'Power, Space and the Body: A Critical Assessment of Foucault's *Discipline and Punish.*' *Environment and Planning D: Society and Space*, 3, 425–46.

Driver, F. (1993a) *Power and Pauperism: The Workhouse System, 1834–1884.* Cambridge: Cambridge University Press.

Driver, F. (1993b) 'Bodies in Space: Foucault's Account of Disciplinary Power.' In C. Jones and R. Porter (Eds) *Reassessing Foucault: Power, Medicine and the Body.* London: Routledge. 113–31.

Elden, S. (2001) *Mapping the Present: Heidegger, Foucault and the Project of a Spatial History.* London: Continuum.

Elden, S. (2002) 'The War of the Races and the Constitution of the State: Foucault's *Il faut defendre la société.*' *boundary 2*, 29, 125–51.

Elden, S. (2003) 'Plague, Panopticon, Police.' *Surveillance and Society*, 3, 240–53.

Elden, S. (2005) 'The Problem of Confession: The Productive Failure of Foucault's *History of Sexuality.*' *Journal of Cultural Research*, 9, 1–41.

Elden, S. (2006) 'Discipline, Health and Madness: Foucault's *Le pouvour psychiatrique.*' Forthcoming in *History of the Human Sciences*, 19(1), 39–66.

Elden, S. (2007) 'Strategies for Waging Peace: Foucault as *Collaborateur.*' In M. Dillon and A. Neal (Eds) *Foucault: Politics, Security and War.* London: Palgrave.

Elden, S., chapter 11 in this volume. 'Strategy, Medicine and Habitat: Foucault in 1976.'

Fontana, A. and M. Bertani (2003) 'Situating the Lectures.' In M. Foucault, *Society Must be Defended: Lectures at the Collège de France, 1975–1976.* New York: Picador. 273–93.

Foucault, M. (1967) *Madness and Civilization: A History of Insanity in the Age of Reason.* London: Tavistock [trans., orig. 1961].

Foucault, M. (1970) *The Order of Things: An Archaeology of the Human Sciences.* London: Tavistock [trans., orig. 1966].

Foucault, M. (1972) *The Archaeology of Knowledge.* London: Tavistock [trans., orig. 1969].

Foucault, M. (1977) *Discipline and Punish: The Birth of the Prison.* London: Allen Lane [trans., orig. 1975].

Foucault, M. (1979a) *The History of Sexuality, Vol.1: The Will to Knowledge.* London: Allen Lane [trans., orig. 1976].

Foucault, M. (1979b) 'Governmentality.' *Ideology and Consciousness*, 6, 5–21 [trans., orig. 1978].

Foucault, M. (1980a) 'Two Lectures.' In C. Gordon (Ed.) *Michel Foucault: Power/ Knowledge: Selected Interviews and Other Writings, 1972–1977.* Brighton, UK: Harvester Press. 78–109 [trans., orig. 1976].

Foucault, M. (1980b) 'Questions on Geography: Interviewers: The Editors of the Journal *Hérodote.*' In C. Gordon (Ed.) *Michel Foucault: Power/Knowledge: Selected Interviews and Other Writings, 1972–1977.* Brighton, UK: Harvester Press. 63–77 [trans., orig. 1976].

Foucault, M. (1986) 'Nietzsche, Genealogy, History.' In P. Rabinow (Ed.) *The Foucault Reader.* Harmondsworth, UK: Peregrine Books. 76–100 [trans., orig. 1971].

Foucault, M. (2003a) *Society Must be Defended: Lectures at the Collège de France, 1975–1976.* New York: Picador [trans., orig. 2001].

Foucault, M. (2003b) *Abnormal: Lectures at the Collège de France, 1974–1975.* London: Verso.

Foucault, M. (2005) *The Hermeneutics of the Subject: Lectures at the Collège de France, 1981–1982.* New York: Picador [trans., orig. 1997].

Foucault, M. (2006) *Psychiatric Power: Lectures at the Collège de France, 1973–1974.* London: Palgrave [trans., orig. 2003].

Geertz, C. (1983) *Local Knowledge: Further Essays in Interpretive Anthropology.* New York: Basic Books.

Gordon, C., Ed. (1980) *Michel Foucault: Power/Knowledge: Selected Interviews and Other Writings, 1972–1977.* Brighton: Harvester Press.

Gregory, D. (1989) 'Areal Differentiation and Postmodern Human Geography.' In D. Gregory and R. Walford (Eds) *Horizons in Human Geography.* London: Methuen. 67–96.

Hamnett, C. (2001) 'The Emperor's New Theoretical Clothes, or Geography Without Origami.' In C. Philo and D. Miller (Eds) *Market Killing: What the Free Market Does and What Social Scientists can Do about It.* Harlow: Pearson Education. 158–69.

Hamnett, C. (2003) 'Contemporary Human Geography: Fiddling While Rome Burns.' *Geoforum,* 34, 1–3.

Hannah, M.G. (1997a) 'Space and the Structuring of Disciplinary Power: An Interpretive Review.' *Geofraska Annaler,* 79B, 171–80.

Hannah, M.G. (1997b) 'Imperfect Panopticism: Envisioning the Construction of Normal Lives.' In G. Benko and U. Strohmayer (Eds) *Space and Social Theory: Interpreting Modernity and Postmodernity.* Oxford: Blackwell. 344–59.

Hannah, M., chapter 12 in this volume, 'Formations of "Foucault" in Anglo–American Geography: An Archaeological Sketch.'

Haraway, D. (1988) 'Situated Knowledges: The Science Question in Feminism and the Privilege of Partial Perspective.' *Feminist Studies,* 14, 575–600.

Hartshorne, R. (1939) *The Nature of Geography: A Critical Survey of Current Thought in the Light of the Past.* Washington, PA: Association of American Geographers.

Harvey, D. (1989) *The Condition of Postmodernity: An Enquiry into the Origins of Cultural Change.* Oxford: Blackwell.

Hussey, A. (2005) 'Book Review of *Society Must be Defended.*' Originally in *New Statesman*, but consulted at http://www/newstatesman.com/Book shop/300000072674 (10/08/05).

Katz, C. (1996) 'Towards Minor Theory.' *Environment and Planning D: Society and Space*, 14, 487–99.

Jones, J. (1996) 'Community-based Mental Health Care in Italy: Are There Lessons for Britain?' *Health and Place*, 2, 125–8.

Jones, J. (2000) 'Mental Health Care Reforms in Britain and Italy Since 1950: A Cross-national Comparative Study.' *Health and Place*, 6, 171–87.

Legg, S. (2005) 'Foucault's Population Geographies: Classifications, Biopolitics and Governmental Spaces.' *Population, Space and Place*, 11, 137–56.

Lemert, C.C. and G. Gillan (1982) *Michel Foucault: Social Theory as Transgression.* New York: Columbia University Press.

Ley, D. (2000) 'Postmodernism.' In R.J. Johnston, D. Gregory, G. Pratt and M. Watts (Eds) *The Dictionary of Human Geography (4ᵗʰ ed.).* Oxford: Blackwell. 620–22.

Ley, D. (2003) 'Forgetting Postmodernism: Recuperating a Social History of Local Knowledge.' *Progress in Human Geography*, 27, 537–60.

Livingstone, D. (1995) 'The Spaces of Knowledge: Contributions Towards a Historical Geography of Science.' *Environment and Planning D: Society and Space*, 13, 5–34.

Livingstone, D.N. and C.W.J. Withers, Eds (1999) *Geography and Enlightenment.* Chicago: University of Chicago Press.

Massey, D. (2005) *For Space.* London: Sage.

Marston, S., J.-P. Jones III and K. Woodward, (2005) 'Human Geography Without Scale.' *Transactions of the Institute of British Geographers*, 30, 416–32.

Merrifield, A. (1995) 'Situated Knowledge through Exploration: Reflections on Bunge's "Geographical Expeditions".' *Antipode*, 27, 49–70.

Natter, W. and J.-P. Jones III (1993) 'Signposts towards a Poststructuralist Geography.' in J.-P. Jones III, W. Natter and T.R. Schatzki (Eds) *Postmodern Contentions: Epochs, Politics, Spaces.* New York: Guilford.165–203.

Ojakangas, M. (2005) 'Impossible Dialogue on Bio-power: Agamben and Foucault.' *Foucault Studies*, 2, 5–28.

Philo, C. 1986 '"The Same and the Other": On Geographies, Madness and Outsiders.' Loughborough University of Technology, Department of Geography, Occasional Paper No.11.

Philo, C. (1989) '"Enough to drive one mad": The Organisation of Space in Nineteenth-century Lunatic Asylums.' In J. Wolch and M. Dear (Eds) *The Power of Geography: How Territory Shapes Social Life.* London: Unwin Hyman. 258–90.

Philo, C. (1992) 'Foucault's Geography.' *Environment and Planning D: Society and Space*, 10, 137–61.

Philo, C. (1999) 'Edinburgh, Enlightenment and the Geographies of Unreason.' In D.N. Livingstone and C.W.J. Withers (Eds) *Geography and Enlightenment.* Chicago: University of Chicago Press. 272–98.

Philo, C. (2001) 'Accumulating Populations: Bodies, Institutions and Space.' *International Journal of Population Geography*, 7, 473–90.

Philo, C. (2004) *A Geographical History of Institutional Provision for the Insane from Medieval Times to the 1860s in England and Wales: 'The Space Reserved for Insanity'.* Lewiston and Queenston, USA, and Lampeter, Wales, UK: Edwin Mellen Press.

Philo, C. (2005) 'Sex, Life, Death, Geography: Fragmentary Remarks inspired by "Foucault"'s Population Geographies.' *Population, Space and Place*, Vol.11. 325–33.

Philo, C. (2006) 'Madness, Memory, Time and Space: A Critical Essay on Thomas Laycock and the Unknown Artist-patient.' Unpublished ms. available from author.

Pratt, G. (2000) 'Poststructuralism.' In R.J. Johnston, D. Gregory, G. Pratt and M. Watts (Eds) *The Dictionary of Human Geography (4th ed.)* Oxford: Blackwell. 625–7.

Robinson, J. (1999) 'Power as Friendship: Spatiality, Femininity and "Noisy Surveillance".' In J.P. Sharp, P. Routledge, C. Philo and R. Paddison (Eds) *Entanglements of Power: Geographies of Domination/Resistance.* London: Routledge. 136–60.

Rose, G. (1997) 'Situating Knowledges: Positionality, Reflexivities and Other Tactics.' *Progress in Human Geography*, 21, 305–20.

Routledge, P. (2003) 'Convergence Space: Process Geographies of Grassroots Globalisation Networks.' *Transactions of the Institute of British Geographers*, 28, 333–66.

Sharp, J.P., P. Routledge, C. Philo and R. Paddison (2000) 'Entanglements of Power: Geographies of Domination/Resistance.' In J.P. Sharp, P. Routledge, C. Philo and R. Paddison, (Eds) *Entanglements of Power: Geographies of Domination/ Resistance.* London: Routledge. 1–42.

Smith, N. and C. Katz (1993) 'Grounding Metaphor: Towards a Spatialised Politics.' In M. Keith and S. Pile (Eds) *Place and the Politics of Identity.* London: Routledge. 67–83.

Soja, E.W. (1989) *Postmodern Geographies: The Reassertion of Space in Critical Social Theory.* London: Verso.

Stone, B.E. (2004) 'Defending Society from the Abnormal: The Archaeology of Biopower.' *Foucault Studies*, 1, 77–91.

Strohmayer, U. and M. Hannah (1992) 'Domesticating Postmodernism.' *Antipode*, 24, 29–55.

Surowiecki, J. (2005) 'Book Review of *Society Must be Defended.*' Originally in *Bookforum*, but consulted at: http://www.holizbrinckpublishers.com./academic/ book/BookDisplay.asp?bookKey.

... and perspectives' *Royal Institution Society*, May 3(2) ... pp. 361.

Pabla, C. (2003) *Accumulating Publications*, Vienna: Institute ... 2nd Stage International Seminar on Reproduction Programme, ... pp. 1–100.

Pabla, C. (2001) 'Retrospective Histories of the ... Protection for the Nazing from Heritage Conservation in the 1980s in Britain and Wales', 7th Symposia: Preserved Settlement', L.ac.k.a. and Queens, in B.s.A. and I. amputer, Watios, UK: Elwin, W.H. in Press.

Pabla, C. (2005) 'And the Deung Coram ... brief enquiry Research inspired by ... Worldwide', *Population Geographies Textbook, Singapore*, pp. Time, Vol. 11, 195–22.

Page, C. (2004) 'Marnings Memory Time and Soul of A Brief Essay on Topous ... Layouts from the Unknown Artist-pattern', Birmingham: [publisher] not available from author.

Paul, G. (2003) 'Participatory data', Ali, R.J. Johnston, D. Gregory, Castree and M.J. Watts (eds), *The 5th ed.*, Oxford, Oxford Geographic: Oxford: Oxford, Blackwell, ... 635.

Thompson, J. (2004) 'Power as Inheritable: Kept by a Detailed layout in Joint Surveillance', in N. Lyon, a Foucaultion ... Ellis, and R.J. Radburn (eds), *Edited Essays on Foucault: Geographies of Domination*, Routledge, London, pp. 1–20.

Rose, G. (1997) 'Situating Knowledges: Positionality, Reflexivities and Other Tactics', *Progress in Human Geography*, 21, 305–320.

Roulier, ... (2002) 'Investigating Square Spaces: Geographies of Leaseholds: Globalisation and Production', *The Journal of the History of Britain's Properties*, 28, 115–38.

Sauer, B.J. Rothberg, C. and ... R. Rudge et al. (2002) 'T Jump Image at Proven Geographies of Foucaultionists ... in B.L. Sher, J.R. Bonk ... (eds), *Production* and R. Davidson (eds), *Transformation of Power in Geographies of Geographications*, ... Relations continued.

Scull, ... and L. Kirk. 1993. *Boundaries: Territory No ... and Importance, Spaces*, in M.J. Cerca and F.J. Crack, *Geographies* ... Planning Limited, Routledge, ... 75.

Sola, E.W. (1990) *Pace: A Re-assertion of ... Preservation of Space in Critical Social Theory*, London: Verso.

Stone, B.L. (2005) *Deconstructing Society from the ... math: The Epistemology of Setting up Labs?*, London: ... 573–74.

Stratview ... and M. Mairch (2002) 'Anti ... Geography and Practice', pp. 57–89.

Stratview, J.L. (2002) *Basic Range of Settlements ... Critical Geography in Employment continued and Important with ... not publishers, an accessible booklet, not ..., England, UK).

Index

Printed in the United States
by Baker & Taylor Publisher Services

Printed in the United States
by Baker & Taylor Publisher Services